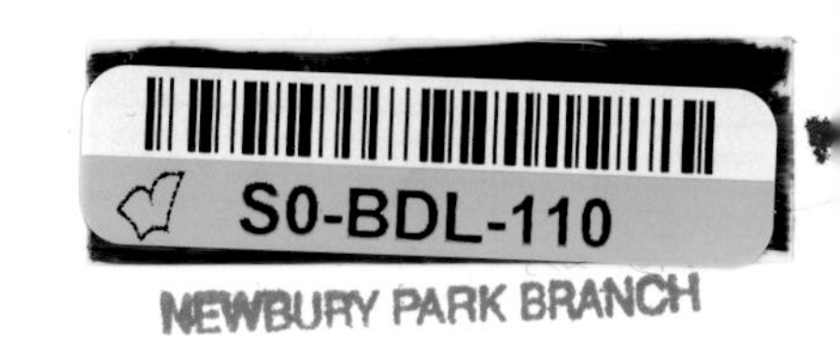
S0-BDL-110

Walker's

Marsupials of the World

Walker's

Marsupials of the World

Ronald M. Nowak

Introduction by
Christopher R. Dickman

The Johns Hopkins University Press
Baltimore

Printed in the United States of America on acid-free paper
9 8 7 6 5 4 3 2 1

The Johns Hopkins University Press
2715 North Charles Street
Baltimore, Maryland 21218-4363
www.press.jhu.edu

Library of Congress Cataloging-in-Publication Data

Nowak, Ronald M.
Walker's marsupials of the world / Ronald M. Nowak ; introduction by Christopher R. Dickman.
p. cm.
Portions of this book have been adapted from Walker's mammals of the world, 6th ed., by Ronald M. Nowak.
Includes bibliographical references and index.
ISBN 0-8018-8222-2 (hardcover : alk. paper) — ISBN 0-8018-8211-7 (pbk. : alk. paper)
1. Marsupials. 2. Marsupials—Classification. I. Title.
QL737.M3N68 2005
599.2—dc22

2004028272

A catalog record for this book is available from the British Library.

Contents

Walker's

Marsupials of the World

Marsupials of the World: An Introduction

Christopher R. Dickman

Marsupials stand apart from all other mammals in producing miniscule young and then suckling them outside the mother's body for extended periods. This ability enthralled the earliest European observers, including Queen Isabella of Spain. In 1500, she was so moved at the sight of a Brazilian opossum with young that she pronounced it an "incredible mother" (Archer 1982). Marsupials are also ecologically diverse, collectively eating almost all kinds of foods and exploiting an extraordinary array of other biotic and physical resources in the many environments they inhabit. In his review of marsupials of just the Australian region, the respected mammalogist Ellis Troughton considered this group to be "the greatest phylogenetic deployment of a single mammalian Order that the World can ever know" (1959, 69). Despite a touch of hyperbole in Troughton's comment, he was not far off the mark.

Constituting almost 7 percent of the world's extant mammals, marsupials predominate in the Australasian region (defined here to include Australia, New Guinea, and islands east to the Solomons and west to the Makassar and Lombok Straits, or Wallace's Line) and are also represented strongly in the Americas. In the latter region all species are referred to generally as *opossums,* but in Australasia use of the term *possum* is restricted to certain small and medium-sized marsupials. The 300 or so species of living marsupials show astonishing variation in size, from just 4.5 grams for the long-tailed planigale *(Planigale ingrami)* to more than 80 kilograms for adult male red kangaroos *(Macropus rufus)*. Diprotodontian marsupials, comprising possums and kangaroos, span the greatest size range of any mammalian order. As with many other groups of mammals, marsupials were represented in the Pleistocene by still larger beasts. In Australia, at least two dozen taxa exceeded 100 kilograms, including kangaroos, diprotodontids, and bizarre, long-snouted palorchestids, with the carnivorous marsupial "lion" *Thylacoleo carnifex* averaging about 100 kilograms and the rhinoceros-sized *Diprotodon optatum* tipping the scales at a likely 2,800 kilograms (Long et al. 2002; Wroe et al. 2003, 2004*a*).

Living marsupials exploit virtually all terrestrial environments in the Americas and Australasia. In the latter region there are marsupial moles that ascend rarely to the surface and slow-moving cuscuses that spend their lives in trees; there have also been significant radiations of gliding forms. Several species are associated with freshwater habitats, with the yapok *(Chironectes minimus)* of South America being a semi-aquatic specialist. The diets of marsupials reflect the diversity of habitats that they occupy. Thus, many of the forest-dwelling small marsupials of the Americas eat a wide variety of fruits, nectar, invertebrates, and small vertebrates, with some tending toward more specialized frugivory (e.g.,

Eastern gray kangaroo *(Macropus giganteus)*. Photograph by Pavel German/Wildlife Images.

woolly opossums, *Caluromys* spp.), insectivory (e.g., short-tailed opossums, *Monodelphis* spp.), or carnivory (e.g., lutrine opossum, *Lutreolina crassicaudata*) (Emmons and Feer 1997; Vieira and Astúa de Moraes 2003). In contrast, marsupials of the Australasian region include both dietary generalists and specialists, with different species eating from almost any organic material (e.g., bandicoots) to just one or two kinds of prey (e.g., numbats, *Myrmecobius fasciatus,* which specialize on termites; greater gliders, *Petauroides volans,* which eat the leaves of selected *Eucalyptus* trees) (Dickman and Vieira 2005).

Despite the astounding diversity in size, distribution, ecology, and behavior of marsupials, all are united by their distinctive mode of reproduction and by dental, cranial, skeletal, and physiological similarities. The marsupials differ profoundly from other mammals in these respects and seem to have diverged at a very early date from modern placental mammals. Unfortunately, in both the Americas and Australasia, this alternative experiment in mammalian evolution is subject to stresses imposed by the proliferation of human populations and the diversion of resources for human activities; some 99 species and subspecies are threatened, and at least 10 species have become extinct in the last two hundred years (Hilton-Taylor 2000).

Taxonomy and Evolution

Recent studies on the phylogenetic systematics of marsupials have done much to clarify the relationships of both extant and extinct forms (Amrine-Madsen et al. 2003; Cordillo et al. 2004; Horovitz and Sánchez-Villagra 2003; Kirsch et al. 1997; Lapointe and Kirsch 2001; Luo et al. 2002; Nilsson et al. 2003), but several controversial problems remain. At the highest level, for example, Marsupialia has been variously given the rank of order, cohort, supercohort, and even class, with the categories Marsupialia and Metatheria often being used interchangeably (see McKenna and Bell 1997 for a review). Here I follow the emerging convention that Marsupialia occupies at least superordinal rank and encompasses the group consisting of the most recent common ancestor of living marsupials and all its descendants, whereas Metatheria includes both Marsupialia and several basal fossil taxa that appear most closely related to marsupials (Horovitz and Sánchez-Villagra 2003; Rougier et al. 1998). Detailed discussions of some of these marsupial-like forms, including *Deltatheridium, Andinodelphys, Mayulestes,* and *Pucadelphys,* are given by Kielan-Jaworowska and Nessov (1990), Marshall and Muizon (1995), Marshall and Sigogneau-Russell (1995), Muizon et al. (1997), Muizon (1998), and Argot (2002, 2003).

Following the above conventions, extant marsupials are usually grouped into seven orders containing some 19 families and 90 genera (Table 1). A further five orders and 37 families of marsupials and marsupial-like taxa are known from the fossil record (Archer et al. 1999; Kielan-Jaworowska et al. 2004; Long et al. 2002; Luo et al. 2002, 2003; Marshall et al. 1990; McKenna and Bell 1997; Szalay 1994). One formerly diverse lineage, that containing the Tasmanian tiger, or thylacine *(Thylacinus cynocephalus),* became extinct as recently as the last century. The last known individual died at the city zoo in Hobart, Tasmania, in 1936 (Owen 2003).

In addition to contention about the higher-level relationships among marsupials, there is much debate about the origin and early radiation of the group. Some molecular evidence suggests that metatherians diverged from the ancestors of placental mammals in the early Cretaceous, perhaps 120 million years ago (Gemmell and Westerman 1994). Other evidence, based on analyses of multigene sequences and BRCA1 and IGF2 receptor protein sequences, indicates a split as much as 70 million years earlier (Kumar and Hedges 1998; Woodburne et al. 2003). Fossil evidence is similarly contradictory. Based on an ostensibly "primitive" pattern of replacement of cheek teeth in *Slaughteria eruptens,* a small mammal from early Cretaceous deposits in Texas, Kobayashi et al. (2002) proposed that metatherian-eutherian separation could have occurred as recently as 110 million years ago. Very recent finds of clearly differentiated eutherian and metatherian mammals, however, from other early Cretaceous formations, most notably from Liaoning Province in China, confirm a separation date of at least 125 million years ago (Ji et al. 2002; Luo et al. 2003) and thus provide a degree of congruence with estimates from some molecular studies.

Whatever the precise time of divergence of the marsupials, there is a growing consensus that the group emerged on the great northern landmass, Laurasia (Luo et al. 2001, 2002, but see Woodburne et al. 2003 for an alternative view), perhaps somewhere in present-day Asia. Dramatic support for this view has come from the recent discovery of an ancient and apparently basal metatherian from the 125 million-year-old Yixian Formation in China (Luo et al. 2003). This taxon, *Sinodelphys szalayi,* is represented by a superbly preserved skull, skeleton, and even fur. The fossil exhibits structures of the fore and hind feet suggesting that the living animal was an agile climber that would have been at home both on the ground and in trees (Cifelli and Davis 2003). If this interpretation is correct, it supports the early but visionary proposal of Huxley (1880) that the first marsupials were tree-dwellers. At least 20 genera of slightly younger Cretaceous marsupial fossils have been discovered at sites in North America (Kielan-Jaworowska et al. 2004; Wroe and Archer 2004) and indicate a striking and rapid radiation in this part of Laurasia. Unfortunately, many of these early marsupial-like taxa, such as *Kokopellia juddi* and *Holoclemensia texana* (Cifelli and Muizon 1997; Slaughter 1968; Springer et al. 1997), are known only from teeth or mandibular remains and, hence, throw little further light on how early marsupials lived.

Before the break-up of Laurasia, early marsupials had dispersed south across

TABLE 1. DISTRIBUTION AND DIVERSITY OF LIVING MARSUPIALS

Order	Family	Common Names	Distribution	No. of Genera	No. of Species
		Cohort Ameridelphia			
Didelphimorphia[1]	Didelphidae	American, four-eyed, and lutrine opossums; yapok; murine, short-tailed, Patagonian, bushy-tailed, black-shouldered, and woolly opossums	North, South, and Central America	15 (+1)	66 (+5)
Paucituberculata	Caenolestidae	shrew opossums	South America	2	7
		Cohort Australidelphia			
Microbiotheria	Microbiotheriidae	monito del monte	South America	1	1
Dasyuromorphia	Dasyuridae	quolls, Tasmanian devil, dunnarts, antechinuses, planigales, ningauis, phascogales, and many others	Australia and New Guinea	17 (+3)	62 (+9)
	Myrmecobiidae	numbat	Australia	1	1
	Thylacinidae	thylacine, Tasmanian tiger	Australia	1	1

Note: The arrangement of orders follows Wilson and Reeder (1993). Numbers represent the arrangement of genera and species followed in the detailed species and generic accounts in this book. The numbers in parentheses represent additions and deletions following recent taxonomic revisions, including taxa that have been recognized but not necessarily named yet (Blacket et al. 2000; Bowyer et al. 2003; Cooper et al. 2000; Díaz et al. 2002; Dickman et al. 1998; Eldridge et al. 2001; Flannery and Boeadi 1995; Helgen and Flannery 2004; Lemos and Cerqueira 2002; Lindenmayer et al. 2002; Norris and Musser 2001; Van Dyck 2002; Van Dyck and Crowther 2000; Voss and Jansa 2003). All living marsupials are included, as well as 10 species from Australia that have become extinct since European settlement.

[1]Didelphimorphia is sometimes split into Didelphidae and Caluromyidae (Kirsch and Palma 1995) or into these two families plus Marmosidae and Glironiidae (Hershkovitz 1992*b*). Following Gardner (1993), current usage places all living didelphimorphians into one family, Didelphidae.

land bridges into South America, as well as to Europe and other parts of Asia (Archer and Kirsch 2005). Now-extinct marsupials have been described from early Eocene to Miocene deposits in Europe, from Oligocene deposits in central Asia and north Africa, and as recently as the middle Miocene (16.4–11 million years ago) from southern Asia (Szalay 1994). The recent discovery of a putative late Cretaceous marsupial from Madagascar suggests that dispersal to some distant regions might have occurred at a particularly early date (Krause 2001, but see Averianov et al. 2003). Marsupial hegemony in these regions seems to have been limited, however, and dwindled with the ascendancy of placental and other forms of mammals, such as multituberculates. In contrast, dispersal of marsupials to South America had occurred by the late Cretaceous, and diversity in that continent remained high from the middle Paleocene (60 million years ago) until the Great American Biotic Interchange about 3 million years ago. Indeed, during this period South America was effectively isolated from other landmasses, and its marsupial fauna underwent an impressive radiation that produced a myriad of forms (e.g., Flynn and Wyss 1999; Goin 2003; Marshall 1978*a*, 1981, 1982, 1987; Pascual and Carlini 1987). In the Pliocene, however, the Central American land bridge reformed, and the southern continent was once again joined to its northern neighbor.

North America was also an important region of diversification for marsupials, but these mammals did not persist there and became extinct by 20–15 million

TABLE 1 *(continued)*

Order	Family	Common Names	Distribution	No. of Genera	No. of Species
Peramelemorphia	Peramelidae[2]	bandicoots, bilbies, pig-footed bandicoot	Australia and New Guinea	4	11 (−1)
	Peroryctidae	spiny bandicoots	Australia, New Guinea, Seram Island	4	11
Notoryctemorphia	Notoryctidae	marsupial moles	Australia	1	1 (+1)
Diprotodontia	Phalangeridae	cuscuses, brush-tailed and scaly-tailed possums	Australia and New Guinea	6	22 (+2)
	Burramyidae	pygmy possums	Australia and New Guinea	2	5
	Acrobatidae	feathertail glider, feathertail possum	Australia and New Guinea	2	2
	Pseudocheiridae	ringtail possums	Australia and New Guinea	6	16 (+1)
	Petauridae	gliders and trioks	Australia and New Guinea	3	10 (+1)
	Macropodidae	kangaroos and wallabies	Australia and New Guinea	12	61 (+1)
	Potoroidae[3]	rat-kangaroos, musky rat-kangaroo	Australia	5	10 (+2)
	Phascolarctidae	koala	Australia	1	1
	Vombatidae	wombats	Australia	2	3
	Tarsipedidae	honey possum	Australia	1	1

[2]This family is split into Peramelidae and Thylacomyidae by many authorities, the latter family including two species of *Macrotis* and monotypic *Chaeropus* (Westerman et al. 1999).
[3]This family is now usually split into Potoroidae and Hypsiprymnodontidae (Archer et al. 1999; Burk et al. 1998), the latter family containing the monotypic *Hypsiprymnodon* and several fossil taxa.

years ago. After establishment of the land bridge, South America was invaded by a host of placental mammals, including raccoons, bears, and cats. The invasion coincided with a sharp reduction in marsupial diversity; the loss of larger South American marsupial carnivores, such as borhyaenids and thylacosmilids, was particularly severe, and the legacy is a marsupial fauna now dominated by small (<5 kg), predominantly omnivorous forms (Dickman and Vieira 2005; Van Valkenburgh 1999; Wroe et al. 2004*b*). In contrast to the surge of invaders moving south, the only marsupial successfully to colonize North America above Mexico is the Virginia opossum *(Didelphis virginiana)*, which probably advanced north within the last million years. There is, however, a rich marsupial fauna of at least 13 species in southeastern Mexico and on the Central American isthmus, including four taxa that occur nowhere else (Gardner 1993; Reid 1997; Wainwright 2002).

As the fortunes of marsupials waxed and waned in most parts of the world after their exodus from Asia and North America, immigrants to eastern regions of the southern supercontinent, Gondwana, fared quite differently. In particular, the marsupials that reached the Australian part of Gondwana subsequently underwent a spectacular radiation comprising seven orders and at least 31 families (Archer 1984*a*, 1984*b*; Long et al. 2002). There is some agreement that marsupials first entered Australia in the Paleocene around 65–60 million years ago, when marsupial diversity in South America was high (Archer et al. 1999; Godthelp et al. 1999). The earliest marsupial fossils discovered to date in Australia are of late Paleocene–early Eocene age (55–50 million years old) from Murgon in south-eastern Queensland, and some of these have affinities with South American taxa (Archer et al. 1998, 1999).

Reconstructions of major geological events also provide support for a South American point of entry for Australian marsupials (Cox 2000). During the Paleocene, Gondwana comprised South America, Antarctica, and Australia and thus provided potentially continuous land for dispersal. This great landmass began to break up in the Eocene, with the final separation of Australia from Antarctica taking place between 45 and 38 million years ago and the sundering of Antarctica and South America 5 million years later. The climate of the early Eocene was warm and humid and favored the development of extensive forests of southern beech (*Nothofagus* spp.) over much of the Antarctic (Arroyo et al. 1996; Woodburne and Case 1996). Confirmation that conditions were indeed conducive to dispersal came in 1982, with the dramatic recovery of the fossilized remains of a now-extinct South American polydolopid marsupial in deposits of middle Eocene age from Seymour Island off the northern Antarctic Peninsula (Woodburne and Zinsmeister 1982, 1984). This and subsequent discoveries of four further species (Case et al. 1988; Goin and Carlini 1995; Goin et al. 1999) provide unambiguous confirmation of marsupial presence on the ancient Gondwanan land bridge.

Despite evidence of marsupial movement across the southern continents, there is deep and continuing debate about the origin and constitution of the Australian marsupial radiation and its relationship to the American marsupials. For example, ground-breaking work by Szalay (1982) showed that, on the basis of detailed similarities in the anatomy and structure of the ankle bones, the extant South American microbiotheriid marsupial *Dromiciops gliroides*, or monito del monte, has much stronger affinities with Australian than with American marsupials. Szalay suggested further that *Dromiciops* was, in fact, basal to the Aus-

White-eared opossum *(Didelphis albiventris).* Photograph by Hugh Tyndale-Biscoe.

tralian radiation, and this hypothesis has received some support from comparative analyses of phosphoglycerate kinase DNA sequences (Colgan 1999) and of the 12S rRNA mitochondrial gene (Palma and Spotorno 1999). If recent discoveries of putative microbiotheriid fossils from late Paleocene deposits at Murgon are confirmed (Archer et al. 1999), this would establish the intercontinental distribution of this group and lend support to the idea that microbiotheriids provided the stock for Australia's extant marsupials. Other research suggests, however, closer affinities between *Dromiciops* and particular orders of Australian marsupials (e.g., Diprotodontia, Burk et al. 1999 and Kirsch et al. 1991, 1997; Dasyuromorphia, Szalay 1994; Notoryctemorphia and Dasyuromorphia, Retief et al. 1995). These latter studies imply that Australian marsupials were founded by multiple dispersal events or even that there was some reverse dispersal from Australia to South America.

An alternative hypothesis advanced by Hershkovitz (1992*a*) holds that *Dromiciops* is the last survivor of a separate cohort, the Microbiotheriomorphia, that is basal to all other marsupials. Certainly, *Dromiciops* has characters, such as a basicaudal cloaca, that are suggestive of an early divergence from other marsupials, as well as cranial and dental characters associated with placental mammals (Hershkovitz 1999). Detailed morphological and molecular studies, however, have found little support for Hershkovitz's proposal (Horovitz and Sánchez-Villagra 2003; Springer et al. 1998), and most authors agree that *Dromiciops* is nested within the Australian radiation (see Amrine-Madsen et al. 2003, Asher et al. 2004, and Palma 2003 for reviews).

A final complication in resolving the early zoogeography of marsupials has been the recent discovery of dental material from a small, very generalized marsupial *(Djarthia murgonensis)* from the late Paleocene–early Eocene of Australia (Godthelp et al. 1999; Wroe et al. 2000). This taxon has so few derived features that it cannot yet be placed with confidence in either the American or the Australasian faunas, and it raises the intriguing possibility that marsupial diversity during the earliest phases of the Australasian radiation was far greater than expected.

Except for the position of *Dromiciops,* the interordinal relationships of marsupials remain generally uncertain. Szalay (1982) produced some stability by including the order Microbiotheria (as order Dromiciopsia) and extant orders confined to the Australasian region within the cohort Australidelphia and by placing American marsupials within the cohort Ameridelphia. This concept was later expanded by others, most notably Woodburne (1984) and Aplin and Archer (1987) (Table 1). Subsequent studies, however, have found contradictory evidence in support of Ameridelphia. For example, although ameridelphians are unique in showing sperm-pairing in the epididymis (Biggers and DeLamater 1965), molecular comparisons fail to show a close relationship between the orders (Burk et al. 1999), and there is recent evidence that sperm-pairing evolved independently in each of the two ameridelphian orders (Jansa and Voss 2000). Surprisingly, morphological comparisons place paucituberculates closer to Australian marsupials than to their putative sister group, Didelphimorphia, within Ameridelphia (Horovitz and Sánchez-Villagra 2003).

Interordinal relationships within the Australidelphia are similarly ambiguous. Longstanding conundrums such as the affinities of the marsupial moles (order Notoryctemorphia) appear no closer to resolution, despite much recent research; moles group with Dasyuromorphia in most molecular comparisons but with Peramelemorphia in morphological analyses (Amrine-Madsen et al. 2003; Horovitz and Sánchez-Villagra 2003). In addition, new conundrums are emerging. Perhaps the most intriguing is the possibility that peramelemorphians have evolved separately from all other Australasian marsupials and are close to ameridelphians (Burk et al. 1999). This is supported by some molecular and biochemical evidence (Asher et al. 2004; Kirsch et al. 1997; Springer et al. 1994; but cf. Nilsson et al. 2003; Phillips et al. 2001). In addition, a putative bandicoot hails from the late Paleocene–early Eocene Tingamarra Local Fauna in Queensland; as noted by Long et al. (2002), this predates the final separation of Australia from Antarctica, making it possible that the group arose elsewhere in Gondwana before entering Australia. Resolution of such puzzles awaits the discovery of further fossil material, especially from the late Cretaceous to Oligocene in Australia, the integration of larger molecular and morphological character sets, and, perhaps, the discovery of new investigative techniques.

Irrespective of the affinities of the Australidelphia, it is clear that representatives of at least three orders within the cohort (Dasyuromorphia, Peramelemorphia, and Diprotodontia) escaped the confines of Australia to occupy New Guinea and many surrounding islands. Several lines of evidence, derived from studies of fossils, molecular phylogenies, and plate tectonics, combine to produce a plausible scenario of these colonization events.

In the first instance, although there is a 30-million-year "dark age" up to the latest Oligocene that has frustratingly yielded no fossils, by the early Miocene rich fossil deposits in Queensland, South Australia, and Tasmania show that marsupials were astonishingly diverse and represented by carnivorous, insectivorous, omnivorous, grazing, and browsing forms (Archer et al. 1999; Rich 1991). During the Oligocene, orogenic activity at the northern edge of the Australian continental plate resulted in the emergence of a landmass across northern Australia that was similar in extent to the present island of New Guinea (Dow 1977; Dow and Ṣukamto 1984). During "greenhouse" conditions in the early Miocene some 23–20 million years ago, land with continuous forest cover probably allowed the ancestors of New Guinea's living old endemic marsupials to colonize from the south (Flannery 1989). This was a vicariant event in that the land connection was broken abruptly later in the Miocene by marine flooding, stranding marsupials on the young New Guinean landmass and on a series of emergent islands. It is not clear how often land bridges reformed until rapid eustatic fluctuations in sea level began some 2.5 million years ago. However, construction of molecular "clocks" suggests that episodes of overland dispersal (or possibly waif dispersal) took place around 12–8 million and 4.7–2.7 million years ago (Aplin et al. 1993; Kirsch and Springer 1993; Krajewski et al. 1993). According to this scenario, the earliest colonists of New Guinea included bandicoots and representatives of two or three families of diprotodontians (Acrobatidae, Phalangeridae, and possibly Petauridae). Later colonists included dasyurids, pseudocheirids, and macropodids, with further representatives of the two latter families, plus more dasyurids, arriving in the last dispersal episode.

After the early Miocene dispersal event, conditions in Australia began to cool and dry out (Martin 1998; McGowran and Li 1994), and the formerly rich continental fauna of arboreal marsupials began a slow but inexorable decline in diversity and distribution. This period, beginning about 15–16 million years ago, saw the demise of several marsupial families, including the enigmatic miralinids, wynyardiids, and yingabalanarids (Archer et al. 1999). Dry conditions intensified over the next 10 million years, eventually favoring vast areas of grassland and savanna woodland that allowed the explosive radiation of now-familiar grazers such as kangaroos and wombats. Despite the severity and length of this "icehouse" period and even more intense drying during the Pleistocene, pockets of wet forest have persisted in northeastern Queensland and throughout New Guinea. Archer (1981) has previously alluded to the plesiomorphic, or archaic, nature of many living New Guinean marsupials, such as forest wallabies of the genus *Dorcopsis* and mouse bandicoots of the genus *Microperoryctes,* and noted also the striking similarity that many show to now-extinct taxa from the Miocene of central Australia. Such forest specialists may thus provide an intriguing glimpse of the Australian Miocene environment, when marsupial diversity approximated or even exceeded that of mammals in the present-day rainforests of Amazonia and Borneo (Archer et al. 1999).

Distinctive Characteristics

MORPHOLOGY

Marsupials differ from all other mammals in several aspects of their development, as well as in the structure of their teeth, skulls, skeletons, and soft tissues. The distinctive nature of the hard body parts is crucial in identifying marsupials in fossil deposits and in inferring the size, diet, degree of arboreality, and general lifestyle of their owners. The morphology of marsupials has been the subject of considerable study, and the following account is summarized largely from the works of Pocock (1921, 1926), Wood Jones (1923–24, 1949), Grassé (1955), Barbour (1977), and Dawson et al. (1989). Further accounts of features that distinguish marsupials from other mammals have been provided by Archer (1984*a*), Sánchez-Villagra (2002), Smith (1997), and Wroe and Archer (2004).

With respect to dentition, marsupials show a unique pattern of tooth eruption and replacement, with just the third upper and lower premolars being replaced by secondary teeth. Archer (1974, 1978) argued that the replacement teeth are actually the first molars, but this view has been abandoned following Luckett (1993). Except for wombats (family Vombatidae), which have just 24 teeth, most marsupials have 40 to 50 teeth, and some individual numbats *(Myrmecobius fasciatus)* have up to 52. The basic dental formula is I 5/4, C 1/1, PM 3/3, and M 4/4, as exemplified by unspecialized members of the Didelphimorphia. There is considerable variation, however, with paucituberculates having four upper and three or four lower incisors in each half of the jaw and no extant Australian marsupials having more than three lower incisors.

The presence of multiple incisors is termed *polyprotodonty* and is a characteristic of omnivorous, insectivorous, and carnivorous species. In contrast, diprotodontian marsupials have one to three upper incisors but only one lower incisor in each half of the jaw. The diprotodont condition is characteristic of grazers and browsers, such as wombats and kangaroos and wallabies (family Macropodidae). In these animals, the sharp, forward-pointing lower incisors oppose their smaller, upper counterparts and allow delicate selection and snipping of plant leaves. Other diprotodontians include possums and gliders, which eat nectar, sap, other plant exudates, and insects, as well as the honey possum *(Tarsipes rostratus),* which specializes on a diet of nectar and pollen (Bradshaw and Bradshaw 2001; Richardson et al. 1986). Extinct diprotodontians such as propleopine kangaroos probably also included some meat in the diet (Ride et al. 1997), while the formidable marsupial "lion" *Thylacoleo carnifex* was almost certainly a pure carnivore (Wroe 2003).

Canine teeth are well developed in all except extant diprotodontian marsupials, in which the lower and often upper canines are lost. In more carnivorous forms there is a tendency for the cheek teeth to be narrow and aligned for cutting, whereas in more herbivorous taxa the molars are flattened and provide a broad, ridged surface that allows efficient crushing and grinding of plant material. In some macropodids, an extraordinary adaptation known as molar progression has been demonstrated. As the early erupting anterior cheek teeth become worn down, they are shed; fresh teeth erupt at the back of the toothrow and the whole row moves forward. This phenomenon, known since the nineteenth century (Thomas 1888), has been demonstrated primarily in species that eat abrasive foods such as grass (Lentle et al. 2003*a*, 2003*b*; Ride 1957; Sanson 1980). Wombats and koalas meet the problem of tooth wear differently. The teeth of wombats are open-rooted and grow continuously through life to provide a constant grinding surface. In contrast, koalas compensate for tooth wear by eating more and by increasing the digestibility of food by regurgitating and remasticating it (Logan 2003).

Marsupial skulls can be distinguished from those of placental mammals by several features in addition to the unique pattern of tooth replacement. The nasal bones are large and expand posteriorly; the auditory bullae, when present, are formed principally from the alisphenoid; the lachrymal bone extends beyond the orbit and onto the face; the jugal bone extends back along the zygomatic arch to the glenoid articulation of the jaw; and there are usually at least two pairs of openings, or fenestrae, in the palate. Except in *Tarsipes rostratus* and the koala *(Phascolarctos cinereus)*, the dentary or lower jaw bone also is directed inward to form an internal angular process (Abbie 1939; Russell and Renfree 1989). This inward inflection allows the insertion of the internal pterygoid muscle and is particularly obvious in terrestrial browsers and grazers that need to generate large bite forces to process their food (Sánchez-Villagra and Smith 1997).

Relative to similar-sized and ecologically equivalent placental mammals, the facial area of the skull of marsupials is large, but the cranial cavity is small. The cerebral hemispheres of the brain tend to have simple convolutions and lack the connection provided by the corpus callosum that is a characteristic of placental brains. The functional significance of these differences remains unclear. A pervasive stereotype is that marsupials are less intelligent than placental mammals because they do not have highly evolved social systems or exhibit complex behaviors such as play, group cohesion, advanced antipredator tactics, or highly developed modes of communication (see Vaughan et al. 2000 for a useful summary). Several recent studies, however, have uncovered considerable complexity in marsupial behavior (e.g., Byers 1999; Charles-Dominique 1983; Colagross and Cockburn 1993; Croft and Eisenberg 2005; Croft and Ganslosser 1996; Fadem 1986) as well as in learning ability (Griffin et al. 2000; McLean et al. 2000). The development of different brain components also is associated strongly with ecology and sociality (Lapointe and Legendre 1996). Eisenberg and Wilson (1981) showed, furthermore, that the cranial capacities of many didelphids are similar to those of placentals of equivalent size.

It has long been known that marsupials are well endowed with specialized skin glands and that olfaction is a major channel of communication (Ewer 1968; Russell 1985). In addition, all marsupials studied to date possess a vomeronasal organ (Sánchez-Villagra 2001), which probably allows the interpretation of pheromonal signals (Jackson and Harder 1996). In light of this research, there are intriguing suggestions that marsupials make particularly extensive use of chemical communication (Toftegaard and Bradley 2003). If this is so, it is possible that the true complexity of marsupial behavior remains to be discovered.

Behind the skull, the arrangement of bones in the marsupial skeleton follows the basic mammalian pattern; however, two features are worthy of note. First, with the exceptions of *Thylacinus cynocephalus*, in which they are present as cartilaginous remnants, and notoryctids, in which they are small and not ossified (Warburton et al. 2003), all living marsupials bear forward-projecting epipubic bones on the pelvic girdle. Epipubic bones were found also in now-extinct multituberculate mammals and two early placentals (Kielan-Jaworowska 1975; Novacek et al. 1997); they are present in modern monotremes but are otherwise characteristic of marsupials. These bones have long been considered as important sites of attachment for abdominal muscles that support the testes and pouch. Recent research suggests, however, that they form part of a complex muscular linkage between the pelvis and hind limbs that stiffens the trunk of the body during movement (Reilly and White 2003).

Second, while many marsupials retain five digits on the hind feet and may use the first digit as an opposable big toe to assist climbing (e.g., *Didelphis, Acrobates, Trichosurus* spp.), foot and toe structure are highly variable. In many small, terrestrial marsupials, such as some dasyurids and peramelemorphians, the first toe is greatly reduced, while in kangaroos it is entirely absent. The second and third toes of peramelemorphians and diprotodontians are slender and bound closely together by an envelope of skin, with only the claws protruding. This condition, termed *syndactyly* (cf. didactyly, where the second and third toes are separate), provides a fur-grooming comb for the owner. In the peramelemorphs, kangaroos, and rat-kangaroos, the fourth toe is enlarged and elongated and forms the major axis of the foot. These adaptations reduce the lateral mobility of the foot but serve to increase forward speed, most notably in the hopping motion of kangaroos. Modifications to the front foot are less remarkable, although the third and fourth fingers of *Notoryctes* spp. are greatly expanded and equipped with stout claws for digging, while the fourth finger of the striped possum and trioks (*Dactylopsila* spp.) is slender, elongated, and used for winkling larvae from bark and rotten wood.

REPRODUCTION

Of all the features that separate marsupials from other mammals, the most obvious is the mode of reproduction. Except when growth of the embryo is arrested during diapause (see below), marsupials generally have short gestations of only 9 to 42 days and give birth to poorly developed young that weigh less than 1 gram. In some of the smallest marsupials, such as *Planigale* spp. (adults weigh 4.5–20 grams, depending on the species), females suckle litters of 8 to 12 young, with individual newborns weighing just 5 milligrams (Tyndale-Biscoe and Renfree 1987). Marsupial litters at birth weigh less than 1 percent of the mass of the mother; young then experience long but variable periods of growth that depend on provision of milk from the female.

In diprotodontians, peramelemorphians, some didelphids, and *Dromiciops gliroides*, much growth takes place while the young are protected in the mother's pouch. This part of the reproductive cycle is so conspicuous that it was remarked upon by the first European observers of marsupials, including Queen Isabella of Spain; the word *marsupial* is in fact derived from the Latin word *marsupium*, meaning purse or pouch. In many marsupials the pouch opens anteriorly, but in *Chironectes minimus*, peramelemorphians, notoryctids, wombats, and koalas it opens toward the rear (Marshall et al. 1990). In burrowers this orientation presumably reduces the amount of soil that gets into the pouch, but in the arboreal koala this explanation clearly does not hold. If it is not selectively disadvantageous, it is possible that the koala's posterior-opening pouch simply reflects shared ancestry with wombats (Murray 1998). Alternatively, rearward pouch orientation may allow the young access to the mother's feces, hence facilitating inoculation of the gut with microbes that will be needed later for digestion of leaf material (M. Renfree, in Springer et al. 1997).

Despite their name, not all marsupials have pouches. Caenolestids, some didelphids, most dasyurids, and the numbat lack obvious pouch structures (Marshall et al. 1990; Woolley 1974, 2003), instead having rings of muscle in the skin surrounding the nipples that contract to provide temporary cover when the young are present. Young fuse to the mother's nipples and remain firmly attached for 4 to 10 weeks but are then left alone in a nest for several more weeks until weaned (Lee et al. 1982). Litters of some small insect-eating and omnivorous marsupials, such as dunnarts (*Sminthopsis* spp.) and antechinuses (*Antechinus* spp.) contain 8 to 12 or more young; by the time they are weaned, their combined mass can be more than three times that of their mother (Russell 1982). This is an extraordinary investment of energy by females at this time and is often followed by high mortality of mothers whose body reserves have been depleted (Dickman 1986*a*; Wood 1970).

In large kangaroos, single young are carried in the pouch for more than 230 days and then begin to venture out progressively until weaned some five to eight months later (Frith and Calaby 1969). Remarkably, as soon as the firstborn begins to stray from the pouch, its mother will often give birth to another young that has been held for several months in an inactive state in her body, and she will then suckle the two different-aged young until the older offspring, or joey, is weaned. During this time, the mother simultaneously produces two different kinds of milk from her teats: dilute milk containing mostly carbohydrates and a little protein nourishes the newborn young, whereas fat-laden milk is provided for the older sibling (Green 1984; Green and Merchant 1988). These marsupials are unique among mammals in being able to adjust the content of their milk so exquisitely to the needs of the young (Krockenberger 2005).

The phenomenon of diapause, alluded to above, is similar to the process of delayed implantation in some placental mammals and occurs in all kangaroos studied thus far (except the western gray kangaroo, *Macropus fuliginosus*), potoroids, *Tarsipes rostratus*, acrobatids, and some burramyids (Renfree 1981; Shaw 2005). Except in the anomalous swamp wallaby *(Wallabia bicolor)*, in which the gestation period lasts longer than the estrous cycle (Calaby and Poole 1971), pregnancy in these marsupials coincides roughly with the length of the estrous cycle but does not affect it. Thus, at about the same time that the female gives birth, she also becomes receptive and mates. The resultant embryo develops to an early stage, the blastocyst, and then enters a quiescent state termed *embryonic diapause*. In macropodids the process of suckling suppresses growth of the blastocyst. With the diminishment of this stimulus as the pouch young begins to take solid food, the blastocyst resumes development and is born usually within four weeks of permanent pouch vacation or loss of the older young (Tyndale-Biscoe and Renfree 1987).

The period of quiescence of the blastocyst varies within and between species. For example, in seasonally breeding populations of the Tasmanian red-necked wallaby *(Macropus rufogriseus)*, diapause may extend for 8 to 12 months depending on the survival of the existing pouch young (Merchant and Calaby 1981). In *Tarsipes rostratus*, by contrast, suckling does not suppress growth of the blastocyst as completely as in macropodids, and termination of suckling does not always stimulate embryonic development (Renfree et al. 1984). Environmental factors such as food or time of year may instead influence the resumption of development (Russell and Renfree 1989), and as a consequence diapause dur-

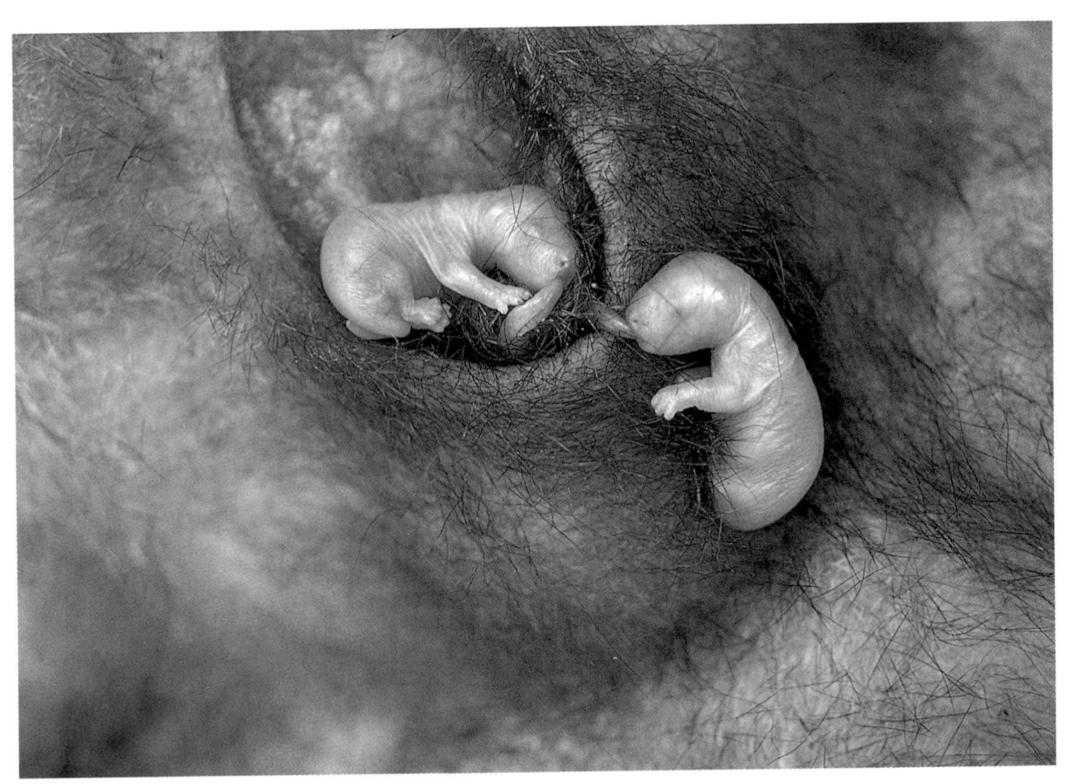

Neonates feeding in pouch. Note nipples connected to mouths. Photograph by Hugh Tyndale-Biscoe.

ing gestation in *T. rostratus* may vary from three to five months (Bradshaw et al. 2000). It is likely that diapause was initially advantageous in preventing a second young from being born when the pouch was already occupied. It also, however, allows rapid replacement of lost young when conditions improve, even in the absence of males.

Because marsupials produce such tiny neonates, it has often been assumed that there is no opportunity for embryos to be nourished via a maternal placenta, as occurs in all placental mammals. This supposed difference between these two great groups of mammals has often been used as a point of demarcation between them. Marsupials do, however, develop a functional and well-vascularized yolk-sac placenta (Dawson et al. 1989). For the short time that the embryo is in the uterus, it is nourished by transfer of nutrients from inside the uterus across the yolk sac that is in loose contact with the uterine wall (Hughes et al. 1990; Krause 1998). In the Phascolarctidae, Vombatidae, and especially the peramelemorphians, a second, chorioallantoic placenta allows much more intimate exchange between the embryo and the circulation of the mother. This placenta resembles that of the "true" placental mammals in both structure and function (Sharman 1965).

The small size of marsupials at birth means that they are quite undeveloped and have limited sensory and motor capabilities. The hind limbs are little more than undifferentiated buds, the lungs lack alveoli, and the ventricles of the heart are not yet separate; overall organ development may not be complete until 30 days after birth (Nelson et al. 2003). Thus, the question of how newborn young get to the pouch has been an enduring mystery. Early naturalists assumed that the young must grow directly out of the nipples, were sneezed into the pouch after nasal copulation (O'Connell 2001), or, if born via the birth canal, became glued to the pouch area by sticky secretions in the milk. Direct observations and detailed neurological studies in recent years have dispelled these fanciful notions, but the passage of the young from the cloaca to the pouch remains an extraordinary journey.

Newborn marsupials have relatively well-developed heads and forelimbs and move toward the pouch using swimming movements of the arms (Hughes and Hall 1988; Pflieger and Cabana 1996). In several species of kangaroos, possums, and opossums, the mother sits upright with her legs extended forward during and after birth; if she changes her orientation at this critical time, the young continue to move upward even if the pouch position has changed. This suggests that the newborn possesses a gravitational (or antigravitational) sense well before smell, vision, or other sensory capabilities develop (Cannon et al. 1976; but see also Gemmell and Rose 1989; Reynolds 1952). In some small marsupials such as *Antechinus* spp., females stand on all four legs with their hips raised at birth, and young climb rapidly down to the teats. Many small marsupials produce two to three times more young at birth than can be accommodated on the available nipples. Tiny neonates are energetically cheap to produce, and it is presumably advantageous to mothers to overproduce young at this stage to ensure that all teats become occupied.

Like other mammals, marsupials have two ovaries and two oviducts. The uteri are separate, however, unlike the single, fused uterus of most placental mammals, and the vagina comprises a single central and two lateral canals. The lateral vaginae serve to transport sperm and are matched in many male marsupials by a penis that has a bilobed tip (Nogueira et al. 1999; Woolley 1982). The two canals are curved and allow the urinary ducts to pass between them. The central, or median, vagina is the birth canal. In Didelphimorphia, Dasyuromorphia, Peramelemorphia, and many diprotodontians, this canal is transitory and forms in the mass of connective tissue between the ureters before each birth event. After birth, it closes, reforming just prior to each subsequent parturition (Tyndale-Biscoe 1966). In the Macropodidae and Tarsipedidae, by contrast, the median vagina becomes lined with epithelium and remains open after the first birth (Sharman 1970). The vaginal apparatus in all species terminates with the ureters in a urogenital sinus; together with the hind gut, this opens caudally into the urogenital opening, or cloaca. Male marsupials differ from their placental counterparts in that the scrotum is situated anterior to the penis (Tyndale-Biscoe 1973). As in females, there is a single cloacal opening that receives the products of the urinary and alimentary systems; the penis also is extruded through the cloaca during copulation.

One final point of interest is the possession, in some didelphids, of bony tubercles on the wrists that probably allow males to maintain a secure grip on females during copulation (Lunde and Schutt 1999). In Robinson's mouse opossum *(Marmosa robinsoni)*, for example, copulation takes place when the pair is suspended precariously above ground, secured only by the prehensile tail of the male (Barnes and Barthold 1969); immobilization of the female in this situation is presumably advantageous. Tubercles have been found in 14 of 16 didelphid species studied thus far (Lunde and Schutt 1999). They are present only in males and represent the only sexually dimorphic limb bone structures known in mammals.

Distribution and Patterns of Diversity

THE AMERICAS

The American opossums range from southwestern Canada at 49° 41′ N to 47° 6′ S in Argentina and from 0 to 4,500 meters in altitude; they are also represented on some islands of the Lesser Antilles (Brown 2004; Eisenberg 1989; Nagorsen 1996; Redford and Eisenberg 1992). At both ends of their latitudinal distribution, marsupials are represented by only one species; in the north it is the Virginia opossum *(Didelphis virginiana)* and in the south it is the Patagonian opossum *(Lestodelphys halli)*. When European settlers first arrived in North America, *D. virginiana* apparently extended no farther north than the modern states of Ohio and Virginia. After land modification by the early settlers, this opossum moved north to the Great Lakes. Its northward dispersal on the Pacific coast to southern Canada was assisted by changes in land use and also by introductions to California in 1890 (O'Connell 2001). It lays down fat throughout the body before the cold northern winter and may spend several days in a nest during particularly harsh periods. In contrast to its northern relative, *L. halli* seems to occupy restricted areas of pampas or patches of other mesic vegetation and may be declining. This species has strong feet with relatively long, stout claws and may escape seasonal cold conditions by burrowing (O'Connell 2001). The tail can store fat, which acts as a temporary food store during severe conditions.

Opossum diversity increases at lower latitudes in the Americas and peaks in the temperate and tropical forests (Kaufman 1995). The richest latitudinal zone, containing 37 species, lies between 10° and 20° S, with 28 and 29 species occurring in the 5° bands south and north of this, respectively (Birney and Monjeau 2003; Brown 2004). Species diversity correlates strongly and positively with mean annual temperature but negatively with the annual temperature range (Birney and Monjeau 2003), indicating that opossums occur preferentially in consistently warm forest conditions (Voss and Emmons 1996; Voss et al. 2001). Detailed analyses of nine genera of rainforest didelphids by Patton and Costa (2003) confirmed the long-term nature of this preference. These authors documented large differences between species in DNA sequences from the mitochondrial cytochrome *b* gene, indicating that most species are "old" and have occupied stable, geographically separate, closed-canopy forests for long periods.

Although many American opossums are arboreal and obtain food and shelter from the forest canopy, most also spend time on the ground. Some of these species, such as gracile and slender mouse opossums (*Gracilinanus* spp. and *Marmosops* spp., respectively), are small and highly agile with long, prehensile tails that assist climbing (Emmons and Feer 1997; Voss et al. 2001). Although the precise habitat requirements have been elucidated for relatively few of these species, some are restricted to small areas of forest that provide particular resources. For example, the Andean slender mouse opossum *(Marmosops impavidus)* occupies moist evergreen forest at elevations above 1,400 meters, while the São Paulo slender mouse opossum *(M. paulensis)* occurs only in montane and cloud forests in the coastal mountains of southeastern Brazil above 800 meters (Gardner 1993).

Among the larger forest dwellers, all members of the Caluromyinae are primarily arboreal, with the black-shouldered opossum *(Caluromysiops irrupta)* seldom descending below the subcanopy (Emmons and Feer 1997; Fonseca et al. 1996; Hunsaker 1977). In contrast, four-eyed opossums *(Metachirus nudicaudatus, Philander andersoni,* and *Philander opossum)* are largely terrestrial, obtaining food resources from the ground and sheltering in logs or the lower branches of trees (Handley 1976; Miles et al. 1981).

Outside the tropical forests, several species of southern mouse opossums (genus *Thylamys*) occur in open and semi-arid areas on the eastern and western slopes of the Andes, with *T. pusillus* extending east into the chaco and other dryland habitats of southern Bolivia, Paraguay, and Argentina (Solari 2003) and *T. macrurus* occupying moist forest east of the Paraguay River (Palma 1995). Common shrew opossums (*Caenolestes* spp.) also have Andean distributions, occupying dense scrub or woodland near meadows in the high Andean paramo (Albuja and Patterson 1996; Kirsch and Waller 1979; Osgood 1921). Finally, with the possible exception of *Didelphis marsupialis,* which has been reported to prefer moist, broad-leaved forest (Cerqueira 1985, but see Pérez-Hernández 1989), other members of this genus occupy a broad range of habitats. *Didelphis virginiana* uses timbered and open country and is conspicuous in disturbed suburban environments (Hall 1981).

Some recent studies have recognized that species diversity in several genera of opossums is higher than current taxonomy would suggest and that this will affect our understanding and interpretation of species' distributions (Mares and Braun 2000; Patterson 2000; Patton et al. 2000; Patton and Costa 2003; Solari 2003). A useful synopsis of major patterns of distribution and diversity is provided by Birney and Monjeau (2003). These authors recognized five "life form" groups of opossums.

Group 1: Small (<100 grams), arboreal species with long prehensile tails

These are mostly frugivores or insectivore-omnivores that occupy tropical rainforest with high minimum temperatures and predictable rainfall. The group includes representatives of *Gracilinanus, Marmosa, Marmosops, Micoureus,* and probably *Hyladelphys kalinowskii* (formerly *Gracilinanus kalinowskii* but included in the new genus *Hyladelphys* by Voss et al. 2001) (Vieira and Monteiro-Filho 2003; A. Monjeau, personal communication, 2004). The distribution of these agile, climbing species seems to be related to the distributional range of lianas that facilitate access to the upper strata and canopy of the forest.

Group 2: Small (<100 grams), terrestrial species with short prehensile tails

These opossums are more insectivorous and less influenced by annual variability in temperature and rainfall than members of group 1. They occur in both forested and open habitats and obtain resources chiefly from ground level. This group comprises *Monodelphis* spp.

Group 3: Medium-sized (100–500 grams) species with prehensile tails

Included in this group are omnivores such as *Caluromys, Caluromysiops, Glironia,* and *Philander* spp. These species tend to be scansorial or arboreal; some are restricted to dense forest, while others occupy a wide range of habitat types. They are less affected by temperature than are members of group 1 but are limited by cold at higher latitudes.

Group 4: Large (>500 grams) species with nonprehensile tails

As titled, this group would include only the semi-aquatic *Chironectes minimus* and *Lutreolina crassicaudata,* which have extensive ranges in South America. The tails of both species are weakly prehensile but too thick to assist in climbing or carrying nest material. Ecologically, *Metachirus nudicaudatus* may also be placed in this group (Vieira and Monteiro-Filho 2003), although only a few individuals exceed 500 grams. The other large opossums are the species of *Didelphis,* which do have prehensile tails. These species are among the most opportunistic of the American marsupials, taking a very broad range of foods and occupying most terrestrial habitats over a large area. Although the distributions of *D. al-*

biventris, D. aurita, and *D. marsupialis* encompass tropical latitudes, *D. virginiana* tolerates a wide thermal range and is apparently less constrained by climate than is any other American marsupial.

Group 5: Small species (<100 grams) with incrassated tails

Species in this group are among the most insectivorous and carnivorous of the opossums. Their small body size and ability to store fat in the tail allow them to exploit harsh, seasonal environments such as the colder, more southerly latitudes of South America, as well as arid puna, desert, and high cloud forest. This group includes *Dromiciops gliroides, Lestodelphys halli,* both species of *Rhyncholestes,* and most *Thylamys.* The enigmatic common shrew opossums (*Caenolestes* spp.), which occur at altitudes of 1,500 to 4,000 meters in the Andes, may also be placed in group 5, although tail incrassation is slight or absent.

South American marsupials are seen occasionally in North America and even Europe, the victims of accidental shipment in cargoes of bananas and other exported fruits. No species are known to have established feral populations outside the Americas.

AUSTRALIA

Australian marsupials occur in all parts of the continent and on many offshore islands, with two species (dusky antechinus, *Antechinus swainsonii,* and mountain pygmy-possum, *Burramys parvus*) occurring at altitudes up to 2,227 meters (Mt. Kosciuszko, the highest peak) in the Australian Alps. The distributions of individual species are related to different suites of factors, but, as with American marsupials, a primary influence on general patterns of diversity is habitat. In the Dasyuridae, for example, the largest family of Australian marsupials with 53 species, diversity is least in woodland and heath (usually only one or two species coexist) and highest in arid regions dominated by hummock grasses and other desert-complex vegetation (up to eight species overlap). For these insectivorous and carnivorous species, the structural complexity of the vegetation and nonliving components of the environment are critical determinants of diversity (Dickman 1989, 2003). Complex habitats allow species access to different niches for foraging, and this, in turn, reduces overlap in their diets and permits more species to coexist.

Habitat complexity also seems to determine the numbers of bandicoot species (family Peramelidae) that occur together (Dickman 1984). In contrast to the dasyurids, however, bandicoots are now confined largely to heathland and wooded habitats on the continental periphery since the regional extinction of four peramelid species from the arid zone after European settlement (Baynes and Johnson 1996).

Possums, gliders, the koala *(Phascolarctos cinereus),* and other arboreal marsupials such as the tuan *(Phascogale tapoatafa)* and antechinuses occur only in forest or woodland habitats where they can access food and shelter resources. Hence, their distributions are centered on the more heavily timbered coastal and subcoastal regions of eastern, western, and northern Australia. On a local scale, patterns of occurrence may be dictated by the presence of particular species of trees that provide preferred foods either directly (e.g., certain *Eucalyptus* species, which are eaten almost exclusively by koalas and greater gliders, *Petauroides volans;* Kavanagh and Lambert 1990; Lee and Martin 1988), or indirectly (e.g., old trees or those with fissured bark that provide foraging substrates for antechinuses and phascogales; Dickman 1991). For gliders that eat insects, nectar,

Grizzled tree kangaroo *(Dendrolagus inustus)*. Photograph by Pavel German / Wildlife Images.

pollen, and plant exudates, sites must contain wattle trees, grevilleas, melaleucas, banksias, and other suitable food plants to sustain local populations.

Intriguingly, research by Braithwaite et al. (1983, 1984) has shown that aggregations of gliders and possums can be found at local "hot spots" in forests where soil and foliar nutrients are concentrated; away from these localized sites, the abundance of arboreal marsupials falls by some 95 percent. For terrestrial forest-dwellers such as potoroos and some bettongs, the habitat components selected by individuals remain difficult to identify. However, these species are known to take substantial amounts of fungus as part of the diet (Claridge and Barry 2000; Claridge et al. 1993), and it is possible that animals are cueing on the scent of fungal fruiting bodies that remains elusive for human observers. For forest generalists, such as the common brushtail possum *(Trichosurus vulpecula)*, a wide variety of lightly and heavily timbered habitats can be used; *T. vulpecula* even occupies arid environments with very sparse tree cover and persists in suburban areas that provide buildings for shelter (Kerle 2001).

Several species of larger Australian marsupials are also restricted to wooded habitats. In the case of tree-kangaroos *(Dendrolagus bennettianus* and *D. lumholtzi)*, rainforest trees provide leaves for food and shelter from inclement weather and predators (Newell 1999; Procter-Gray 1984). Terrestrial browsers such as swamp wallabies *(Wallabia bicolor)* and red-necked wallabies *(Macropus rufogriseus)* forage on low-growing, broad-leaved shrubs and grasses, using scrapes under shrubs or debris on the forest floor for shelter. The largest kangaroos, including the eastern gray *(Macropus giganteus)*, western gray *(M. fuliginosus)*, and red kangaroo *(M. rufus)*, also take refuge in forest patches or wooded creek lines to escape danger and to seek shade or respite from very hot, cold, or wet weather (Frith and Calaby 1969; Newsome 1965*a*). These species and the euro *(M. robustus)* are primarily grazers, however, and are often found in pastures far from the nearest woodland cover. Except for these four large kangaroos and the rock wallabies, kangaroo and rat-kangaroo diversity is highest in woodland and forest habitats in eastern Australia, where up to 12 species (three rat-kangaroo and nine species of larger macropodids) may coexist (Fox 1989; Kaufmann 1974; Seebeck et al. 1989).

A further aspect of habitat that determines the distribution of many Australian marsupials is substrate. This is most obvious for species such as the marsupial moles *(Notoryctes caurinus* and *N. typhlops)*, which spend most of their active time burrowing 20–100 cm below the soil surface and occur only in sandy soils that allow subterranean movement (Benshemesh and Johnson 2003). At least two species of planigales *(Planigale gilesi* and *P. tenuirostris)* are tied closely to the presence of deeply cracking soils that provide below-ground refugia (Andrew and Settle 1982), while kowaris *(Dasyuroides byrnei)* and mulgaras *(Dasycercus cristicauda)* show strong preferences, respectively, for stony gibber and sandridge desert, which facilitate burrow construction (Lim 1992; Masters 1993; Masters et al. 2003).

Hard, rocky substrates are selected by other marsupials, such as the mountain pygmy-possum *(Burramys parvus)*, fat-tailed antechinus *(Pseudantechinus macdonnellensis)*, and at least 16 species of rock wallabies (*Petrogale* spp.). For the former two species, rock surfaces are used extensively as foraging substrates, while crevices between and under rocks are used for shelter and nest sites (Gilfillan 2001; Mansergh and Broome 1994). Rock wallabies use rock surfaces for movement to and from feeding sites, for evasion of predators, and as the template for patterns of social organization (Jarman and Bayne 1997). They display remarkable speed and agility even on steep and smooth, worn rocks, using their long, cylindrical tails for balance and their rough-soled, padded hind feet to gain purchase (Hume et al. 1989).

While the availability of suitable habitat in any locality determines whether species can potentially live there, other factors such as the presence of competitors, predators, or facilitators often influence whether they actually do occur. For example, competition between related species affects the local distributions of some dasyurids, possums, and bandicoots (Dickman 1984, 1986*a*, 1986*b*) and is suspected to be a cause of the already-small and declining distribution of the Proserpine rock wallaby *(Petrogale persephone)* in Queensland (Sharman et al. 1995). Predation by the dingo *(Canis familiaris dingo)* and interference from humans almost certainly resulted in the extermination of the thylacine *(Thylacinus cynocephalus)* and devil *(Sarcophilus harrisii)* from the Australian mainland within the last three thousand years (Johnson and Wroe 2003), and the depredations of introduced red foxes *(Vulpes vulpes)* and cats *(Felis catus)* have had catastrophic effects on many contemporary species of small and medium-sized marsupials (Dickman 1996*a;* Saunders et al. 1995). Conversely, the burrowing activities of other vertebrates, large spiders, and scorpions have beneficial effects on small dasyurids in arid areas. Small dasyurids do not excavate their own burrows; hence, the diggings of "engineer" species provide crucial subterranean refugia for dasyurids and underpin the presence of the rich dasyurid communities that characterize arid areas (Dickman 1996*b*).

Finally, chance and historical events also have important influences on present-day patterns of diversity and distribution. For example, related species are often found on either side of a current or former barrier to dispersal and provide potential examples of allopatric speciation. Several groups of closely related marsupials have been identified in Tasmania and on the adjacent mainland, as well as either side of the Nullarbor in southern Australia and the Macleay-McPherson overlap zone in northern New South Wales (Crowther and Blacket 2003; Hope 1973; Kitchener et al. 1984; Spencer et al. 2001). As with speciation in Amazonian opossums (Patton and Costa 2003), most divergences reflect separations that occurred pre-Pleistocene, more than 1.75 million years ago; those among dasyurids are even more ancient, having occurred pre-Pliocene, more than 5 million years ago (Krajewski et al. 2000).

In a further example, in the Wet Tropics of northern Queensland four species of ringtail possums, several species of dasyurids, macropodids, and other marsupials represent a relictual fauna that was once more widespread but is now confined to patches of rainforest (Williams et al. 1996). Despite climatically driven expansions and contractions in size, these patches remained large enough over time to sustain and allow divergence of their remnant marsupial populations. No endemic species now occur in rainforest remnants in more southerly parts of Queensland and New South Wales, despite providing apparently suitable habitat. Here, rainforest patches seem to have fluctuated dramatically in size during Quaternary glacial-interglacial cycles, reducing marsupial populations during the driest periods to the point that numbers were too low to be viable (Winter 1988, 1997).

In the northern rainforests, declining rainforest endemics were also more likely to be "rescued" or replaced by immigrants from across Torres Strait. Thus, cuscuses disappeared from northern Australia between 4 million and 1 million years ago but have been replaced by a subsequent reinvasion of two species from New Guinea (Archer et al. 1999). Small, isolated areas of habitat also account for the paucity of specialist marsupials in monsoon rainforest across Kimberley and the "Top End" of Australia (Bowman and Woinarski 1994; Friend et al. 1991).

As detailed below, human activity has had dramatic and often deleterious effects on marsupial abundance and diversity. In addition to these obvious effects, Europeans have also translocated marsupials both within and outside Australia and have created many distortions to species' natural distributions. Sugar gliders *(Petaurus breviceps)* were probably introduced to Tasmania in the 1830s (Green 1973) and have since spread to cover almost 20 percent of that island (Rounsevell et al. 1991). Koalas and three other marsupials have been similarly translocated to Kangaroo Island off the South Australian coast (Inns et al. 1979). Outside Australia, brush-tailed rock wallabies *(Petrogale penicillata)* and parma *(Macropus parma)*, red-necked *(M. rufogriseus)*, and tammar *(M. eugenii)* wallabies have become established on Kawau Island, New Zealand; red-necked wallabies *(M. rufogriseus)* have become established in the Peak District of Great Britain; and eastern gray kangaroos *(M. giganteus)* have made themselves at home in the Rambouillet forest west of Paris (Long 2003; Yalden 1988). Perhaps the most notorious translocation took place in 1858, when common brushtail possums *(Trichosurus vulpecula)* were taken to New Zealand to start a fur industry. The possums have expanded to occupy virtually all forest habitats in New Zealand, causing severe damage to native trees and to agricultural interests (Montague 2000).

NEW GUINEA REGION

Despite being founded by Australian immigrants, the modern marsupial fauna of the New Guinea region is strikingly different from that of the southern continent. Just 11 of the collective number of Australasian marsupial species occur in both regions, and 16 of 31 marsupial genera occurring in New Guinea and surrounding islands are endemic. The 11 species that occur in both regions include two species each of dasyurids, macropodids, petaurids, and phalangerids and one species of peramelid, peroryctid, and burramyid. In addition to the long period available for separate evolution, New Guinea has provided marsupials with a relatively stable climatic environment and a forest-cloaked landmass that has favored arboreal and scansorial adaptations. Thus, while some 30 of Australia's 155 marsupial species are arboreal, at least 50 of New Guinea's present 83 species are restricted to the forest canopy (Dickman and Russell 2001).

Small and medium-sized browsers such as cuscuses and striped and ringtail possums exploit resources in the upper levels of the forest, with individuals of most species seldom descending to the ground. Larger browsers, such as tree-kangaroos (*Dendrolagus* spp.), also spend much time above ground, but the less agile and heavier-bodied members of the genus either avoid the canopy or make increased use of the lower strata and forest floor (Flannery 1995*a*). Other forest-dwellers, such as bandicoots and echymiperas (*Echymipera* spp.), are entirely terrestrial (Anderson et al. 1988), while many dasyures seem to move efficiently among all levels in the forest (Flannery 1995*a*; Woolley 1989).

The habitat preferences of most New Guinea marsupials have been little studied, but the highly localized distributions of some species suggest preferences for particular forest types. For example, the white-striped dorcopsis wallaby *(Dorcopsis hageni)* is confined to mixed alluvial forest at altitudes less than 400 meters, while the speckled dasyure *(Neophascogale lorentzii)* occurs primarily in moss forest above 2,000 meters (Flannery 1995*a*). Most of the eight New Guinean species of *Dendrolagus* are restricted to rugged mountain slopes carrying dense rainforest, with different species occurring at different elevations up to 4,000 meters (Flannery et al. 1996; Menzies 1991). Interestingly, the more derived or "advanced" species in this genus occupy the central montane areas of New Guinea, while the "primitive" species are consigned to lower-lying peripheral regions (Groves 1990). Several species of cuscus also show altitudinal replacements in occurrence, and this has been interpreted as being a response, in part, to interspecific competition (Dickman 1984; George 1979). Altitudinal zonation among other marsupial species has been documented by Gressitt and Nadkarni (1978).

A few New Guinean marsupials occur in grassland and woodland, including the bronze quoll *(Dasyurus spartacus)* and planigale *(Planigale novaeguineae)*, as well as species such as the northern brown bandicoot *(Isoodon macrourus)* and agile wallaby *(Macropus agilis)* that occupy similar habitats in Australia (Dickman and Vieira 2005). As in both the Americas and Australia, some species show little evident habitat preference. The common echymipera *(Echymipera kalubu)* occurs in plantations, gardens, and other disturbed habitats, as well as in forest; other echymiperas and bandicoots are also broadly distributed (Flannery 1995*a*).

Beyond the mainland of New Guinea, marsupials have a broad but scattered distribution. One species, the northern common cuscus *(Phalanger orientalis)*, occurs throughout the Solomon Islands; other endemics are scattered on islands in the Banda, Timor, Coral, Arafura, and Solomon Seas. Many of these taxa remain poorly known, despite great efforts in recent years to survey and study them (e.g., Flannery 1995*b*; Flannery and White 1991; Heinsohn and Hope 2004; Kitchener and Suyanto 1996; Macdonald et al. 1993; Musser 1987). For example, the highly distinctive Seram bandicoot *(Rhynchomeles prattorum)* is known from just seven specimens collected in 1920 from rugged and heavily forested limestone terrain at high elevation on Seram Island (Flannery 1995*b*). Two species occur as far west as Sulawesi, which abuts Wallace's Line: the small Sulawesi cuscus *(Strigocuscus celebensis)* and the bear cuscus *(Ailurops ursinus)*. Both of these cuscuses are endemic, with recent DNA-hybridization experiments indicating that *A. ursinus* diverged from its phalangerid relatives some 16 million years ago (Kirsch and Wolman 2001).

Extensive human movements in the New Guinea region during the Holocene (last 10,000 years) and earlier have resulted in translocations of many of the region's marsupials, complicating attempts to resolve their former distributions. Among the most commonly moved marsupials are cuscuses, which are captured live and carried for food, trade and ceremonial purposes, or as pets. The common spotted cuscus *(Spilocuscus maculatus)*, for example, has been transported to dozens of islands, including Selayar Island, south of Sulawesi, and to the Saint Matthias Group more than 3,000 km to the east (Heinsohn 2002*a*, 2002*b*). *Spilocuscus kraemeri, Phalanger orientalis, P. ornatus,* and *P. lullulae* are among other cuscus species known or suspected to have been moved by human agency (Flannery 1995*b*). Other translocated marsupials include the petaurid glider *Petaurus breviceps,* the bandicoots *Echymipera kalubu* and *E. rufescens,* and five species of macropodids (Heinsohn 2003).

Diets and Resource Requirements

Given the diverse habitats used by marsupials, it is not surprising that dietary diversity within the group is also very broad. In the Americas most living species are omnivorous and eat invertebrates, small vertebrates, eggs, fruits, and other plant products in different amounts, but in the Australasian region marsupials may be either dietary generalists or specialists that take particular foods from across the dietary spectrum. If diets are categorized as carnivorous (including diets that comprise mostly invertebrates or vertebrates), omnivorous, and herbivorous (Hume 1999), strikingly different patterns of feeding behavior can be identified between and within the American and Australasian radiations that correlate with body size, habitat, and phylogeny (Eisenberg 1981; Lee and Cockburn 1985).

CARNIVORY

Carnivorous marsupials that specialize on invertebrates predominate in Australia and New Guinea; most are dasyurids weighing less than 100 grams. Larger insect-eaters include highly arboreal species such as the tuan *(Phascogale tapoatafa)* and some of the dasyures of New Guinea, perhaps suggesting that tree trunks and the upper strata of forests are rich sources of invertebrate food (Dickman and Vieira 2005). Although *P. tapoatafa* is nocturnal, as are most of the smaller dasyurids, there is a tendency for larger members of the group, such as *Antechinus swainsonii*, *Antechinus minimus*, and *Parantechinus apicalis* (all up to 100 grams) and *Neophascogale lorentzii* (200 grams), to be increasingly active by day. In absolute terms, large bodies are energetically more expensive to maintain than are smaller ones, and some diurnal activity is presumably necessary to obtain sufficient food. The numbat *(Myrmecobius fasciatus)* is the largest insect-eater at 500 grams; it specializes on termites and is active by day or night in accordance with the rhythms of its prey (Cooper 2003).

In South America, short-tailed opossums (*Monodelphis* spp.), at least some species of southern mouse opossums (*Thylamys* spp.), the monito del monte *(Dromiciops gliroides)*, and the Chilean shrew opossum *(Rhyncholestes raphanurus)* take invertebrates as a major part of the diet (Busch and Kravetz 1991; Vieira and Palma 1996), with *R. raphanurus* consuming large numbers of earthworms (Meserve et al. 1988). As with their Australasian counterparts, there is evidence that some of the larger insectivores, such as *Monodelphis brevicaudata*, *M. dimidiata*, and *M. domestica* (all $\geqslant$40 grams), are partly or wholly diurnal; *M. americana*, while small (30 grams), may also be day-active (Emmons and Feer 1997; Gardner 1993; Streilein 1982). Recent studies of *Metachirus nudicaudatus*, however, confound this trend. This larger (300–450 grams) species is more insectivorous than previously thought (Astúa de Moraes et al. 2003; Santori et al. 1995), but there is little evidence that it is active by day (Emmons and Feer 1997).

Carnivorous marsupials that eat chiefly vertebrate flesh usually exceed 100 grams and are represented in all regions. In Australia this group includes four species of quolls (*Dasyurus* spp.), the Tasmanian devil *(Sarcophilus harrisii)*, and the recently extinct *Thylacinus cynocephalus;* two further species of quolls occur in New Guinea. Small quolls such as *D. viverrinus* can eat large quantities of invertebrates if warm-blooded prey are not available (Blackhall 1980), but the larger members of the group tend to be more obligate flesh-eaters that hunt by stalking and pouncing (Jones and Barmuta 1998, 2000). *Sarcophilus harrisii* is equipped with bone-crunching molars and includes much carrion in its diet (Guiler 1970), whereas *T. cynocephalus* is believed to have been a pursuit predator that preyed upon wallabies, kangaroos, and smaller mammals and birds (Guiler and Godard 1998; Paddle 2000).

Three South American marsupials, all didelphimorphians, can be considered serious flesh-eaters. *Chironectes minimus* includes fish and frogs in its diet, as well as crabs and other aquatic invertebrates, which it catches with its fore feet while paddling in streams and ponds (Fish 1993; Marshall 1978*b*). *Lutreolina crassicaudata* is also common in many riparian environments; in addition to frogs, this ferocious hunter eats other small mammals, lagomorphs, birds, snakes, and invertebrates (Hunsaker 1977; Monteiro-Filho and Dias 1990). The third didelphimorphian carnivore is *Lestodelphys halli*. Although there are few observations of it feeding in the wild, observations at traps and in captivity suggest that this small species is an aggressive hunter capable of killing and entirely consuming a variety of small mammals and birds (Birney et al. 1996*a*, 1996*b*). In addition to these species, some caenolestids may turn out to be effective flesh-eaters. Captive *Caenolestes* readily take young rats, holding them down with their fore feet while using their bayonet-like lower incisors to stab them. Caenolestids have unusual lip flaps that may help to prevent the sensory hairs and fur at the side of the mouth from becoming clogged with blood and soil during feeding (Kirsch and

Waller 1979). Hume (2003) has commented that meat-based diets are high in protein, vitamins, minerals, and water and are generally highly digestible. Carnivores, however, have relatively short and simple guts and physiological specializations that limit their ability to process other foods; hence, strict carnivory can be viewed as representing a particularly narrow nutritional niche (Hume 2003).

OMNIVORY

Marsupials with broad-based diets are well represented in both the American and Australasian radiations, although none exceeds 5,000 grams in weight. It is convenient to break this large and diverse dietary category into at least four groups (Dickman and Vieira 2005). The first, containing *insectivore-omnivores,* consists of species for which invertebrates compose up to about half the diet by volume, with a variety of other foods making up the remainder. Species in this group predominate in South America, where they are represented by at least 26 of 46 species (57%) of mostly small didelphids for which dietary information is available (Dickman and Vieira 2005). Fruits and other plant materials, including small seeds, make up much of the noninvertebrate diet for these species; they are often picked directly from branches near tree tops but can also be retrieved from the forest floor (Carvalho et al. 1999; Charles-Dominique et al. 1981; Meserve et al. 1988). Australasian insectivore-omnivores include burramyids (*Burramys* and *Cercartetus* spp.), trioks (*Dactylopsila* spp.), some gliders (*Acrobates pygmaeus* and *Petaurus* spp.), and the feather-tailed possum *(Distoechurus pennatus).* In addition to invertebrates, these species take variable amounts of fruit, nectar, pollen, flowers, and plant leaves and exudates in their diets (Goldingay and Jackson 2004; Huang et al. 1987; Smith 1982; Turner and McKay 1989; van Tets 1998).

The second omnivore group, comprising *frugivore-omnivores,* is largely confined to South America. It includes the three species of *Caluromys* and probably the poorly known *Caluromysiops irrupta* and *Glironia venusta* (Izor and Pine 1987; Vieira and Astúa de Moraes 2003). Frugivore-omnivores are poorly represented in Australasia. The single species confirmed to fall in this group is the musky rat-kangaroo *(Hypsiprymnodon moschatus),* which includes much fallen fruit from the forest floor in its diet (Dennis 2002; Dennis and Marsh 1997). Highly arboreal cuscuses within at least three genera *(Phalanger, Spilocuscus,* and *Strigocuscus)* also eat small amounts of fruit, although quantitative observations are limited. In the most detailed study to date, Salas (2002) recorded New Guinean montane cuscuses and the coppery ringtail possum *(Pseudochirops cupreus)* eating fruits quite commonly. He noted further that individuals would move rapidly at dusk to certain trees with ripe fruit and, at least in the case of *Pandanus* fruit racemes, defend them against conspecifics of the same sex (Hide et al. 1984; Menzies 1991; Winter and Leung 1995*a,* 1995*b).*

Fungivore-omnivores are the third omnivore group, and these are confined to Australia. All are rat-kangaroos of the family Potoroidae. Fungal fruiting bodies and hyphae compose 50–80 percent of the diets of the four species that have been subject to most study (*Bettongia penicillata, B. tropica, Potorous longipes* and *P. tridactylus;* see Claridge et al. 1993; Claridge and May 1994; McIlwee and Johnson 1998), with invertebrates, fruits, seeds, plant leaves, and tubers making up most of the remainder. All potoroids are terrestrial and excavate subterranean fungi using powerful forelimbs and digits that are equipped with stout claws (Seebeck and Rose 1989).

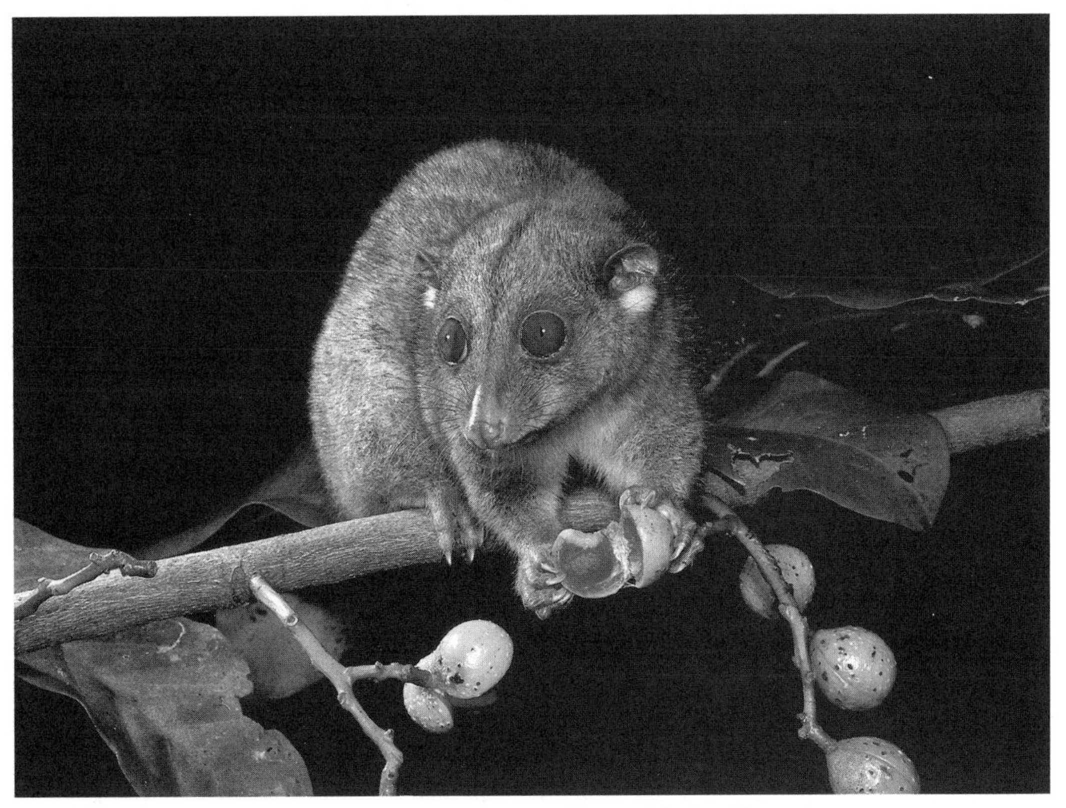

Southern common cuscus *(Phalanger intercastellanus)* eating seeds. Photograph by Pavel German/ Wildlife Images.

The final group contains *generalist-omnivores* and is represented by species that are able to take virtually any foods that are available within their immediate environments. The four large, scansorial species of *Didelphis* from North and South America fall within this group (Cordero and Nicolas 1987; Santori et al. 1995), as do peramelemorphians in Australasia (Gibson 2001; Jackson 2003). When environmental conditions are unfavorable, it is worth noting that many omnivores in the above categories can take a broader range of foods than usual and could be categorized, temporarily, as generalist omnivores. These include smaller species of *Dasyurus* (Oakwood 1997) and probably both species of *Philander* (Emmons and Feer 1997; Santori et al. 1997). In addition, dietary differences within species may also occur between localities (e.g., *Didelphis marsupialis,* Cordero and Nicolas 1987).

In view of the ability of many marsupials to switch diets between times and places, several authors have cautioned against forcing species into artificial dietary categories and have suggested that a dietary continuum from frugivory to carnivory should be recognized (e.g., Astúa de Moraes et al. 2003; Vieira and Astúa de Moraes 2003). At least for Neotropical marsupials, which are predominantly omnivorous, this suggestion should assist in guiding future dietary research.

HERBIVORY

Herbivorous marsupials form the largest dietary category in the faunas of both Australia and New Guinea but are conspicuously absent from the Americas. Ecologically, we can group them into browsers and grazers. These groups do not necessarily contain species that belong to separate families or exhibit consistent differences in dental morphology or digestive physiology (Sanson 1978), but they do show certain similarities in diet and way of life.

Browsers eat the shoots and leaves of shrubs and trees and include the koala *(Phascolarctos cinereus)*, brushtail and scaly-tailed possums, ringtail possums, greater glider *(Petauroides volans)*, swamp wallaby *(Wallabia bicolor)*, and tree kangaroos (Kerle 2001; Lindenmayer 2002). The larger species of cuscus also fall into this group, as did many species of now-extinct megafauna (Johnson and Prideaux 2004). Except for the ground-dwelling *W. bicolor*, extant browsers are largely arboreal and use trees for food and shelter. They either exploit foliage or hollows for nests or use branches and tree forks as rest sites (Russell 1980); however, most browsers spend some time on the ground to move between trees or feed on low shrubs. The common brushtail possum *(T. vulpecula)* is astonishingly versatile in this regard, being almost completely arboreal in heavily timbered areas, where predators such as the red fox are present, but terrestrial in savanna or rocky habitats where trees are sparse (Kerle 2001) and predators absent (Inns et al. 1979). Its diet is accordingly very varied (Evans 1992; Kerle 1984).

Browsers usually also include small amounts of plant material other than leaves in their diet, including blossoms, fruits, and occasionally invertebrates. The smallest browser, the pygmy ringtail *(Pseudochirulus mayeri)* (150 grams) of the New Guinea highlands, is unusual in including some fungus and pollen in its otherwise leafy diet and is the only marsupial known to consume substantial amounts of moss and lichen (Hume et al. 1993). The ground cuscus *(Phalanger gymnotis)* also is unusual in eating vertebrate flesh, at least in captivity, and a wide range of fruits, ferns, and even bark in the wild (George 1982; Majnep and Bulmer 1990). At the opposite extreme, *Phascolarctos cinereus* and *Petauroides volans* of eastern Australia usually eat the leaves of only one or two species of *Eucalyptus* in any locality, although the diets are more varied over the species' entire ranges.

Grazing marsupials include wallabies, kangaroos, and wombats. This group shows considerable variation in size, from the monjon *(Petrogale burbidgei)* and nabarlek *(P. concinna)* at 1,250–1,350 grams to red kangaroos that weigh 50–60 times as much. Grass, forbs, and browse form the staple diet of all grazers. The smaller species in general prefer to eat more succulent, less fibrous plant material than do the wombats and large kangaroos, with some fruit and browse being taken in addition by hare-wallabies (*Lagorchestes* spp.) and forest-dwelling pademelons (*Thylogale* spp.) (Dawson 1989, 1995; Jarman 1984). During periods of food shortage or drought, drier foods such as *Acacia* leaf litter may be eaten (Irlbeck and Hume 2003). Grazers are primarily terrestrial. Wombats, however, shelter in deep subterranean burrows (Triggs 1988), and rock wallabies show remarkable agility in negotiating steep-sided rock outcrops. The yellow-footed rock wallaby *(P. xanthopus)* has even been observed foraging in the lower limbs of trees, but this behavior is exceptional and probably a response to food shortage at ground level (Allen 2001).

Green plant material seems readily available to herbivores, but in reality this food source poses many problems for animals that ingest it. Although plant foods contain nutritious proteins, carbohydrates, and lipids, these constituents are locked within leaf and stem cells with tough cellulose walls. Many plants also repel herbivores by lacing cellular contents with distasteful or toxic compounds, by deploying external deterrents such as spines, or by embedding shards of silica within the leaf tissue. Marsupials partially reduce these problems by choosing to eat only the most nutritious and energetically profitable parts of plants, such as buds or young leaves (e.g., Dwiyahreni et al. 1999). Smaller species, especially, also supplement their diets with high-quality foods such as invertebrates and even birds' eggs when available (Kerle 2001). Once plants have been selected and gathered, the grinding action of the molars fragments the cell walls and releases the cell contents. In kangaroos, the cellular fragments are subjected to further digestion in a highly elaborated forestomach that houses dense populations of cellulose-fermenting microbes (Hume 1982; Moir et al. 1956). In the koala, possums, and wombats, by contrast, microbial populations are housed in a greatly enlarged cecum and/or colon. Some 25–50 percent of the cell wall contents are digested in these specialized fermentation chambers and provide a vital contribution to the overall energy budgets of these marsupial herbivores (Dawson et al. 1989; Hume 1982).

Although marsupials display some common solutions to the problem of digesting green plants, different species exhibit an astonishing array of individual solutions. For example, the common ringtail *(Pseudocheirus peregrinus)* and green ringtail *(Pseudochirops archeri)* possums pass food through their guts twice to maximize absorption of nutrients; soft feces excreted by day are reingested and passed as hard pellets after full digestion at night (Chilcott 1984). These and other possums produce enzymes that neutralize toxic chemicals in leaves (Kerle 2001), while differences in digestive efficiency among larger kangaroo species arise in part from differences in the production and composition of saliva (Beal 1989). Full accounts of the digestive adaptations of herbivorous marsupials are provided by Hume (1982, 1999, 2005) and Dawson et al. (1989).

Life Histories

Marsupials exhibit striking variations in life history, reflecting their enormous diversity in body size, behavior, and ecology. Large kangaroos can occasionally live up to 27 years in the wild (Bailey and Best 1992), but males of several species of small dasyurids never survive more than 12 months (Lee et al. 1982). Litter size is one in many herbivorous species but 13 to 15 in some didelphids, and breeding can be seasonal or continuous, with up to five litters produced in a year. Males are the larger sex in most marsupials; in some dasyurids and macropodids, males can weigh more than twice as much as females. In the honey possum *(Tarsipes rostratus)*, by contrast, females are 25 percent heavier than males (Renfree et al. 1984).

To understand why life history traits diverge so greatly among marsupials, we must consider the effects of body size, diet, and the seasonality of food resources. This topic has received much attention, and the account below draws heavily on several works: Cockburn (1997), Dickman and Vieira (2005), Eisenberg (1981, 1989), Eisenberg and Redford (1999), Fisher et al. (2001), Flannery (1995*a*), Harder (1992), Harder and Fleck (1997), Lee and Cockburn (1985), McAllan (2003), Redford and Eisenberg (1992), Russell (1982), Strahan (1995), Taggart et al. (1998), Tyndale-Biscoe (1973, 1979, 1984), Tyndale-Biscoe and Renfree (1987), and Ward (1998).

CARNIVORES

Marsupials that specialize on invertebrates invariably rear litters of two or more young and produce one to three litters per year, but very few individuals survive for more than three years. This "live fast, die young" strategy is exem-

Wongai ningaui *(Ningaui ridei)* eating a grasshopper. Photograph by Pavel German/Wildlife Images.

plified by many species of small dasyurids, in which litter size is 4 to 12, and marmosids, which frequently bear litters of 12 or more young. In contrast, larger flesh-eaters such as the Tasmanian devil tend to carry smaller litters (average size, four) but live longer; devils may survive up to six years in the wild and eight years in captivity (Slater 1993).

For all carnivorous marsupials, decisions about when to breed and how much effort to expend depend critically on the seasonal availability of food. Because of the high energetic costs of suckling large litters for two to four months until the young are weaned (up to 240 days in the Tasmanian devil), strong selective advantages accrue to females that time lactation to coincide with peaks in food abundance (Braithwaite and Lee 1979). In environments where invertebrates and other food resources are available for a restricted period each year, females usually achieve sexual maturity at 11 months, are monestrous, and have time to produce only a single litter. In these situations, most or all females breed and litter sizes are maximal. The males of short-season species also expend much effort in finding and copulating with females during the annual rut, and in several species the breeding season terminates dramatically with their abrupt and total death. All members of the Australian genera *Antechinus* and *Phascogale* exhibit this remarkable semelparous strategy, as do populations of some species of *Dasyurus* and *Parantechinus* (Dickman 1993; Oakwood et al. 2001). In South America, male and female semelparity has been inferred for both the short-tailed opossum *(Monodelphis dimidiata)* (Pine et al. 1985) and the gray slender mouse opossum *(Marmosops incanus)* (Lorini et al. 1994).

In environments where food resources are available for longer periods each year, females can mature in as little as five to six months and potentially breed in the season of their birth. Species of *Planigale* and *Sminthopsis* provide examples. Females are polyestrous and produce two (occasionally three) litters during the breeding season but reduce their reproductive effort on each occasion by rearing litters of only five to eight young. Because reproductive effort is also less for males, iteroparity, or repeated breeding, is observed. If food resources are available consistently throughout the year, temporal constraints on reproduction are likely to be absent and year-round breeding could be expected. This seems to be the case for several species of dasyurids in tropical regions of New Guinea and northern Australia (Taylor et al. 1982; Woolley 2003), as well as for South American *Monodelphis domestica* in moist refugia around rock outcrops and tropical species of *Gracilinanus* (Hunsaker and Shupe 1977; Streilein 1982).

The timing of breeding in some carnivorous marsupials seems to be influenced by changes in temperature or rainfall (Fleming 1973; O'Connell 1979, 1989), but in most species photoperiod is probably the major trigger (McAllan 2003). In several species of dunnarts (*Sminthopsis* spp.), for example, reproductive activity begins in spring when daylength is increasing and slows or ceases after the summer solstice. Testicular enlargement in males can be stimulated at any time of year in captivity if artificial light is increased, thus confirming the daylength effect (Godfrey 1969; Holloway and Geiser 1996). Field populations of *Monodelphis domestica* in semi-arid Caatinga may also respond to photoperiod (Bergallo and Cerqueira 1994), although animals appear unresponsive to environmental cues in captivity (Fadem and Rayve 1985). Perhaps the most extraordinary response to photoperiod is that shown by several species of *Antechinus*. In *A. agilis*, *A. flavipes*, and *A. stuartii*, at least, females ovulate when a critical threshold is reached in the rate of change of daylength in spring (McAllan and Dickman 1986; McAllan et al. 2005). This threshold, ranging from 77 to 137 seconds per day, allows populations of each species to breed progressively earlier at higher latitudes in eastern Australia and, hence, to achieve weaning in late spring and summer when invertebrate abundance is greatest (Van Dyck 1982).

Of all life history traits exhibited by carnivorous marsupials, that which has attracted most attention is the abrupt and synchronous postmating death of males in *Antechinus* and *Phascogale* species (see, e.g., reviews by Bradley 1997, 2003; Lee and Cockburn 1985; Lee and McDonald 1985). Although ovulations are precisely triggered by a threshold in the rate of change of daylength, both sexes appear primed for reproduction in the lead-up to the breeding season by an internal clock. This annual rhythm prompts increases in body condition before breeding and, in males, stimulates an increase in circulating stress hormones. Cortisol is the most potent of these hormones; it has the effects of suppressing appetite and promoting conversion of protein into sugars, thus allowing animals to engage in sustained activity (Lee and McDonald 1985).

During the breeding season, males often interact aggressively with each other, search for mates by day and night, and copulate for extended periods with several females over several days (Shimmin et al. 2002). Elevated cortisol levels can be tolerated for only a limited time, however, because the hormone suppresses immune and inflammatory responses and predisposes animals to increased risk of disease. Death follows swiftly and inevitably. Autopsy of males after the mating period reveals massive hemorrhaging from ulcers in the stomach and gut, organ failure, and elevated loads of parasites that have taken advantage of immune system failure. From this period in the year until the young become independent in summer, field sampling reveals only females. Hence, populations of all species are low in spring and do not reach peaks until late summer and autumn, when young emerge from maternal nests (Lee and Cockburn 1985; Wood 1970).

OMNIVORES

Because they do not rely on a single kind of food that may be seasonally limited, many omnivores are able to breed over extended periods or even year-round. Most are polyestrous. Litter size varies from one in potoroids to 14 or 15 in omnivorous didelphids such as *Marmosa robinsoni* (O'Connell 1983; Seebeck and Rose 1989), and longevity usually exceeds one year. Among the slow breeders is the mountain pygmy-possum *(Burramys parvus)*, a small (40–60 gram) burramyid that is restricted to boulder fields in the Australian Alps. Females produce just one litter of four young during the short alpine summer but compensate for the lack of annual fecundity by exceptional longevity; wild animals have been recorded to live more than 11 years (Mansergh and Broome 1994) and presumably produce a litter each year after achieving sexual maturity at 12 months. Longevity in *B. parvus* exceeds that of other mammals of comparable size except some bats. At the other extreme, *M. robinsoni* can produce two litters of 14 in a season but is unlikely to survive beyond two years (O'Connell 1979, 1983).

In the large American opossums (*Didelphis* spp.), reproduction is tuned finely to environmental conditions and varies greatly throughout the enormous distributional range of the genus. At higher latitudes, the North American opossum *(D. virginiana)* begins to breed sometime in January, February, or March, when daylength and temperature increase and food supplies become more assured. Reproductive activity ceases in midyear before the sharp autumnal decline in food resources, usually allowing females to raise just one litter (Harder 1992). At equatorial latitudes, by contrast, the related black-eared opossum *(D. marsupialis)* begins breeding in January and often weans two litters during the wetter months from May to September, when fruits are readily available (Fleming 1973). In years when rainfall and fruit production are relatively constant, females can raise three litters (Harder 1992). Similar rainfall-related shifts in patterns of reproduction have been observed in other broadly omnivorous marsupials such as *Caluromys philander* and *Philander opossum* (Charles-Dominique et al. 1981; Julien-Laferrière and Atramentowicz 1990).

Production of one to three litters per year might imply a threefold variation in reproductive effort between geographic locations in *Didelphis*. However, females produce larger litters at high than at lower latitudes (Tyndale-Biscoe 1973; Gardner 1982), suggesting a trade-off between litter frequency and litter size when the growing season is short. Intriguingly, while food supply is, as expected, an important determinant of litter growth and survival (Atramentowicz 1992), it seems also to influence the sex of young that females produce. In both *Didelphis virginiana* and *Didelphis marsupialis*, females in good body condition or provided with extra food raise male-biased litters (Austad and Sunquist 1986; Sunquist and Eisenberg 1993), perhaps because the sons of well-fed mothers get a head start in growth and are disproportionately successful subsequently in competition for mates. The mechanism, or mechanisms, contributing to sex bias remain unknown but probably operate before fertilization or during embryonic stages rather than after birth (Cockburn 1990*a;* Ward 2003).

In the Australasian region, peramelemorphians exhibit reproductive flexibility similar to that of the large American opossums. At equatorial latitudes, the Papuan bandicoot *(Microperoryctes papuensis)* and common echymipera *(Echymipera kalubu)*, at least, can breed year-round (Flannery 1995*a;* Hide et al. 1984). At higher latitudes in Australia, most bandicoots are capable of producing young in all months in captivity but exhibit a hiatus in reproduction from late summer to winter (Gemmell 1990; Jackson 2003). Daylength probably initiates breeding in the northern brown bandicoot *(Isoodon macrourus)*, but temperature and rainfall are also important in stimulating new flushes of food for this and other species (Gemmell 1990; Scott et al. 1999).

Reproductive output is high in all peramelemorphians. In *I. macrourus* and the long-nosed bandicoot *(Perameles nasuta)*, gestation lasts 12½ days, young are weaned at 60–70 days, and females can breed at 3–3½ months of age (Lyne 1964, 1976). Although litter sizes average only two to four, four (occasionally five) litters can be produced in years when conditions are favorable. Despite this high output, recruitment of young into the population is usually in the order of 10–15 percent, with most juveniles presumably dying or dispersing soon after weaning (Friend 1990; Scott et al. 1999). As bandicoots often occur in vegetation that is prone to fire or other disturbances, however, their almost continuous production of young allows rapid colonization of new habitats as they become available (Cockburn 1990*b*). Few other marsupials exploit such regeneration niches as successfully.

HERBIVORES

Because of the apparently ready availability of green plant food, it might be expected that herbivorous marsupials would show much commonality in their life histories. There are, in fact, considerable similarities in traits such as litter size (litter size is usually one but may occasionally be two or more), numbers of nipples (two to four), and even longevity (individuals in most species studied have been recorded to survive five years or more). There is, however, also striking variation in traits between species, such as seasonality of breeding, length of gestation, presence of postpartum estrus and diapause, rates of growth, and social organization. For simplicity, we will consider the life histories of selected browsers and grazers.

Browsers

Perhaps the most familiar of the browsing marsupials is the koala *(Phascolarctos cinereus)*. This distinctive species is associated most strongly with *Eucalyptus* forest in eastern Australia, but it can be found occasionally in areas dominated by *Melaleuca, Callitris, Casuarina,* and other species of tall trees. Although *Eucalyptus* leaves form the great bulk of the diet, they are low in protein and other nutrients and high in indigestible secondary compounds and toxic chemicals (Hume 1982, 1999). To maintain their body weight, individual koalas need to eat about half a kilogram of fresh leaf per day but have to work hard to extract sufficient energy and nutrients (Cork et al. 1983; Lee et al. 1991). The energetic costs of eating a poor-quality diet have selected for a suite of adaptations, including high-ridged grinding teeth that allow efficient mastication of leaves, a very extensive cecum that acts as a fermentation chamber for plant fibers, an enlarged liver that inactivates leaf toxins, and a slow metabolic rate that reduces energy requirements. The reduced metabolic rate, which is about 70 percent that expected for a marsupial of this size (Degabriele and Dawson 1979), allows koalas to rest for some twenty hours a day but also constrains the rate of reproduction.

Koalas mate in spring and early summer and give birth to a single young (occasionally twins) after a gestation of 35 days. Cubs are weaned at about a year, emerging to forage independently the following spring or summer when *Eucalyptus* leaves are fresh and most nutritious, and become sexually mature at two to three years (Eberhard 1978). Individual females do not enter estrus until the previous year's cub has been fully weaned and thus cannot produce more than one young a year. Moreover, if the previous year's young is suckled until after the summer solstice, there is some evidence that decreasing daylength inhibits further estrus (K. Handasyde, in Lee and Martin 1988). These constraints result in an average reproductive rate, or number of young produced annually by females, of just 0.65–0.70 (Eberhard 1978; Lee et al. 1991; Martin 1981). This low output is balanced, however, by the survival of some 80 percent of young to weaning and by a productive life span for females of at least 10 to 12 years (Martin and Handasyde 1991).

Among other browsers, the short-eared possum *(Trichosurus caninus)* and greater glider *(Petauroides volans)* have reproductive rates similarly low to that of the koala (0.68–0.73). These species also breed seasonally, achieve sexual maturity at two to three years, wean single young once a year, and, at least in the short-eared possum, have life spans up to 17 years (How 1981; Kerle 2001). In contrast, two other species of browser, the common brushtail possum *(Trichosurus vulpecula)* and common ringtail possum *(Pseudocheirus peregrinus)*, achieve much higher reproductive outputs. Both can breed twice each year, achieve sexual maturity at 12 to 14 months, and wean their young after only six or seven months (How 1978; Kerle 2001). The annual reproductive rate of the brushtail possum is 0.9–1.4, with females living 10 years or more (Kerle 2001). Female ringtail possums produce 1.8–2.4 young on average each year; the increase is achieved by the production of at least two young per litter. Longevity, which is six years at most, seems to be compromised as a consequence of the elevated reproductive output (Kerle 2001).

The faster-breeding possums in general are better able to exploit disturbed habitats, including suburban backyards. Conversely, the greater glider, short-eared possum, and koala are more restricted to stable forests and colonize new habitats slowly. Their populations often vary little over long periods. These species are particularly susceptible to both human-induced disturbances, such as logging, and drought, fire, or forest die-back, although paradoxically they can survive in small patches of intact forest, at least for short periods (Lindenmayer 1997; Melzer et al. 2000).

Grazers

Kangaroos, wallabies, and wombats produce single joeys after relatively long gestations of 27 to 37 days, retain them in the pouch for six months or more, and wean them after periods of 7 to 18 months (Lee and Ward 1989; Tyndale-Biscoe and Renfree 1987). In general, reproductive rates range from 0.5 to 1.8, slightly lower than those of browsing marsupials. Some variation in reproductive rate is due to differences among species in the rate of growth and duration of lactation, but much arises also from the degree of seasonality of reproduction. Following Tyndale-Biscoe (1989), three main strategies can be distinguished.

In the first, exemplified by *obligate seasonal breeders*, most or all mature females give birth over a single restricted period each year. This strategy is exhibited by the tammar wallaby *(Macropus eugenii)* and western gray kangaroo *(M. fuliginosus)*, as well as by the Tasmanian subspecies of the red-necked wallaby *(M. rufogriseus rufogriseus)*. In the latter subspecies, most births occur from January to March and young remain in the pouch for some 270 days before being weaned at 12 to 17 months. Females enter estrus and mate again within hours of birth but then hold the resulting embryo in a quiescent state for up to a year. It is reactivated and born 16 to 29 days after its older sibling has stopped suckling (Merchant and Calaby 1981). Because of the short window of opportunity for reproduction, the maximum reproductive rate of obligate seasonal breeders is 1.0. Species exhibiting this strategy predominate in temperate regions of southern Australia. Here, fresh grasses and forbs are predictably abundant in spring and summer, thus providing females with nutritious food when their young are emerging from the pouch and the demands of lactation are high.

The second strategy is exemplified by *facultative seasonal breeders* such as the western brush wallaby *(Macropus irma)*, banded hare-wallaby *(Lagostrophus fasciatus)*, and quokka *(Setonix brachyurus)*. Breeding in these smaller grazers takes place largely in the austral summer and autumn, but births can also occur under certain conditions in the second half of the year (Sharman 1955; Tyndale-Biscoe 1965). In populations of the quokka on Rottnest Island, for example, females provided with extra food or that are otherwise in good condition may enter estrus and produce young during winter or spring. As joeys remain in the pouch for only 190 days and are weaned less than two months later (Shield 1968), two young can be produced within a year and elevate the annual reproductive rate of quokkas to 1.8. The predominantly seasonal breeding of these grazers suggests that they occupy largely seasonal environments, but their ability to produce out-of-season young indicates further that they are attuned to improved conditions at any time of year.

Grazers are mostly *continuous breeders* and can produce young at any time of year. This third life history strategy is exhibited by many of the larger species of kangaroos *(Macropus* spp., including mainland populations of *M. rufogriseus banksianus)*, some rock wallabies (*Petrogale* spp.), and mainland populations of the quokka (Ealey 1967; Merchant 1976; Tyndale-Biscoe 1989). Continuous breeders occupy mesic forest environments or protected situations such as rock outcrops where resources are available year-round and achieve annual reproductive rates up to 1.8. They occur also in the central deserts of Australia, but in these arid habitats they can achieve continuous reproduction only when green food resources are plentiful (Newsome 1965*b*, 1966). During drought, breeding slows or ceases, although females with existing young usually continue to suckle. As drought deepens, females either become anestrous and stop breeding altogether or continue to produce young that have little chance of survival (Newsome 1966). When rains return, however, nonbreeding females enter estrus and ovulate within two weeks, while some 90 percent of females with pouch young will wean them successfully (Newsome 1966).

Economic and Ecological Importance

For indigenous peoples of the Americas and Australasia, marsupials are an integral and often very important part of the local economy. In both regions marsupials are used as food, pets, or items for trade, while in Australasia, in particular, skins and fur of the larger species such as kangaroos and possums are used for dress, ornamental, and ceremonial purposes (Dixon and Huxley 1985; Dwyer 1974; Heinsohn 2003). Parts of these marsupials are used also to construct water bags, pouches, weapons, or percussion instruments (Allen 1983; Gould 1969). Irrespective of their practical utility, however, marsupials usually are well known to indigenous peoples and placed into local classificatory and knowledge systems (see, e.g., Bulmer and Menzies 1972, 1973; Finlayson 1961; Flannery 1995*a*, 1995*b*; Patton et al. 1982). Collaborative studies using both traditional knowledge and scientific approaches have provided unusually deep insights into marsupial behavior, ecology, and distributions (Baker et al. 1993; Burbidge et al. 1988; Flannery et al. 1996; Harder and Fleck 1997; Newsome 1997).

The first European explorers in the Americas and Australasia viewed marsupials with a mixture of fascination, fear, and disgust, describing them variously and endearingly in terms of mammals familiar to them such as rats and ferrets or, more ominously, as "monstrous beasts," "vampires," "devils," or chimeras constructed from the body parts of unrelated placentals (Austad 1988; Strahan 1995). A detailed account of reactions to one particularly maligned species, the thylacine or Tasmanian tiger, has been provided by Paddle (2000).

European perceptions of the economic value of marsupials also were divergent and continue to be so. On the one hand, large and charismatic species such as woolly opossums, kangaroos, and the Tasmanian tiger were much sought after by menageries and museums in Europe (Fisher 1984) and provided a small contribution to the settler economies. In Australia, possums, koalas, and bilbies formed the basis of an export fur trade (Marshall 1966), while kangaroos were shot for their meat and skins and hunted for sport using hounds. Koalas and bilbies are now protected, but brushtail possums are still used for fur (Callister 1991), and some 5.8 million kilograms of kangaroo meat are exported annually for human consumption (Grigg 2002). Marsupials are also being used increasingly as companion animals, tourist attractions, commercial symbols, and models in biological and biomedical research (Lunney and Dickman 2002; VandeBerg and Robinson 1997). In the United States, introduced sugar gliders *(Petaurus breviceps)* have become increasingly popular companion animals in recent years (J. A. W. Kirsch, personal communication, 2004).

On the other hand, marsupials have long been vilified for their damage to agricultural, pastoral, and silvicultural interests. This is particularly so in Australia because of the abundance of large, conspicuous herbivorous and carnivorous species, although American opossums sometimes cause local damage in orchards and other crops. In Queensland, Marsupial Destruction Acts in force between 1877 and 1930 placed bounties on the heads of many native species and led to the destruction of 27 million macropodids and peramelids (Hrdina 1997); in Tasmania over much the same period, the thylacine was persecuted to reduce its perceived depredations on sheep (Paddle 2000). Current costs to overall agricultural production from kangaroo grazing have been estimated at over Aust$100 million a year and used to justify annual culls of 2–3 million animals; the true cost to agriculture is, however, much disputed (Dawson 1995; Grigg 2002).

The common brushtail possum, introduced to New Zealand in 1858, causes extensive damage to native vegetation and fauna as well as to pastoral, horticultural, forestry, and apicultural industries; it also acts as a reservoir for bovine tuberculosis (Butcher 2000; Payton 2000; Sadleir 2000). Returns on 58 million possum skins exported up to 1997 amounted to a total of NZ$830 million, thus providing some offset to the species' more negative effects (Warburton et al. 2000).

The ecological importance of marsupials is more difficult to specify, in part because they have been subject to insufficient study, but also because we fail generally to place value on ecological services (Kinzig et al. 2001; Randall 1991). Marsupials are, of course, integral components of ecological communities and play important but mostly unquantified roles as competitors, mutualists, predators, and prey. Patterns of resource use and overlap have been described within communities in diverse habitats in both the Americas (e.g., Crespo 1982; Fonseca and Kierulff 1989; Glanz 1990; Peres 1999; Smythe 1986) and Australia (e.g., Dick-

Bilby *(Macrotis lagotis)*. Photograph by Pavel German/Wildlife Images.

man 1989; Fox 1989; Henry et al. 1989). There is clear evidence of partitioning of resources by body size, habitat, diet, and, to a lesser extent, time of activity, much as observed in ecological communities generally (Fox 1999; Jones 1997; Schoener 1974).

Marsupials provide ecological services directly by moving pollen, plant seeds, and fungal spores and perhaps also by limiting outbreaks of certain pest invertebrates. In both the Americas and Australasia, marsupials are frequently seen visiting flowers to obtain floral resources, such as nectar, or invertebrates, and they pick up pollen in the fur while foraging (Carthew and Goldingay 1997; Charles-Dominique et al. 1981; Gribel 1988; Janson et al. 1981; Steiner 1981). If pollen is moved to the stigmas of other flowers, pollination may be achieved. There is some doubt that marsupials are efficient pollinators in comparison with more mobile taxa such as birds or bats (Hopper and Burbidge 1982) or that any plant species depend specifically on marsupials for pollen transfer (Turner 1982). However, experimental evidence has confirmed that some dasyurid and burramyid species are effective pollinators (Cunningham 1991; Goldingay et al. 1991), and there seems little doubt that more species will be shown to play important pollination roles when appropriate work is carried out (Goldingay 2000).

In addition to moving pollen, many marsupials act as vectors for plant seeds. In the Americas, frugivore-omnivores and other didelphids ingest fleshy fruits and arils, excreting the seeds later at some distance from the parent plants. Seed dispersal can be effected also if fruits have sticky appendages or hooks that lodge in fur. In one particularly detailed study in tropical forest in French Guyana, Charles-Dominique et al. (1981) showed that marsupials eat mostly fruits with seeds less than 20 mm in diameter and eat the fruit pulp but not the seeds of larger species. Intriguingly, at least five plant species produce seeds that seem to be dispersed exclusively by marsupials; fruits bearing these seeds are available year-round.

In another remarkable study in southern Argentina, Amico and Aizen (2000) demonstrated that *Dromiciops gliroides* is the sole dispersal agent for seed of the loranthaceous mistletoe *Tristerix corymbosus*. Fruits eaten by *D. gliroides* are excreted undamaged above ground and must pass through the gut of the marsupial for germination and adherence to the host tree to take place. In view of the great antiquity of both microbiotheriids and Loranthaceae, Amico and Aizen (2000) postulated that co-evolutionary relationships between the groups could have developed some 60–70 million years ago.

Australasian marsupials are generally less frugivorous than their American counterparts and so may play a lesser role in seed dispersal. Two ground-dwelling species, however, are exceptional. The brush-tailed bettong *(Bettongia penicillata)* has long been suspected of dispersing a range of shrub seeds but has now been shown to be a major vector for the large seeds of Western Australian sandalwood *(Santalum spicatum)*. Bettongs pick up fallen seeds and place them singly in caches up to 80 meters distant, apparently increasing the chance of both local germination and sandalwood regeneration over large areas (M. Murphy and M. Garkaklis, personal communications, 2003). Seed caching occurs also in the diminutive musky rat-kangaroo *(Hypsiprymnodon moschatus)* of northern Queensland; as with the bettong, seed burial probably facilitates germination away from parent plants. Detailed studies by Dennis (2003) have shown that musky rat-kangaroos favor large, fleshy fruits and move them 17 meters on average from the point of discovery. Animals move up to 900 fruits per hectare each month, burying over three-quarters of these just below the soil surface in a scatter-hoard pattern.

In New Zealand, the introduced brushtail possum also has been suggested to be a major vector for large tree seeds in degraded forest where large-gaped birds and other native seed dispersers have been extirpated (Dungan et al. 2002). This suggestion has been disputed, however, and the precise role of possums remains to be resolved (Williams 2003). Some fruit-eating species of cuscus are suspected to play a role in seed dispersal and possibly even in seed caching (Flannery 1995*a*), but confirmation awaits further research.

Fungal spores are dispersed in much the same way as seeds: the sporocarps, or fruiting bodies, are eaten and the intact spores excreted elsewhere. The fungi that are eaten most frequently by marsupials are underground species that form ectomycorrhizal associations with plant roots (Johnson 1996). These associations are extremely beneficial to plants because the fungi assist in concentrating important nutrients from the soil. If mycophagous marsupials are present in an area after a disturbance, such as a fire, they could assist recovery by disseminating spores and speeding the regeneration of plants. Both *Bettongia penicillata* and partly mycophagous bandicoots *(Isoodon obesulus, Perameles nasuta)* are active in fragmented and newly burnt areas (Christensen 1980; Scott et al. 1999; Stoddart and Braithwaite 1979) and could thus be important agents of vegetation regeneration.

While some organisms benefit from marsupials directly by being dispersed, others may benefit indirectly. For example, if plant recovery in a burnt area is speedy because bandicoots import ectomycorrhizae, this will in turn facilitate fast recovery of small marsupials, rodents, and other animals that require dense vegetative cover (Wilson and Friend 1999). The feeding activities of some marsupials can also have beneficial indirect effects. Thus, yellow-bellied gliders *(Petaurus australis)* gouge V-shaped notches in the trunks of some two dozen species of *Eucalyptus* in eastern Australia to obtain sap; the notches provide feeding opportunities for several species of smaller gliders and possums that would otherwise be unable to access this food source (Fleay 1947; Russell 1980). Alternatively, the ingestion of herbivorous insects such as lerps, locusts, and leaf and scarab beetles by dasyurids and petaurids (Strahan 1995) reduces the chance of plague development and hence assists in the maintenance of native vegetation.

The indirect effects of marsupials on other organisms are not limited to those created by their feeding behavior and include a suite of additional effects caused by their interactions with the nonliving environment. Such effects result from "ecosystem engineering" (Jones et al. 1994, 1997) and can be profound. In Western Australia, elegant studies by Garkaklis et al. (1998, 2000, 2003, 2004) have shown that the foraging digs of brush-tailed bettongs dramatically alter water repellency of the soil. At the surface, water infiltration and leaching of available nitrate, ammonium, and sulfur increase, but at depths below 10 cm repellency can become severe. In the absence of bettongs, soils become so hard-packed that within three to four years water cannot penetrate and plant establishment is greatly reduced. Individual bettongs make 38–114 holes during nightly foraging bouts and displace some 4.8 tons of soil annually (Garkaklis et al. 2004), clearly having great influence on soil structure and the maintenance of nutrient and hydrological cycles.

The engineering credentials of the burrowing bettong *(Bettongia lesueur)* are even more impressive. This species disappeared from mainland Australia in the late nineteenth and early twentieth centuries, but enduring structures seem to represent old bettong warrens (Noble 1993). The structures are typically circular or subcircular, 20–100 meters in diameter, with 50-cm-high walls that surround a central lens of calcrete (Noble 1999). Surviving warrens reduce erosion by trapping large amounts of organic matter and topsoil and greatly increase variability in soil structure and the composition of plant communities at the landscape scale. On Barrow Island, northwestern Australia, where burrowing bettongs still persist, warrens are used as shelters by brushtail possums and golden bandicoots *(Isoodon auratus)*; evidence from indigenous Australian informants indicates that they were used also by western quolls *(Dasyurus geoffroii)* in the central deserts (Burbidge et al. 1988).

The engineering effects of bettongs, other deep-digging marsupials such as wombats and bilbies, and shallower diggers such as bandicoots, some dasyurids, and didelphids have been little appreciated. Unfortunately, many members of this underground set have declined or become extinct, especially in arid parts of Australia. It is becoming increasingly apparent that their losses have not just impoverished the regional marsupial faunas but also contributed to massive simplification of ecological processes, landscape function, and resilience (Dickman 2004).

Conservation

In Australasia, marsupials have generally fared poorly in the presence of people. In Australia, at least 36 species of large marsupials that were not part of the modern fauna survived into the late Pleistocene (Johnson and Prideaux 2004; Long et al. 2002; Murray 1984). Many of these were present after the arrival of humans, probably between 60,000 and 53,000 years ago (Roberts et al. 1994; Thorne et al. 1999), but disappeared subsequently and are known only from fossil remains. In New Guinea, where human presence can be traced back at least 40,000 years (Groube et al. 1986) and agricultural development to 10,000 years ago (Denham 2004), six species of megafauna vanished sometime in the late Pleistocene, three smaller species became extinct in the last 10,000 years, and many others have declined in range (Flannery 1994, 1995*a*). Another four species disappeared from elsewhere in Australasia (Aplin et al. 1999; Flannery 1995*b*) over the same period.

It is important to note that these numbers are minima; they are tallied from the relatively few fossil sites that have been studied and include few small species that are easy to overlook. It is not possible to distinguish the influence of human hunting and modification of habitat from the effects of climate change on the marsupial losses; this is at present the subject of vigorous debate (Roberts et al. 2001; Wroe et al. 2004*c*). Nonetheless, it is clear that, for Australasia, even before the disastrous arrival of Europeans, the legacy of the recent past is a substantially impoverished modern fauna.

In the Americas, by contrast, humans arrived at least 13,000 years ago (Meltzer 2004*a*, 2004*b*) and became established shortly thereafter (Bray 1988). Although these early colonists seem to have had little effect on marsupials (Redford and Eisenberg 1992), all species then present were small, and any negative effects on them of human settlement would be difficult to discern.

The arrival of Europeans in both regions has had dramatic and deleterious effects on marsupials. Assessments of status carried out by the World Conservation Union (IUCN) using standardized methods indicate that 52 of 189 extant species and subspecies of Australian marsupials are at some risk of extinction, as are 22 of 78 marsupials of the New Guinea region and 23 of 78 from Central and South America (Hilton-Taylor 2000). In addition to these numbers, 10 species and six subspecies of Australian marsupials are listed as extinct (Maxwell et al. 1996). And there may be more. One species of potoroid *(Bettongia pusilla)* is known from subfossil material in South and Western Australia and was probably present when Europeans first arrived (McNamara 1997).

The IUCN assessments are worth further consideration. In the first instance, "risk of extinction," as used above, covers three categories that express different degrees of risk. The most extreme category is "critically endangered." As defined by the IUCN, this includes species with populations declining at the rate of 90 percent in just 10 years or three generations, geographic distributions covering less than 100 km^2, small populations containing fewer than 250 mature individuals, or a 50 percent probability of extinction in the wild within 10 years or three generations, as predicted in quantitative analyses (IUCN 2001). Other criteria are used, too, such as whether the species is known to fluctuate or to aggregate in a single place and whether the species faces known threats. At present, seven marsupials fall within this most worrying of categories, four in Australia and three in South America.

The next most extreme IUCN category, "endangered," includes species with populations declining at a rate of 70 percent in 10 years or three generations, geographic distributions covering less than 5,000 km^2, populations with fewer than 2,500 mature individuals, or a 20 percent probability of extinction in the wild within 20 years or five generations. Listed in this category are 17 species in Australia, eight in the New Guinea region, and three in South America.

The third risk category is "vulnerable" and includes species with populations declining at a rate of 50 percent in 10 years or three generations, geographic distributions covering less than 20,000 km^2, populations with fewer than 10,000 mature individuals, or a 10 percent probability of extinction within 100 years. More marsupials are included in this formal risk category than in any other: 31 species and subspecies in Australia, 14 in the New Guinea region, and 17 in Central and South America.

A further IUCN category, "near threatened," allows for placement of species that do not meet the minimal requirements for listing as vulnerable but nonetheless exhibit trends in population size, distribution, or other indicators of status that warrant concern. This category contains an extraordinary 41 species and subspecies from Australia, in comparison with none from New Guinea and 18 from Central and South America. The most positive category, that of "least concern," contains just under half (39–48%) the marsupials in the three regions where they occur. Even in this category, however, there is little room for complacency. A recent review of the status of New World marsupials identified 9 of 33 species in the least concern category that had highly restricted ranges in two regions undergoing extensive habitat fragmentation, the tropical Andes and the Brazilian Atlantic forest (Fonseca et al. 2003). The review concluded that these species require timely attention.

In addition to the delineated categories of risk, 33 species of marsupials are considered too poorly known to be assigned reliably to a status category. Of these "data deficient" species, 25 occur in the New Guinea region, 3 in Australia, and 5 in South America. Much of the New Guinean fauna is genuinely poorly known because of difficulties of access and limited opportunities for field survey. For example, the ringtail possums *Pseudochirulus caroli, P. canescens,* and *P. schlegeli* are each known from a handful of specimens collected in a limited number of localities (Flannery 1995*a*). There is not enough information on population size, distribution, or threats to be confident that they are secure or at risk. Similarly, little can be said of the Seram bandicoot *(Rhynchomeles prattorum)*, an enigmatic species collected in 1920 in rugged rainforest on Mount Manusela, Seram (George and Maynes 1990). Surveys carried out in the second half of the twentieth century have failed to record it, but it is not clear whether inappropriate habitats were explored or whether the species has disappeared (Flannery 1995*b*). Data-deficient marsupials in Australia and South America also are species that have proven difficult or unreliable to find, even during targeted surveys.

The IUCN assessments show that many marsupials have declined in abundance and distribution and provide a gloomy prognosis for the future. Why have marsupials fared so poorly? In all regions, large areas of native vegetation have been fragmented and removed for agriculture and original faunas have been transformed by the introduction of exotic species.

In the Americas, European colonization began in earnest in the sixteenth century, especially in temperate portions of the southern cone and in high-altitude plateau areas of the tropics. Redford and Eisenberg (1992) chronicled the fluxes in settlement that took place at this time and drew attention to the vast numbers of horses, mules, cattle, and, later, sheep and pigs that were liberated from both abandoned and successful colonies. By 1699 the pampayan grasslands and other habitats were teeming with cattle and feral horses, allowing Charles Darwin to comment, in 1832, that the herds had "altered the whole aspect of the vegetation" (Darwin 1845, 121). Additional changes to vegetation, and the marsupials associated with it, were probably caused by introductions of deer, goats, and lagomorphs (Grigera and Rapoport 1983; Ramírez et al. 1981). Introductions of house mice and black and brown rats and the establishment of feral populations of cats and dogs may have had further effects on marsupials, although these have been little documented.

In the tropical regions of South America, large tracts of forest have been logged for timber or cleared for production of cattle or of cash crops such as coffee, bananas, soy, and sugar cane. Sharp increases in human populations in the Amazon basin are now being accompanied by increased frequency of fire and more extensive fragmentation, with some 3–4 million hectares of forest being destroyed each year (Cochrane 2001; Laurance et al. 2001). As fragmentation proceeds, the remaining forest becomes more susceptible to drought and climate changes (Laurance and Williamson 2001). Despite this onslaught on the tropical rainforests,

Burbidge and Eisenberg (2005) have noted that much forest still remains and that this habitat is likely to constitute a major repository of marsupial diversity for decades to come. Marsupials are likely to be at more immediate risk in the less extensive tropical dry forests and the temperate southerly rainforests, where fragmentation effects are acute and several species have small populations and reduced ranges (Chiarello 1999; Fonseca et al. 2003).

In Central America, field surveys and habitat modeling suggest that marsupials are faring variably because of regional differences in human population density and environmental disturbance (Cuarón 2000). Large areas of intact forest remain, although many are under threat from expanding agriculture, bioprospecting, and oil and mineral extraction (Medellín 1994). There is encouraging evidence, however, that some marsupials, at least, can persist in forest fragments amid agricultural land (Daily et al. 2003), especially the smaller, more generalist species.

Europeans visited Australia sporadically from at least the early seventeenth century, variously describing the small wallabies that they encountered as strange-looking rats, cats, or even "civet-cats" (Finney 1984). Sustained colonization began in 1788 and saw rapid but localized settlement of southeastern and then southwestern coastal districts, with Van Diemen's Land (Tasmania) established as a penal colony in 1803. Early coastal settlement was accompanied by limited clearing of land for building materials and for agricultural produce, but this was also a period when domestic cats, dogs, and pigs made their first forays into the bush and game animals such as deer were liberated (Rolls 1969). For several decades after the 1820s, there was a sustained quest to find new grazing lands and a bountiful (but mythical) inland sea. Since this period there has been a national obsession with maintaining large herds of grazing animals, especially sheep, and this has undoubtedly been the single most important factor driving declines and extinctions of Australian marsupials (Fisher et al. 2003; Lunney 2001).

Early warnings about the effects of the expanding pastoral industry on native mammals were made by Gould (1863) and Krefft (1866) and echoed by later commentators (e.g., Wood Jones 1923–24; Troughton 1932, 1944). The effects of overgrazing by sheep include degradation of native vegetation, removal of food and shelter resources, and damage to topsoil and hence the shelters of burrowing species. In addition, vast areas of woodland and scrub have been deliberately cleared to provide improved pasture for stock; indeed, until 2003, some 400,000 hectares of vegetation continued to be removed each year in Queensland and a further 60,000 hectares in New South Wales. Most Australian states also mandated the elimination of marsupials at some point to support pastoral interests. The extinction of the Tasmanian tiger is the most familiar example of this (Paddle 2000), but populations of many species of bandicoots, bettongs, and larger marsupials have been depleted in concerted poisoning, trapping, and shooting campaigns. During the 53-year tenure (1877–1930) of Marsupial Destruction Acts in Queensland, for example, some 27 million medium-sized terrestrial marsupials were killed (Hrdina 1997).

The effects of sheep and cattle on Australian marsupials are obvious because of the abundance and ubiquity of these introduced grazing animals in the landscape. Several other threats, however, are likely to affect particular marsupial species at certain times or places. These include loss or fragmentation of habitat (Maxwell et al. 1996; Morton 1990); overgrazing by nondomestic introduced species such as rabbits and hares (Jarman and Johnson 1977); predation, competition, and disease transmission from carnivores such as foxes, cats, and other in-

Golden bandicoot *(Isoodon auratus)*. Photograph by Elizabeth Tasker/Wildlife Images.

troduced mammals (Dickman 1996*a;* Smith and Quin 1996); changes in vegetation due to altered fire regimes (Allen 1983); inappropriate use of pesticides and poisons (Dickman 2004); and climate change (Brereton et al. 1995). In addition, populations of geographically restricted species such as island endemics and some rock wallabies often have limited genetic variation (Eldridge et al. 1999; Frankham 1998) and hence retain little potential to adapt to future environmental change. The Tasmanian devil may provide an unfortunate but dramatic example of this. This species is characterized by very limited genetic diversity, and, in the decade since the early 1990s, more than one-third of Tasmania's devil population has succumbed to a debilitating facial tumor disease (Mooney 2004). The relative importance of these various threats has been much debated, and it is likely that many marsupials are at risk because two or more threatening processes act at the same time and have synergistic effects (Allen 1983; Burbidge and McKenzie 1989).

Despite the difficulties of identifying particular threats, several studies have managed to elucidate the attributes of Australian marsupials that place them at risk. In general, threatened species tend to be specialists with limited geographic ranges, especially if they occupy nonwooded environments (Jones et al. 2003; Wilson et al. 2003). Species of intermediate body weight (35–5,500 grams) tend to be at greater risk (Burbidge and McKenzie 1989; Morton 1990), although recent analyses indicate that even large marsupials are vulnerable to decline and extinction (Cardillo and Bromham 2001). Most intriguingly, marsupials also seem to be at greater risk if they belong to genera with few species or that are "old" or phylogenetically distinct. Johnson et al. (2002) raised the possibility that old genera with a long evolutionary history may become more extinction-prone if they evolve toward greater ecological specialization or lower fecundity. Examples would include the marsupial moles, numbat, and Tasmanian tiger.

European interest in the New Guinean–Indonesian region dates back to the sixteenth century, with initial settlements being established in Micronesia and the Moluccas. The first written account of an Australasian marsupial can also be traced to this period, with an accurate description being made of the ornate cuscus *(Phalanger ornatus)*, apparently by António Galvão, Portuguese governor of the Moluccas from 1536 to 1540, in 1544 (Calaby 1984). In contrast to Australia, however, European encroachment in the New Guinea region has been relatively limited, and the major threats to marsupials are clearing and alteration of land, overhunting, and the rapid growth of indigenous human populations (Burbidge and Eisenberg 2005). Flannery (1995*a*, 1995*b*) listed 13 species of introduced mammals on the main island of New Guinea and 17 on islands elsewhere throughout the region but commented that he had observed decreases in numbers of the common echymipera *(Echymipera kalubu)* and a native rat only after the local introduction of domestic cats to a village in Sandaun Province. L. Salas has noted that unrestrained hunting dogs kill many marsupials in the vicinity of villages (personal communication, 2005).

Most people in the New Guinea region practice some form of agriculture, and in some areas, such as Simbu Province, most accessible land is used for growing fruit and staple crops of taro, sweet potato, or sago. There is little evidence that local farming has affected marsupials, but recent escalation in logging activities gives greater cause for concern. Saulei (1990) estimated that 90,000 hectares of forest were being logged each year in New Guinea for the trade in tropical timbers and foresaw sharp rises in rates of clearing due to increasing penetration of the region by rapacious multinational corporations. Coupled with overhunting, which has resulted historically in the local extinction of several species of large marsupials (Flannery 1992*a*, 1992*b*, 1994), and human populations that are increasing from in situ growth and Indonesian resettlement schemes, the loss and fragmentation of habitat due to agriculture, burning, and timber-getting are likely to place marsupials of the New Guinea region under increasing pressure in the future.

In the three major regions where marsupials occur, legal protection, limited reservation of habitat, and targeted conservation actions offer some hope that threatened (and common) species will continue to persist (Ceballos and Simonetti 2002; Leary and Mamu 2004; Maxwell et al. 1996). The long lists of threatened species and multiplicity of ongoing threats, however, show that there is no room for complacency. There is a clear need to increase public awareness of the plight of many marsupials and their habitats and of the finality and consequences of extinction. There is also a need to inform different levels of government so that political decisions about the environment are not made in ignorance of marsupials and other natural resources or at the behest only of developers, economic rationalists, and their ilk. The antiquity of the marsupial radiation, the extraordinary diversity of forms, the ecological, behavioral, physiological, and demographic strategies that marsupials exhibit, and the opportunities that they present for us to learn make it imperative that we conserve and manage these "alternative mammals" successfully.

Acknowledgments

This introduction was drafted while I was on sabbatical leave in the School of Animal Science at the University of Western Australia. I thank Dale Roberts for access to facilities, Don and Felicity Bradshaw for helpful discussions on marsupial biology, and especially Phil Withers for the provision of space, equipment, resources, and the camaraderie of his excellent laboratory. Discussions with Patsy Armati, Mike Calver, Chris Cooper, Mathew Crowther, Mark Garkaklis, Ian Hume, Emerson Vieira, Phil Withers, and Steve Wroe clarified my thinking on many points. Alex Baynes, Mathew Crowther, John Kirsch, Peter Meserve, Adrián Monjeau, Bill Poole, Leo Salas, Marcelo Sánchez, Phil Withers, and Steve Wroe provided detailed and critical comments on parts or all of the manuscript; Alex

Common wombat *(Vombatus ursinus)*. Photograph by Pavel German / Wildlife Images.

Baynes alerted me to many new references; and Vince Burke provided editorial assistance. Tim Flannery, Pavel German, Elizabeth Tasker, and Hugh Tyndale-Biscoe kindly provided images of marsupials, while Carol McKechnie provided support throughout the writing process.

Literature Cited

Abbie, A. A. 1939. A masticatory adaptation peculiar to some diprotodont marsupials. Proc. Zool. Soc. Lond. B 109:261–79.

Albuja, L. V., and B. D. Patterson. 1996. A new species of northern shrew-opossum (Paucituberculata: Caenolestidae) from the Cordillera del Cóndor, Ecuador. J. Mammal. 77:41–53.

Allen, C. B. 2001. Analysis of dietary competition between three sympatric herbivores in semi-arid west Queensland. Ph.D. diss., Univ. of Sydney.

Allen, H. 1983. Nineteenth c. faunal change in western NSW and N-W Victoria. Working Papers in Anthropology, Archaeology, Linguistics and Maori Studies, Univ. of Auckland, 69 pp.

Amico, G., and M. A. Aizen. 2000. Mistletoe seed dispersal by a marsupial. Nature 408:929–30.

Amrine-Madsen, H., M. Scally, M. Westerman, M. J. Stanhope, C. Krajewski, and M. S. Springer. 2003. Nuclear gene sequences provide evidence for the monophyly of australidelphian marsupials. Mol. Phyl. Evol. 28:186–96.

Anderson, T. J. C., A. J. Berry, J. N. Amos, and J. M. Cook. 1988. Spool and line tracking of the New Guinea spiny bandicoot, *Echymipera kalubu*. J. Mammal. 69:114–20.

Andrew, D. L., and G. A. Settle. 1982. Observations on the behaviour of species of *Planigale* (Dasyuridae, Marsupialia) with particular reference to the narrow-nosed planigale *(Planigale tenuirostris)*. *In* Archer, M., ed., Carnivorous marsupials, 1:311–24. Royal Zoological Society of New South Wales, Sydney.

Aplin, K. P., and M. Archer. 1987. Recent advances in marsupial systematics with a new syncretic classification. *In* Archer, M., ed., Possums and opossums: studies in evolution, 1:xv–lxxii. Surrey Beatty & Sons and Royal Zoological Society of New South Wales, Sydney.

Aplin, K. P., P. R. Baverstock, and S. C. Donnellan. 1993. Albumin immunological evidence for the time and mode of origin of the New Guinean terrestrial mammal fauna. Sci. New Guinea 19:131–45.

Aplin, K. P., J. M. Pasveer, and W. E. Boles. 1999. Late Quaternary vertebrates from the Bird's Head Peninsula, Irian Jaya, Indonesia, including descriptions of two previously unknown marsupial species. Rec. West. Aust. Mus. Suppl. 57:351–87.

Archer, M. 1974. The development of the cheek-teeth in *Antechinus flavipes* (Marsupialia, Dasyuridae). J. Roy. Soc. West. Aust. 57:54–63.

———. 1978. The nature of the molar-premolar boundary in marsupials and a re-interpretation of the homology of marsupial cheek-teeth. Mem. Qld Mus. 18:157–64.

———. 1981. A review of the origins and radiations of Australian mammals. *In* Keast, A., ed., Ecological biogeography of Australia, 1437–88. Junk, The Hague.

———. 1982. Genesis: and in the beginning there was an incredible carnivorous mother. *In* Archer, M., ed., Carnivorous marsupials, 1:vii–x. Royal Zoological Society of New South Wales, Sydney.

———. 1984*a*. Origins and early radiations of marsupials. *In* Archer, M., and G. Clayton, eds., Vertebrate zoogeography and evolution in Australasia, 585–625. Hesperian Press, Perth.

———. 1984*b*. The Australian marsupial radiation. *In* Archer, M., and G. Clayton, eds., Vertebrate zoogeography and evolution in Australasia, 633–808. Hesperian Press, Perth.

Archer, M., and J. Kirsch. 2005. The evolution and classification of marsupials. *In* Armati, P. J., C. R. Dickman, and I. D. Hume, eds., Marsupials. Cambridge Univ. Press, Cambridge, in press.

Archer, M., I. Burnley, J. Dodson, R. Harding, L. Head, and P. A. Murphy. 1998. From plesiosaurs to people: 100 million years of Australian environmental history. State of the Env. Tech. Pap. Ser. (Portrait of Australia), Dept. of the Env., Canberra, 66 pp.

Archer, M., R. Arena, M. Bassarova, K. Black, J. Brammall, B. Cooke, P. Creaser, K. Crosby, A. Gillespie, H. Godthelp, M. Gott, S. J. Hand, B. Kear, A. Krikmann, B. Mackness, J. Muirhead, A. Musser, T. Myers, N. Pledge, Y. Wang, and S. Wroe. 1999. The evolutionary history and diversity of Australian mammals. Aust. Mammal. 21:1–45.

Argot, C. 2002. Functional-adaptive analysis of the hind limb anatomy of extant marsupials and the paleobiology of the Paleocene marsupials *Mayulestes ferox* and *Pucadelphys andinus*. J. Morph. 253:76–108.

———. 2003. Functional-adaptive anatomy of the axial skeleton of some extant marsupials and the paleobiology of the Paleocene marsupials *Mayulestes ferox* and *Pucadelphys andinus*. J. Morph. 255:279–300.

Arroyo, M. T. K., M. Riveros, A. Peñaloza, L. Cavieres, and A. M. Faggi. 1996. Phytogeographic relationships and regional richness patterns of the cool temperate rainforest flora of southern South America. *In* Lawford, R. G., P. B. Alaback, and E. Fuentes, eds., High-latitude rainforests and associated ecosystems of the west coast of the Americas: climate, hydrology, ecology and conservation, 134–72. Springer-Verlag, Berlin.

Asher, R. J., I. Horovitz, and M. R. Sánchez-Villagra. 2004. First combined cladistic analysis of marsupial mammal interrelationships. Mol. Phyl. Evol. 33:240–50.

Astúa de Moraes, D., R. T. Santori, R. Finotti, and R. Cerqueira. 2003. Nutritional and fibre contents of laboratory-established diets of Neotropical opossums (Didelphidae). *In* Jones, M. E., C. R. Dickman, and M. Archer, eds., Predators with pouches: the biology of carnivorous marsupials, 229–37. CSIRO Publishing, Melbourne.

Atramentowicz, M. 1992. Optimal litter size: does it cost more to raise a large litter in *Caluromys philander?* Can. J. Zool. 70:1511–15.

Austad, S. N. 1988. The adaptable opossum. Sci. Am. 258:54–59.

Austad, S. N., and M. E. Sunquist. 1986. Sex-ratio manipulation in the common opossum. Nature 324:58–60.

Averianov, A. O., J. D. Archibald, and T. Martin. 2003. Placental nature of the alleged marsupial from the Cretaceous of Madagascar. Acta Palaeont. Pol. 48:149–51.

Bailey, P., and L. Best. 1992. A red kangaroo, *Macropus rufus,* recovered 25 years after marking in north-western New South Wales. Aust. Mammal. 15:141.

Baker, L., S. Woenne-Green, and the Mutitjulu Community. 1993. Anangu knowledge of vertebrates and the environment. *In* Reid, J. R. W., J. A. Kerle, and S. R. Morton, eds., Uluru fauna: the distribution and abundance of vertebrate fauna of Uluru (Ayers Rock–Mount Olga) National Park, N.T., 79–132. Australian National Parks and Wildlife Service, Canberra.

Barbour, R. A. 1977. Anatomy of marsupials. *In* Stonehouse, B., and D. Gilmore, eds., The biology of marsupials, 237–62. Macmillan, London.

Barnes, R. D., and S. M. Barthold. 1969. Reproduction and breeding behaviour in an experimental colony of *Marmosa mitis* Bangs (Didelphidae). J. Reprod. Fert. Suppl. 6:477–92.

Baynes, A., and K. A. Johnson. 1996. The contributions of the Horn Expedition and cave deposits to knowledge of the original mammal fauna of central Australia. *In* Morton, S. R., and D. J. Mulvaney, eds., Exploring central Australia: society, the environment and the 1894 Horn Expedition, 168–86. Surrey Beatty & Sons, Sydney.

Beal, A. M. 1989. Differences in salivary flow and composition among kangaroo species: implications for digestive efficiency. *In* Grigg, G. C., P. Jarman, and I. D. Hume, eds., Kangaroos, wallabies and rat-kangaroos, 1:189–95. Surrey Beatty & Sons, Sydney.

Benshemesh, J., and K. Johnson. 2003. Biology and conservation of marsupial moles *(Notoryctes). In* Jones, M. E., C. R. Dickman, and M. Archer, eds., Predators with pouches: the biology of carnivorous marsupials, 464–74. CSIRO Publishing, Melbourne.

Bergallo, H. G., and R. Cerqueira. 1994. Reproduction and growth of the opossum *Monodelphis domestica* (Mammalia: Didelphidae), in northeastern Brazil. J. Zool. Lond. 232:551–63.

Biggers, J. D., and E. D. DeLamater. 1965. Marsupial spermatozoa pairing in the epididymis of American forms. Nature 208:402–4.

Birney, E. C., and J. A. Monjeau. 2003. Latitudinal variation in South American marsupial biology. *In* Jones, M. E., C. R. Dickman, and M. Archer, eds., Predators with pouches: the biology of carnivorous marsupials, 297–317. CSIRO Publishing, Melbourne.

Birney, E. C., J. A. Monjeau, C. J. Phillips, R. S. Sikes, and I. Kim. 1996*a*. *Lestodelphys halli:* new information on a poorly-known Argentine marsupial. Mast. Zool. 3:171–81.

Birney, E. C., R. S. Sikes, J. A. Monjeau, N. Guthmann, and C. J. Phillips. 1996*b*. Comments on Patagonian marsupials from Argentina. *In* Genoways, H. H., and R. J. Baker, eds., Contributions in mammalogy: a memorial volume honoring Dr. J. Knox Jones, Jr., 149–54. Museum of Texas Tech Univ., Lubbock.

Blacket, M. J., M. Adams, C. Krajewski, and M. Westerman. 2000. Genetic variation within the dasyurid marsupial genus *Planigale.* Aust. J. Zool. 48:443–59.

Blackhall, S. 1980. Diet of the eastern native-cat, *Dasyurus viverrinus* (Shaw), in southern Tasmania. Aust. Wildl. Res. 7:191–97.

Bowman, D. M. J. S., and J. C. Z. Woinarski. 1994. Biogeography of Australian monsoon rainforest mammals: implications for the conservation of rainforest mammals. Pac. Cons. Biol. 1:98–106.

Bowyer, J. C., G. R. Newell, C. J. Metcalfe, and M. B. D. Eldridge. 2003. Tree-kangaroos *Dendrolagus* in Australia: are *D. lumholtzi* and *D. bennettianus* sister taxa? Aust. Zool. 32:207–13.

Bradley, A. J. 1997. Reproduction and life-history in the red-tailed phascogale, *Phascogale calura* (Marsupialia: Dasyuridae): the adaptive-stress senescence hypothesis. J. Zool. Lond. 241:739–55.

———. 2003. Stress, hormones and mortality in small carnivorous marsupials. *In* Jones, M. E., C. R. Dickman, and M. Archer, eds., Predators with pouches: the biology of carnivorous marsupials, 254–67. CSIRO Publishing, Melbourne.

Bradshaw, F., L. Everett, and D. Bradshaw. 2000. On the rearing of honey possums. West. Aust. Nat. 22:281–88.

Bradshaw, F. J., and S. D. Bradshaw. 2001. Maintenance nitrogen requirement of an obligate nectarivore, the honey possum, *Tarsipes rostratus*. J. Comp. Physiol. B 171:59–67.

Braithwaite, L. W., M. L. Dudziński, and J. Turner. 1983. Studies on the arboreal marsupial fauna of eucalypt forests being harvested for woodpulp at Eden, N.S.W. II. Relationship between the fauna density, richness and diversity, and measured variables of the habitat. Aust. Wildl. Res. 10:231–47.

Braithwaite, L. W., J. Turner, and J. Kelly. 1984. Studies on the arboreal marsupial fauna of eucalypt forests being harvested for woodpulp at Eden, N.S.W. III. Relationships between faunal densities, eucalypt occurrence and foliage nutrients, and soil parent materials. Aust. Wildl. Res. 11:41–48.

Braithwaite, R. W., and A. K. Lee. 1979. A mammalian example of semelparity. Am. Nat. 113:151–56.

Bray, W. 1988. How old are the Americans? Américas 40:50–55.

Brereton, R., S. Bennett, and I. Mansergh. 1995. Enhanced greenhouse climate change and its potential effect on selected fauna of south-eastern Australia: a trend analysis. Biol. Cons. 72:339–54.

Brown, B. E. 2004. Atlas of New World marsupials. Fieldiana: Zool., New Series 102:1–308.

Bulmer, R. N. H., and J. I. Menzies. 1972. Karam classification of marsupials and rodents. Part 1. J. Polynesian Soc. 81:472–99.

———. 1973. Karam classification of marsupials and rodents. Part 2. J. Polynesian Soc. 82:86–107.

Burbidge, A. A., and J. F. Eisenberg. 2005. Conservation and management. *In* Armati, P. J., C. R. Dickman, and I. D. Hume, eds., Marsupials. Cambridge Univ. Press, Cambridge, in press.

Burbidge, A. A., and N. L. McKenzie. 1989. Patterns in the modern decline of Western Australia's vertebrate fauna: causes and conservation implications. Biol. Cons. 50:143–98.

Burbidge, A. A., K. A. Johnson, P. J. Fuller, and R. I. Southgate. 1988. Aboriginal knowledge of the mammals of the central deserts of Australia. Aust. Wildl. Res. 15:9–39.

Burk, A., M. Westerman, and M. S. Springer. 1998. The phylogenetic position of the musky rat-kangaroo and the evolution of bipedal hopping in kangaroos (Macropodidae: Diprotodontia). Syst. Biol. 47:457–74.

Burk, A., M. Westerman, D. J. Kao, J. R. Kavanagh, and M. S. Springer. 1999. An analysis of marsupial interordinal relationships based on 12S rRNA, tRNA, valine, 16S rRNA, and cytochrome *b* sequences. J. Mammal. Evol. 6:317–34.

Busch, M., and F. O. Kravetz. 1991. Diet composition of *Monodelphis dimidiata* (Marsupialia, Didelphidae). Mammalia 55:619–21.

Butcher, S. 2000. Impact of possums on primary production. *In* Montague, T. L., ed., The brushtail possum: biology, impact and management of an introduced marsupial, 105–10. Manaaki Whenua Press, Lincoln.

Byers, J. A. 1999. The distribution of play behaviour among Australian marsupials. J. Zool. Lond. 247:349–56.

Calaby, J. H. 1984. Foreword. *In* Smith, A. P., and I. D. Hume, eds., Possums and gliders, iii–iv. Surrey Beatty & Sons, and Australian Mammal Society, Sydney.

Calaby, J. H., and W. E. Poole. 1971. Keeping kangaroos in captivity. Int. Zoo Yearbook 11:5–12.

Callister, D. J. 1991. A review of the Tasmanian brushtail possum industry. Traffic Bull. 12:49–58.

Cannon, J. R., H. R. Bakker, S. D. Bradshaw, and I. R. McDonald. 1976. Gravity as the sole navigational aid to the newborn quokka. Nature 259:42.

Cardillo, M., and L. Bromham. 2001. Body size and risk of extinction in Australian mammals. Cons. Biol. 15:1435–40.

Cardillo, M., O. R. P. Bininda-Emonds, E. Boakes, and A. Purvis. 2004. A species-level phylogenetic supertree of marsupials. J. Zool. Lond. 264:11–31.

Carthew, S. M., and R. L. Goldingay. 1997. Non-flying mammals as pollinators. Trends Ecol. Evol. 12:104–8.

Carvalho, F. M. V., P. S. Pinheiro, F. A. S. Fernandez, and J. L. Nessimian. 1999. Diet of small mammals in Atlantic forest fragments in south-eastern Brazil. Rev. Brasil. Zoociências 1:91–101.

Case, J. A., M. O. Woodburne, and D. S. Chaney. 1988. A new genus and species of polydolopid marsupial from the La Meseta formation, late Eocene, Seymour Island, Antarctic Peninsula. *In* Feldmann, R. M., and M. O. Woodburne, eds., Geology and paleontology of Seymour Island, Geol. Soc. Am. Mem. No. 169, Boulder, 505–21.

Ceballos, G., and J. A. Simonetti, eds. 2002. Diversidad y conservación de los mamíferos neotropicales. Comisión Nacional para el Conocimiento y Uso de la Biodiversidad, Mexico, Distrito Federal, Mexico, 582 pp.

Cerqueira, R. 1985. The distribution of *Didelphis* in South America (Polyprotodontia, Didelphidae). J. Biogeog. 12:135–45.

Charles-Dominique, P. 1983. Ecology and social adaptations in didelphid marsupials: comparison with eutherians of similar ecology. *In* Eisenberg, J. F., and D. G. Kleiman, eds., Advances in the study of mammalian behavior, 395–422. American Society of Mammalogists, Shippensburg, Special Publ. No. 7.

Charles-Dominique, P., M. Atramentowicz, M. Charles-Dominique, H. Gérard, A. Hladik, C. M. Hladik, and M. F. Prévost. 1981. Les mammifères frugivores arboricoles nocturnes d'une forêt guyanaise: inter-relations plantes-animaux. Rev. Ecol. (La Terre et la Vie) 35:341–435.

Chiarello, A. G. 1999. Effects of fragmentation of the Atlantic forest on mammal communities in south-eastern Brazil. Biol. Cons. 89:71–82.

Chilcott, M. J. 1984. Coprophagy in the common ringtail possum, *Pseudocheirus peregrinus* (Marsupialia: Petauridae). Aust. Mammal. 7:107–10.

Christensen, P. E. S. 1980. The biology of *Bettongia penicillata* Gray 1837, and *Macropus eugenii* (Desmarest, 1817), in relation to fire. For. Dept. West. Aust. Bull. 91:1–90.

Cifelli, R. L., and B. M. Davis. 2003. Marsupial origins. Science 302:1899–1900.

Cifelli, R. L., and C. de Muizon. 1997. Dentition and jaw of *Kokopellia juddi*, a primitive marsupial or near-marsupial from the medial Cretaceous of Utah. J. Mammal. Evol. 4:241–58.

Claridge, A. W., and S. C. Barry. 2000. Factors influencing the distribution of medium-sized ground-dwelling mammals in southeastern mainland Australia. Austral Ecol. 25:676–88.

Claridge, A. W., and T. W. May. 1994. Mycophagy among Australian mammals. Aust. J. Ecol. 19:251–75.

Claridge, A. W., M. T. Tanton, and R. B. Cunningham. 1993. Hypogeal fungi in the diet of the long-nosed potoroo *(Potorous tridactylus)* in mixed-species and regrowth eucalypt forest stands in south-eastern Australia. Wildl. Res. 20:321–37.

Cochrane, M. A. 2001. Synergistic interactions between habitat fragmentation and fire in evergreen tropical forests. Cons. Biol. 15:1515–21.

Cockburn, A. 1990*a*. Sex ratio variation in marsupials. Aust. J. Zool. 37:467–79.

———. 1990*b*. Life history of the bandicoots: developmental rigidity and phenotypic plasticity. *In* Seebeck, J. H., P. R. Brown, R. L. Wallis, and C. M. Kemper, eds., Bandicoots and bilbies, 285–92. Surrey Beatty & Sons, and Australian Mammal Society, Sydney.

———. 1997. Living slow and dying young: senescence in marsupials. *In* Saunders, N. R., and L. A. Hinds, eds., Marsupial biology: recent research, new perspectives, 163–74. Univ. New South Wales Press, Sydney.

Colagross, A. M. L., and A. Cockburn. 1993. Vigilance and grouping in the eastern grey kangaroo, *Macropus giganteus*. Aust. J. Zool. 41:325–34.

Colgan, D. J. 1999. Phylogenetic studies of marsupials based on phosphoglycerate kinase DNA sequences. Mol. Phyl. Evol. 11:13–26.

Cooper, C. 2003. Physiological specialisations of the numbat, *Myrmecobius fasciatus* Waterhouse 1832 (Marsupialia: Myrmecobiidae): a unique termitivorous marsupial. Ph.D. diss., Univ. of Western Australia, Perth.

Cooper, N. K., K. P. Aplin, and M. Adams. 2000. A new species of false antechinus (Marsupialia: Dasyuromorphia: Dasyuridae) from the Pilbara region, Western Australia. Rec. West. Aust. Mus. 20:115–36.

Cordero, G. A., and R. A. Nicolas. 1987. Feeding habits of the opossum *(Didelphis marsupialis)* in northern Venezuela. *In* Patterson, B. D., and R. M. Timms, eds., Studies in Neotropical mammalogy: essays in honor of Philip Hershkovitz. Fieldiana: Zool., New Series 39:125–31.

Cork, S. J., I. D. Hume, and T. J. Dawson. 1983. Digestion and metabolism of a mature foliar diet *(Eucalyptus punctata)* by an arboreal marsupial, the koala *(Phascolarctos cinereus)*. J. Comp. Physiol. B 153:181–90.

Cox, C. B. 2000. Plate tectonics, seaways and climate in the historical biogeography of mammals. Mem. Inst. Oswaldo Cruz, Rio de Janeiro 95:509–16.

Crespo, J. A. 1982. Ecología de la comunidad de mamíferos del Parque Nacional Iguazú, Misiones. Rev. Argent. Cienc. Nat. "Bernardino Rivadavia," Ecol. 3:45–162.

Croft, D. B., and J. F. Eisenberg. 2005. Behaviour of marsupials. *In* Armati, P. J., C. R. Dickman, and I. D. Hume, eds., Marsupials. Cambridge University Press, Cambridge, in press.

Croft, D. B., and U. Ganslosser, eds. 1996. Comparison of marsupial and placental behaviour. Filander, Fuerth, 303 pp.

Crowther, M. S., and M. J. Blacket. 2003. Biogeography and speciation in the Dasyuridae: why are there so many kinds of dasyurids? *In* Jones, M. E., C. R. Dickman, and M. Archer, eds., Predators with pouches: the biology of carnivorous marsupials, 124–30. CSIRO Publishing, Melbourne.

Cuarón, A. D. 2000. Effects of land-cover changes on mammals in a Neotropical region: a modeling approach. Cons. Biol. 14:1676–92.

Cunningham, S. A. 1991. Experimental evidence for pollination of *Banksia* spp. by non-flying mammals. Oecologia 87:86–90.

Daily, G. C., G. Ceballos, J. Pacheco, G. Suzán, and A. Sánchez-Azofeifa. 2003. Countryside biogeography of Neotropical mammals: conservation opportunities in agricultural landscapes of Costa Rica. Cons. Biol. 17:1814–26.

Darwin, C. 1845. The voyage of the "Beagle." Facsimile ed., 1968. Heron Books, Geneva, 551 pp.

Dawson, T. J. 1989. Diets of macropodoid marsupials: general patterns and environmental influences. *In* Grigg, G. C., P. Jarman, and I. D. Hume, eds., Kangaroos, wallabies and rat-kangaroos, 1:129–42. Surrey Beatty & Sons, Sydney.

———. 1995. Kangaroos: biology of the largest marsupials. Univ. of New South Wales Press, Sydney, 162 pp.

Dawson, T. J., E. Finch, L. Freedman, I. D. Hume, M. B. Renfree, and P. D. Temple-Smith. 1989. Morphology and physiology of the Metatheria. *In* Walton, D. W., and B. J. Richardson, eds., Fauna of Australia. Vol. 1B: Mammalia, 451–504. Australian Government Publishing Service, Canberra.

Degabriele, R., and T. J. Dawson. 1979. Metabolism and heat balance in an arboreal marsupial, the koala *Phascolarctos cinereus*. J. Comp. Physiol. B 134:293–301.

Denham, T. 2004. Agriculture's origins in the highlands of New Guinea. Aust. Sci. 25:23–26.

Dennis, A. J. 2002. The diet of the musky rat-kangaroo, *Hypsiprymnodon moschatus*, a rainforest specialist. Wildl. Res. 29:209–19.

———. 2003. Scatter-hoarding by musky rat-kangaroos, *Hypsiprymnodon moschatus*, a tropical rain-forest marsupial from Australia: implications for seed dispersal. J. Tropical Ecol. 19:619–27.

Dennis, A. J., and H. Marsh. 1997. Seasonal reproduction in musky rat-kangaroos, *Hypsiprymnodon moschatus:* a response to changes in resource availability. Wildl. Res. 24:561–78.

Díaz, M. M., D. A. Flores, and R. M. Barquez. 2002. A new species of gracile mouse opossum, genus *Gracilinanus* (Didelphimorphia: Didelphidae), from Argentina. J. Mammal. 83:824–33.

Dickman, C. R. 1984. Competition and coexistence among the small marsupials of Australia and New Guinea. Acta Zool. Fenn. 172:27–31.

———. 1986*a*. An experimental study of competition between two species of dasyurid marsupials. Ecol. Monogr. 56:221–241.

———. 1986*b*. Niche compression: two tests of an hypothesis using narrowly sympatric predator species. Aust. J. Ecol. 11:121–34.

———. 1989. Patterns in the structure and diversity of marsupial carnivore communities. *In* Morris, D. W., Z. Abramsky, B. J. Fox, and M. R. Willig, eds., Patterns in the structure of mammalian communities, 241–51. Texas Tech Univ. Press, Lubbock.

———. 1991. Use of trees by ground-dwelling mammals: implications for management. *In* Lunney, D., ed., Conservation of Australia's forest fauna, 125–36. Royal Zoological Society of New South Wales, Sydney.

———. 1993. Evolution of semelparity in male dasyurid marsupials: a critique, and an hypothesis of sperm competition. *In* Roberts, M., J. Carnio, G. Crawshaw, and M. Hutchins, eds., The biology and management of Australasian carnivorous marsupials, 25–38. Metropolitan Toronto Zoo, Ontario, and American Association of Zoological Parks and Aquariums, Washington, D.C.

———. 1996*a*. Overview of the impacts of feral cats on Australian native fauna. Australian Nature Conservation Agency, Canberra, 92 pp.

———. 1996*b*. Vagrants in the desert. Nature Aust. 25:54–62.

———. 2003. Distributional ecology of dasyurid marsupials. *In* Jones, M. E., C. R. Dickman, and M. Archer, eds., Predators with pouches: the biology of carnivorous marsupials, 318–31. CSIRO Publishing, Melbourne.

———. 2004. Mammals of the Darling River basin. *In* Breckwoldt, R., R. Boden, and J. Andrew, eds., The Darling River, 169–91. Murray-Darling Basin Commission, Canberra.

Dickman, C. R., and E. M. Russell. 2001. Marsupials. *In* Macdonald, D. W., ed., The new encyclopedia of mammals, 802–7. Oxford Univ. Press, Oxford.

Dickman, C. R., and E. M. Vieira. 2005. Ecology and life histories. *In* Armati, P. J., C. R. Dickman, and I. D. Hume, eds., Marsupials. Cambridge University Press, Cambridge, in press.

Dickman, C. R., H. E. Parnaby, M. S. Crowther, and D. H. King. 1998. *Antechinus agilis* (Marsupialia: Dasyuridae), a new species from the *A. stuartii* complex in southeastern Australia. Aust. J. Zool. 46:1–26.

Dixon, J. M., and L. Huxley, eds. 1985. Donald Thomson's mammals and fishes of northern Australia. Thomas Nelson, Melbourne, 210 pp.

Dow, D. B. 1977. A geological synthesis of Papua New Guinea. Aust. Bur. Miner. Res. Geol. Geophys. Bull. 201, 41 pp.

Dow, D. B., and R. Sukamto. 1984. Late Tertiary to Quaternary tectonics of Irian Jaya. Episodes 7:3–9.

Dungan, R. J., M. J. O'Cain, M. L. Lopez, and D. A. Norton. 2002. Contribution by possums to seed rain and subsequent seed germination in successional vegetation, Canterbury, New Zealand. New Zealand J. Ecol. 26:121–28.

Dwiyahreni, A. A., M. F. Kinnaird, T. G. O'Brien, J. Supriatna, and N. Andayani. 1999. Diet and activity of the bear cuscus, *Ailurops ursinus*, in north Sulawesi, Indonesia. J. Mammal. 80:905–12.

Dwyer, P. D. 1974. The price of protein: five hundred hours of hunting in the New Guinea highlands. Oceania 44:278–93.

Ealey, E. H. M. 1967. Ecology of the euro, *Macropus robustus* (Gould) in north-western Australia. CSIRO Wildl. Res. 12:9–80.

Eberhard, I. H. 1978. Ecology of the koala, *Phascolarctos cinereus* (Goldfuss) Marsupialia: Phascolarctidae, in Australia. *In* Montgomery, G. G., ed., The ecology of arboreal folivores, 315–28. Smithsonian Institution Press, Washington, D.C.

Eisenberg, J. F. 1981. The mammalian radiations: an analysis of trends in evolution, adaptation, and behavior. Univ. of Chicago Press, Chicago, 610 pp.

———. 1989. Mammals of the Neotropics. Vol. 1: The northern Neotropics: Panama, Colombia, Venezuela, Guyana, Suriname, French Guiana. Univ. of Chicago Press, Chicago, 449 pp.

Eisenberg, J. F., and K. H. Redford. 1999. Mammals of the Neotropics. Vol. 3: The central Neotropics: Ecuador, Peru, Bolivia, Brazil. Univ. of Chicago Press, Chicago, 609 pp.

Eisenberg, J. F., and D. E. Wilson. 1981. Relative brain size and demographic strategies in didelphid marsupials. Am. Nat. 118:1–15.

Eldridge, M. D. B., J. M. King, A. K. Loupis, P. B. S. Spencer, A. C. Taylor, L. C. Pope, and G. P. Hall. 1999. Unprecedented low levels of genetic variation and inbreeding depression in an island population of the black-footed rock-wallaby. Cons. Biol. 13:531–41.

Eldridge, M. D. B., A. C. C. Wilson, C. J. Metcalfe, A. E. Dollin, J. N. Bell, P. M. Johnson, P. G. Johnston, and R. L. Close. 2001. Taxonomy of rock-wallabies, *Petrogale* (Marsupialia: Macropodidae). III. Molecular data confirm the species status of the purple-necked rock-wallaby *Petrogale purpureicollis* Le Souef 1924. Aust. J. Zool. 49:323–43.

Emmons, L. H., and F. Feer. 1997. Neotropical rainforest mammals: a field guide, 2nd ed. Univ. of Chicago Press, Chicago, 307 pp.

Evans, M. C. 1992. Diet of the brushtail possum *Trichosurus vulpecula* (Marsupialia: Phalangeridae) in central Australia. Aust. Mammal. 15:25–30.

Ewer, R. F. 1968. Ethology of mammals. Elek Science, London, 418 pp.

Fadem, B. H. 1986. Chemical communication in gray short-tailed opossums *(Monodelphis domestica)* with comparisons to other marsupials and with reference to monotremes. *In* Duvall, D., D. Müller-Schwarze, and R. M. Silverstein, eds., Chemical signals in vertebrates, 4:587–607. Plenum Press, New York.

Fadem, B. H., and R. S. Rayve. 1985. Characteristics of the oestrous cycle and influence of social factors in grey short-tailed opossums, *Monodelphis domestica*. J. Reprod. Fert. 73:337–42.

Finlayson, H. H. 1961. On central Australian mammals. Part IV: The distribution and status of central Australian species. Rec. South Aust. Mus. 14:141–91.

Finney, C. M. 1984. To sail beyond the sunset: natural history in Australia, 1699–1829. Rigby Publishers, Adelaide, 206 pp.

Fish, F. E. 1993. Comparison of swimming kinematics between terrestrial and semiaquatic opossums. J. Mammal. 74:275–84.

Fisher, C. T. 1984. Australasian mammal specimens in the collections of Merseyside County Museums. Aust. Mammal. 7:205–13.

Fisher, D. O., I. P. F. Owens, and C. N. Johnson. 2001. The ecological basis of life history variation in marsupials. Ecology 82:3531–40.

Fisher, D. O., S. P. Blomberg, and I. P. F. Owens. 2003. Extrinsic versus intrinsic factors in the decline and extinction of Australian marsupials. Proc. Roy. Soc. Lond. B 270: 1801–8.

Flannery, T. F. 1989. Origins of the Australo-Pacific land mammal fauna. Aust. Zool. Rev. 1:15–24.

———. 1992*a*. Taxonomic revision of the *Thylogale brunii* complex (Macropodidae: Marsupialia) in Melanesia, with description of a new species. Aust. Mammal. 15:7–23.

———. 1992*b*. New Pleistocene marsupials (Macropodidae, Diprotodontidae) from subalpine habitats in Irian Jaya. Alcheringa 16:321–31.

———. 1994. The fossil land mammal record of New Guinea: a review. Sci. New Guinea 20:39–48.

———. 1995*a*. Mammals of New Guinea, 2nd ed. Reed Books, Sydney, 568 pp.

———. 1995*b*. Mammals of the south-west Pacific and Moluccan Islands. Reed Books, Sydney, 464 pp.

Flannery, T. F., and Boeadi. 1995. Systematic revision within the *Phalanger ornatus* complex (Phalangeridae: Marsupialia), with description of a new species and subspecies. Aust. Mammal. 18:35–44.

Flannery, T. F., and J. P. White. 1991. Animal translocation: zoogeography of New Ireland mammals. Nat. Geog. Res. Expl. 7:96–113.

Flannery, T. F., R. Martin, and A. Szalay. 1996. Tree kangaroos: a curious natural history. Reed Books, Sydney, 202 pp.

Fleay, D. 1947. Gliders of the gum trees. Bread and Cheese Club, Melbourne, 113 pp.

Fleming, T. H. 1973. The reproductive cycles of three species of opossums and other mammals in the Panama Canal Zone. J. Mammal. 54:439–55.

Flynn, J. J., and A. R. Wyss. 1999. New marsupials from the Eocene-Oligocene transition of the Andean main range, Chile. J. Vert. Paleo. 19:533–49.

Fonseca, G. A. B., and M. C. M. Kierulff. 1989. Biology and natural history of Brazilian Atlantic forest small mammals. Bull. Florida State Mus. Biol. Sci. 34:99–152.

Fonseca, G. A. B., G. Herrmann, Y. L. R. Leite, R. A. Mittermeier, A. B. Rylands, and J. L. Patton. 1996. Lista anotada dos mamíferos do Brasil. Occ. Pap. No. 4, Conservation International, Washington, D.C., and Fundação Biodiversitas, Belo Horizonte, 38 pp.

Fonseca, G. A. B., A. P. Paglia, J. Sanderson, and R. A. Mittermeier. 2003. Marsupials of the New World: status and conservation. *In* Jones, M. E., C. R. Dickman, and M. Archer, eds., Predators with pouches: the biology of carnivorous marsupials, 399–406. CSIRO Publishing, Melbourne.

Fox, B. J. 1989. Community ecology of macropodoids. *In* Grigg, G. C., P. Jarman, and I. D. Hume, eds., Kangaroos, wallabies and rat-kangaroos, 1:89–104. Surrey Beatty & Sons, Sydney.

———. 1999. The genesis and development of guild assembly rules. *In* Weiher, E., and P. Keddy, eds., Ecological assembly rules: perspectives, advances, retreats, 23–57. Cambridge Univ. Press, Cambridge.

Frankham, R. 1998. Inbreeding and extinction: island populations. Cons. Biol. 12:665–75.

Friend, G. R. 1990. Breeding and population dynamics of *Isoodon macrourus* (Marsupialia: Peramelidae): studies from the wet-dry tropics of northern Australia. *In* Seebeck, J. H., P. R. Brown, R. L. Wallis, and C. M. Kemper, eds., Bandicoots and bilbies, 357–65. Surrey Beatty & Sons and Australian Mammal Society, Sydney.

Friend, G. R., K. D. Morris, and N. L. McKenzie. 1991. The mammal fauna of Kimberley rainforests. *In* McKenzie, N. L., R. B. Johnston, and P. G. Kendrick, eds., Kimberley rainforests of Australia, 393–412. Surrey Beatty & Sons, Sydney, with Dept. of Cons. and Land Mgmt., Perth, and Dept. of Arts, Heritage and Env., Canberra.

Frith, H. J., and J. H. Calaby. 1969. Kangaroos. F. W. Cheshire, Melbourne, 209 pp.

Gardner, A. L. 1982. Virginia opossum. *In* Chapman, J. A., and G. A. Feldhamer, eds., Wild mammals of North America: biology, economics, and management, 3–36. Johns Hopkins Univ. Press, Baltimore.

———. 1993. Order Didelphimorphia. *In* Wilson, D. E., and D. M. Reeder, eds., Mammal species of the world: a taxonomic and geographic reference, 15–23. Smithsonian Institution Press, Washington, D.C.

Garkaklis, M. J., J. S. Bradley, and R. D. Wooller. 1998. The effects of woylie *(Bettongia penicillata)* foraging on soil water repellency and water infiltration in heavy textured soils in southwestern Australia. Aust. J. Ecol. 23:492–96.

———. 2000. Digging by vertebrates as an activity promoting the development of water-repellent patches in sub-surface soil. J. Arid Env. 45:35–42.

———. 2003. The relationship between animal foraging and nutrient patchiness in southwest Australian woodland soils. Aust. J. Soil Res. 41:665–73.

———. 2004. Digging and soil turnover by a mycophagous marsupial. J. Arid Env. 56:569–78.

Gemmell, N. J., and M. Westerman. 1994. Phylogenetic relationships within the Class Mammalia: a study using mitochondrial 12S RNA sequences. J. Mammal. Evol. 2:3–23.

Gemmell, R. T. 1990. The initiation of the breeding season of the northern brown bandicoot *Isoodon macrourus* in captivity. *In* Seebeck, J. H., P. R. Brown, R. L. Wallis, and C. M. Kemper, eds., Bandicoots and bilbies, 205–12. Surrey Beatty & Sons and Australian Mammal Society, Sydney.

Gemmell, R. T., and R. W. Rose. 1989. The senses involved in movement of some newborn Macropodoidea and other marsupials from cloaca to pouch. *In* Grigg, G. C., P. Jarman, and I. D. Hume, eds., Kangaroos, wallabies and rat-kangaroos, 1:339–47. Surrey Beatty & Sons, Sydney.

George, G. G. 1979. The status of endangered Papua New Guinea mammals. *In* Tyler,

M. J., ed., The status of endangered Australasian wildlife, 93–100. Royal Zoological Society of South Australia, Adelaide.

———. 1982. Cuscuses *Phalanger* spp.: their management in captivity. *In* Evans, D. D., ed., The management of Australian mammals in captivity, 67–72. Zoological Board of Victoria, Melbourne.

George, G. G., and G. M. Maynes. 1990. Status of New Guinea bandicoots. *In* Seebeck, J. H., P. R. Brown, R. L. Wallis, and C. M. Kemper, eds., Bandicoots and bilbies, 93–105. Surrey Beatty & Sons and Australian Mammal Society, Sydney.

Gibson, L. A. 2001. Seasonal changes in the diet, food availability and food preference of the greater bilby *(Macrotis lagotis)* in south-western Queensland. Wildl. Res. 28:121–34.

Gilfillan, S. L. 2001. An ecological study of a population of *Pseudantechinus macdonnellensis* (Marsupialia: Dasyuridae) in central Australia. I. Invertebrate food supply, diet and reproductive strategy. Wildl. Res. 28:469–80.

Glanz, W. E. 1990. Neotropical mammal densities: how unusual is the community on Barro Colorado Island, Panama? *In* Gentry, A., ed., Four Neotropical rainforests, 287–311. Yale Univ. Press, New Haven.

Godfrey, G. K. 1969. The influence of increased photoperiod on reproduction in the dasyurid marsupial *Sminthopsis crassicaudata.* J. Mammal. 50:132–33.

Godthelp, H., S. Wroe, and M. Archer. 1999. A new marsupial from the early Eocene Tingamarra Local Fauna of Murgon, southeastern Queensland: a prototypical Australian marsupial? J. Mammal. Evol. 6:289–313.

Goin, F. J. 2003. Early marsupial radiations in South America. *In* Jones, M. E., C. R. Dickman, and M. Archer, eds., Predators with pouches: the biology of carnivorous marsupials, 30–42. CSIRO Publishing, Melbourne.

Goin, F. J., and A. A. Carlini. 1995. An early Tertiary microbiotheriid marsupial from Antarctica. J. Vert. Paleo. 15:205–7.

Goin, F. J., J. Case, M. O. Woodburne, S. F. Vizcaíno, and M. Reguero. 1999. New discoveries of "opossum-like" marsupials from Antarctica (Seymour Island, medial Eocene). J. Mammal. Evol. 6:335–65.

Goldingay, R. L. 2000. Small dasyurid marsupials—are they effective pollinators? Aust. J. Zool. 48:597–606.

Goldingay, R. L., and S. M. Jackson, eds. 2004. The biology of Australian possums and gliders. Surrey Beatty & Sons, Sydney, 573 pp.

Goldingay, R. L., S. M. Carthew, and R. J. Whelan. 1991. The importance of non-flying mammals in pollination. Oikos 61:79–87.

Gould, J. 1863. The mammals of Australia. The Author, London.

Gould, R. A. 1969. Yiwara: foragers of the Australian desert. Collins, London, 239 pp.

Grassé, P.-P. 1955. Ordre des Marsupiaux. Traité Zool. 17:93–185.

Green, B. 1984. Composition of milk and energetics of growth in marsupials. Symp. Zool. Soc. Lond. 51:369–87.

Green, B., and J. C. Merchant. 1988. The composition of marsupial milk. *In* Tyndale-Biscoe, C. H., and P. A. Janssens, eds., The developing marsupial: models for biomedical research, 41–54. Springer-Verlag, Berlin.

Green, R. H. 1973. The mammals of Tasmania. Mary Fisher Bookshop, Launceston, 120 pp.

Gressitt, J. L., and N. Nadkarni. 1978. Guide to Mt. Kaindi: background to montane New Guinea ecology. Wau Ecology Inst., Handbook No. 5, 135 pp.

Gribel, R. 1988. Visits of *Caluromys lanatus* (Didelphidae) to flowers of *Pseudobombax tomentosum* (Bombacaceae): a probable case of pollination by marsupials. Biotropica 20:344–47.

Griffin, A. S., D. T. Blumstein, and C. S. Evans. 2000. Training captive-bred or translocated animals to avoid predators. Cons. Biol. 14:1317–26.

Grigera, D. E., and E. H. Rapoport. 1983. Status and distribution of the European hare in South America. J. Mammal. 64:163–66.

Grigg, G. 2002. Conservation benefit from harvesting kangaroos: status report at the start of a new millennium. A paper to stimulate discussion and research. *In* Lunney, D., and C. R. Dickman, eds., A zoological revolution: using native fauna to assist in its own survival, 53–76. Royal Zoological Society of New South Wales and Australian Museum, Sydney.

Groube, L., J. Chappell, J. Muke, and D. Price. 1986. A 40,000 year-old human occupation site at Huon Peninsula, Papua New Guinea. Nature 324:353–55.

Groves, C. P. 1990. The centrifugal pattern of speciation in Meganesian rainforest mammals. Mem. Qld Mus. 28:325–28.

Guiler, E. R. 1970. Observations on the Tasmanian devil, *Sarcophilus harrisii* (Marsupialia: Dasyuridae) I. Numbers, home range, movements, and food in two populations. Aust. J. Zool. 18:49–62.

Guiler, E. R., and P. Godard. 1998. Tasmanian tiger: a lesson to be learnt. Abrolhos Publishing, Perth, 256 pp.

Hall, E. R. 1981. The mammals of North America, 2nd ed. John Wiley, New York, 1175 pp.

Handley, C. O., Jr. 1976. Mammals of the Smithsonian Venezuelan project. Brigham Young Univ. Sci. Bull. Biol. Ser. 20:1–91.

Harder, J. D. 1992. Reproductive biology of South American marsupials. *In* Hamlett, W. C., ed., Reproductive biology of South American vertebrates, 211–28. Springer-Verlag, New York.

Harder, J. D., and D. W. Fleck. 1997. Reproductive ecology of New World marsupials. *In* Saunders, N. R., and L. A. Hinds, eds., Marsupial biology: recent research, new perspectives, 175–203. Univ. New South Wales Press, Sydney.

Heinsohn, T. E. 2002*a*. Status of the common spotted cuscus *Spilocuscus maculatus* and other wild mammals on Selayar Island, Indonesia, with notes on Quaternary faunal turnover. Aust. Mammal. 24:199–207.

———. 2002*b*. Observations of probable camouflaging behaviour in a semi-commensal common spotted cuscus *Spilocuscus maculatus maculatus* (Marsupialia: Phalangeridae) in New Ireland, Papua New Guinea. Aust. Mammal. 24:243–45.

———. 2003. Animal translocation: long-term human influences on the vertebrate zoogeography of Australasia (natural dispersal versus ethnophoresy). Aust. Zool. 32:351–76.

Heinsohn, T. E., and G. S. Hope. 2004. The Torresian connections: zoogeography of New Guinea. *In* Merrick, J. R., M. Archer, G. M. Hickey, and M. S. Y. Lee, eds., Evolution and zoogeography of Australasian vertebrates, 77–99. Auscipub, Sydney.

Helgen, K. M., and T. F. Flannery. 2004. Notes on the phalangerid marsupial genus *Spilocuscus,* with description of a new species from Papua. J. Mammal. 85:825–833.

Henry, S. R., A. K. Lee, and A. P. Smith. 1989. The trophic structure and species richness of assemblages of arboreal mammals in Australian forests. *In* Morris, D. W., Z. Abramsky, B. J. Fox, and M. R. Willig, eds., Patterns in the structure of mammalian communities, 229–40. Texas Tech Univ. Press, Lubbock.

Hershkovitz, P. 1992*a*. Ankle bones: the Chilean opossum *Dromiciops gliroides* Thomas, and marsupial phylogeny. Bonn. Zool. Beiträge 43:181–213.

———. 1992*b*. The South American gracile mouse opossums, genus *Gracilinanus* Gardner and Creighton, 1989 (Marmosidae, Marsupialia): a taxonomic review with notes on general morphology and relationships. Fieldiana: Zool., New Series 70:1–56.

———. 1999. *Dromiciops gliroides* Thomas, 1894, last of the Microbiotheria (Marsupialia), with a review of the family Microbiotheriidae. Fieldiana: Zool., New Series 93:1–60.

Hide, R. L., J. C. Pernetta, and T. Senabe. 1984. Exploitation of wild animals. *In* The research report of the Simbu land use project. Vol. 4: South Simbu: studies in demography, nutrition, and subsistence, 291–380. I.A.S.E.R., Port Moresby.

Hilton-Taylor, C. 2000. 2000 IUCN Red List of threatened species. World Conservation Union, IUCN, Gland, 61 pp.

Holloway, J. C., and F. Geiser. 1996. Reproductive status and torpor of the marsupial *Sminthopsis crassicaudata:* effect of photoperiod. J. Thermal Biol. 21:373–80.

Hope, J. H. 1973. Mammals of the Bass Strait islands. Proc. Roy. Soc. Vic. 85:163–95.

Hopper, S. D., and A. A. Burbidge. 1982. Feeding behaviour of birds and mammals on flowers of *Banksia grandis* and *Eucalyptus angulosa. In* Armstrong, J. A., J. M. Powell, and A. J. Richards, eds., Pollination and evolution, 67–75. Royal Botanic Gardens, Sydney.

Horovitz, I., and M. R. Sánchez-Villagra. 2003. A morphological analysis of marsupial mammal higher-level phylogenetic relationships. Cladistics 19:181–212.

How, R. A. 1978. Population strategies of four species of Australian "possums." *In* Montgomery, G. G., ed., The ecology of arboreal folivores, 305–13. Smithsonian Institution Press, Washington, D.C.

———. 1981. Population parameters of two congeneric possums, *Trichosurus* spp., in north-eastern New South Wales. Aust. J. Zool. 29:205–15.

Hrdina, F. 1997. Marsupial destruction in Queensland, 1877–1930. Aust. Zool. 30:272–86.

Huang, C., S. Ward, and A. K. Lee. 1987. Comparison of the diets of the feathertail glider, *Acrobates pygmaeus*, and the eastern pygmy-possum, *Cercartetus nanus* (Marsupialia: Burramyidae) in sympatry. Aust. Mammal. 10:47–50.

Hughes, R. L., and L. S. Hall. 1988. Structural adaptations of the newborn marsupial. *In* Tyndale-Biscoe, C. H., and P. A. Janssens, eds., The developing marsupial: models for biomedical research, 8–27. Springer-Verlag, Berlin.

Hughes, R. L., L. S. Hall, M. Archer, and K. Aplin. 1990. Observations on placentation and development in *Echymipera kalubu*. *In* Seebeck, J. H., P. R. Brown, R. L. Wallis, and C. M. Kemper, eds., Bandicoots and bilbies, 259–70. Surrey Beatty & Sons and Australian Mammal Society, Sydney.

Hume, I. D. 1982. Digestive physiology and nutrition of marsupials. Cambridge Univ. Press, Cambridge, 256 pp.

———. 1999. Marsupial nutrition. Cambridge Univ. Press, Cambridge, 434 pp.

———. 2003. Nutrition of carnivorous marsupials. *In* Jones, M. E., C. R. Dickman, and M. Archer, eds., Predators with pouches: the biology of carnivorous marsupials, 221–28. CSIRO Publishing, Melbourne.

———. 2005. Nutrition and digestion. *In* Armati, P. J., C. R. Dickman, and I. D. Hume, eds., Marsupials. Cambridge Univ. Press, Cambridge, in press.

Hume, I. D., P. J. Jarman, M. B. Renfree, and P. D. Temple-Smith. 1989. Macropodidae. *In* Walton, D. W., and B. J. Richardson, eds., Fauna of Australia. Vol. 1B: Mammalia, 679–715. Australian Government Publishing Service, Canberra.

Hume, I. D., E. Jazwinski, and T. F. Flannery. 1993. Morphology and function of the digestive tract in New Guinean possums. Aust. J. Zool. 41:85–100.

Hunsaker, D. 1977. Ecology of New World marsupials. *In* Hunsaker, D., ed., The biology of marsupials, 95–156. Academic Press, New York.

Hunsaker, D., and D. Shupe. 1977. Behavior of New World marsupials. *In* Hunsaker, D., ed., The biology of marsupials, 279–347. Academic Press, New York.

Huxley, T. H. 1880. On the application of the laws of evolution to the arrangement of the Vertebrata, and more particularly of the Mammalia. Proc. Zool. Soc. Lond. 1880:649–62.

Inns, R. W., P. F. Aitken, and J. K. Ling. 1979. Mammals. *In* Tyler, M. J., C. R. Twidale, and J. K. Ling, eds., Natural history of Kangaroo Island, 91–102. Royal Society of South Australia, Adelaide.

Irlbeck, N. A., and I. D. Hume. 2003. The role of *Acacia* in the diets of Australian marsupials—a review. Aust. Mammal. 25:121–34.

IUCN. 2001. IUCN Red List categories and criteria. Version 3.1. IUCN Species Survival Commission, Gland, Switzerland, 30 pp.

Izor, R. K., and R. H. Pine. 1987. Notes on the black-shouldered opossum, *Caluromysiops irrupta*. *In* Patterson, B. D., and R. M. Timms, eds., Studies in Neotropical mammalogy: essays in honor of Philip Hershkovitz. Fieldiana: Zool., New Series 39:117–24.

Jackson, L. M., and J. D. Harder. 1996. Vomeronasal organ removal blocks pheromonal induction of estrus in gray short-tailed opossums *(Monodelphis domestica)*. Biol. Reprod. 54:506–12.

Jackson, S. 2003. Bandicoots. *In* Jackson, S., ed., Australian mammals: biology and captive management, 127–44. CSIRO Publishing, Melbourne.

Jansa, S. A., and R. S. Voss. 2000. Phylogenetic studies on didelphid marsupials. I. Introduction and preliminary results from nuclear IRBP gene sequences. J. Mammal. Evol. 7:43–77.

Janson, C. H., J. Terborgh, and L. H. Emmons. 1981. Non-flying mammals as pollinating agents in the Amazonian forest. Biotropica (Suppl.) 13:1–6.

Jarman, P. J. 1984. The dietary ecology of macropod marsupials. Proc. Nutrition Soc. Aust. 9:82–87.

Jarman, P. J., and P. Bayne. 1997. Behavioural ecology of *Petrogale penicillata* in relation to conservation. Aust. Mammal. 19:219–28.

Jarman, P. J., and K. A. Johnson. 1977. Exotic mammals, indigenous mammals, and land-use. Proc. Ecol. Soc. Aust. 10:146–63.

Ji, Q., Z.-X. Luo, C.-X. Yuan, J. R. Wible, J.-P. Zhang, and J. A. Georgi. 2002. The earliest known eutherian mammal. Nature 416:816–22.

Johnson, C. N. 1996. Interactions between mammals and ectomycorrhizal fungi. Trends Ecol. Evol. 11:503–7.

Johnson, C. N., and G. J. Prideaux. 2004. Extinctions of herbivorous mammals in the late Pleistocene of Australia in relation to their feeding ecology: no evidence for environmental change as cause of extinction. Aust. Ecol. 29:553–57.

Johnson, C. N., and S. Wroe. 2003. Causes of extinction of vertebrates during the Holocene of mainland Australia: arrival of the dingo, or human impact? Holocene 13:941–48.

Johnson, C. N., S. Delean, and A. Balmford. 2002. Phylogeny and the selectivity of extinction in Australian marsupials. Anim. Cons. 5:135–42.

Jones, C. G., J. H. Lawton, and M. Shachak. 1994. Organisms as ecosystem engineers. Oikos 69:373–86.

———. 1997. Positive and negative effects of organisms as physical ecosystem engineers. Ecology 78:1946–57.

Jones, M. E. 1997. Character displacement in Australian dasyurid carnivores: size relationships and prey size patterns. Ecology 78:2569–87.

Jones, M. E., and L. A. Barmuta. 1998. Diet overlap and relative abundance of sympatric dasyurid carnivores: a hypothesis of competition. J. Anim. Ecol. 67:410–21.

———. 2000. Niche differentiation among sympatric Australian dasyurid carnivores. J. Mammal. 81:434–47.

Jones, M. E., M. Oakwood, C. A. Belcher, K. Morris, A. J. Murray, P. A. Woolley, K. B. Firestone, B. Johnson, and S. Burnett. 2003. Carnivore concerns: problems, issues and solutions for conserving Australasia's marsupial carnivores. *In* Jones, M. E., C. R. Dickman, and M. Archer, eds., Predators with pouches: the biology of carnivorous marsupials, 422–34. CSIRO Publishing, Melbourne.

Julien-Laferrière, D., and M. Atramentowicz. 1990. Feeding and reproduction of three didelphid marsupials in two Neotropical forests (French Guiana). Biotropica 22:404–15.

Kaufman, D. W. 1995. Diversity of New World mammals: universality of the latitudinal gradients of species and bauplans. J. Mammal. 76:322–34.

Kaufmann, J. H. 1974. Habitat use and social organization of nine sympatric species of macropod marsupials. J. Mammal. 55:66–80.

Kavanagh, R. P., and M. J. Lambert. 1990. Food selection by the greater glider, *Petauroides volans:* is foliar nitrogen a determinant of habitat quality? Aust. Wildl. Res. 17:285–99.

Kerle, J. A. 1984. Variation in the ecology of *Trichosurus:* its adaptive significance. *In* Smith, A. P., and I. D. Hume, eds., Possums and gliders, 115–28. Surrey Beatty & Sons and Australian Mammal Society, Sydney.

———. 2001. Possums: the brushtails, ringtails and greater glider. Univ. New South Wales Press, Sydney, 128 pp.

Kielan-Jaworowska, Z. 1975. Possible occurrence of marsupial bones in Cretaceous eutherian mammals. Nature 255:698–99.

Kielan-Jaworowska, Z., and L. A. Nessov. 1990. On the metatherian nature of the Deltatheroida, a sister group of the Marsupialia. Lethaia 23:1–10.

Kielan-Jaworowska, Z., R. L. Cifelli, and Z.-X. Luo, eds. 2004. Mammals from the age of dinosaurs: origins, structure, and evolution. Columbia Univ. Press, New York, 700 pp.

Kinzig, A. P., S. W. Pacala, and D. Tilman, eds. 2001. The functional consequences of biodiversity: empirical progress and theoretical extensions. Princeton Univ. Press, Princeton, 365 pp.

Kirsch, J. A. W., and R. E. Palma. 1995. DNA-DNA hybridization studies of carnivorous marsupials. V. A further estimate of relationships among opossums (Marsupialia: Didelphidae). Mammalia 59:403–25.

Kirsch, J. A. W., and M. S. Springer. 1993. Timing of the molecular evolution of New Guinean marsupials. Sci. New Guinea 19:147–56.

Kirsch, J. A. W., and P. F. Waller. 1979. Notes on the trapping and behavior of the Caenolestidae. J. Mammal. 60: 390–95.

Kirsch, J. A. W., and M. A. Wolman. 2001. Molecular relationships of the bear cuscus, *Ailurops ursinus* (Marsupialia: Phalangeridae). Aust. Mammal. 23:23–30.

Kirsch, J. A. W., A. W. Dickerman, O. A. Reig, and M. S. Springer. 1991. DNA hybridization evidence for the Australasian affinity of the American marsupial *Dromiciops australis.* Proc. Natl. Acad. Sci. USA 88:10465–69.

Kirsch, J. A. W., F.-J. Lapointe, and M. S. Springer. 1997. DNA-hybridisation studies of marsupials and their implications for metatherian classification. Aust. J. Zool. 45:211–80.

Kitchener, D. J., and A. Suyanto, eds. 1996. Proceedings of the first international conference on eastern Indonesian-Australian vertebrate fauna, Manado, Indonesia, November 22–26, 1994. Western Australian Museum, Perth, 174 pp.

Kitchener, D. J., J. Stoddart, and J. Henry. 1984. A taxonomic revision of the *Sminthopsis murina* complex (Marsupialia, Dasyuridae) in Australia, including descriptions of four new species. Rec. West. Aust. Mus. 11:201–48.

Kobayashi, Y., D. A. Winkler, and L. L. Jacobs. 2002. Origin of the tooth-replacement pattern in therian mammals: evidence from a 110 myr old fossil. Proc. Roy. Soc. Lond. B 269:369–73.

Krajewski, C., J. Painter, A. C. Driskell, L. Buckley, and M. Westerman. 1993. Molecular systematics of New Guinean dasyurids (Marsupialia: Dasyuridae). Sci. New Guinea 19:157–65.

Krajewski, C., S. Wroe, and M. Westerman. 2000. Molecular evidence for the pattern and timing of cladogenesis in dasyurid marsupials. Zool. J. Linn. Soc. 130:375–404.

Krause, D. W. 2001. Fossil molar from a Madagascan marsupial. Nature 412:497–98.

Krause, W. J. 1998. A review of histogenesis/organogenesis in the developing North American opossum *(Didelphis virginiana).* Springer-Verlag, Berlin, 160 pp.

Krefft, G. 1866. On the vertebrated animals of the lower Murray and Darling, their habits, economy, and geographical distribution. Trans. Phil. Soc. New South Wales 1862–1865:1–33.

Krockenberger, A. 2005. Lactation. *In* Armati, P. J., C. R. Dickman, and I. D. Hume, eds., Marsupials. Cambridge Univ. Press, Cambridge, in press.

Kumar, S., and S. B. Hedges. 1998. A molecular timescale for vertebrate evolution. Nature 392:917–20.

Lapointe, F.-J., and J. A. W. Kirsch. 2001. Construction and verification of a large phylogeny of marsupials. Aust. Mammal. 23:9–22.

Lapointe, F.-J., and P. Legendre. 1996. Evolution of the marsupial brain: does it reflect the evolution of behavior? *In* Cabana, T., ed., Animals in their environment: a tribute to Paul Pirlot, 187–212. Éditions Orbis Publishing, Québec.

Laurance, W. F., and G. B. Williamson. 2001. Positive feedbacks among forest fragmentation, drought, and climate change in the Amazon. Cons. Biol. 15:1529–35.

Laurance, W. F., M. A. Cochrane, S. Bergen, P. M. Fearnside, P. Delamonica, C. Barber, S. D'Angelo, and T. Fernandes. 2001. The future of the Brazilian Amazon. Science 291:438–39.

Leary, T., and T. Mamu. 2004. Conserving Papua New Guinea's forest fauna through community planning. In Lunney, D., ed, Conservation of Australia's forest fauna, 2nd ed., 186–207. Royal Zoological Society of New South Wales, Sydney.

Lee, A. K., and A. Cockburn. 1985. Evolutionary ecology of marsupials. Cambridge Univ. Press, Cambridge, 274 pp.

Lee, A. K., and R. W. Martin. 1988. The koala: a natural history. New South Wales Univ. Press, Sydney, 102 pp.

Lee, A. K., and I. R. McDonald. 1985. Stress and population regulation in small mammals. Oxford Rev. Reprod. Biol. 7:261–304.
Lee, A. K., and S. J. Ward. 1989. Life histories of macropodoid marsupials. *In* Grigg, G. C., P. Jarman, and I. D. Hume, eds., Kangaroos, wallabies and rat-kangaroos, 1:105–15. Surrey Beatty & Sons, Sydney.
Lee, A. K., P. Woolley, and R. W. Braithwaite. 1982. Life history strategies of dasyurid marsupials. *In* Archer, M., ed., Carnivorous marsupials, 1:1–11. Royal Zoological Society of New South Wales, Sydney.
Lee, A. K., K. A. Handasyde, and G. D. Sanson, eds. 1991. Biology of the koala. Surrey Beatty & Sons, Sydney, 336 pp.
Lemos, B., and R. Cerqueira. 2002. Morphological differentiation in the white-eared opossum group (Didelphidae: *Didelphis*). J. Mammal. 83:354–69.
Lentle, R. G., I. D. Hume, K. J. Stafford, M. Kennedy, S. Haslett, and B. P. Springett. 2003*a*. Molar progression and tooth wear in tammar *(Macropus eugenii)* and parma *(Macropus parma)* wallabies. Aust. J. Zool. 51:137–51.
———. 2003*b*. Comparisons of indices of molar progression and dental function of brush-tailed rock-wallabies *(Petrogale penicillata)* with tammar *(Macropus eugenii)* and parma *(Macropus parma)* wallabies. Aust. J. Zool. 51:259–69.
Lim, L. 1992. Recovery plan for the kowari *Dasyuroides byrnei* Spencer, 1896 (Marsupialia: Dasyuridae). Australian National Parks and Wildlife Service, Canberra, 46 pp.
Lindenmayer, D. B. 1997. Differences in the biology and ecology of arboreal marsupials in forests of southeastern Australia. J. Mammal. 78:1117–27.
———. 2002. Gliders of Australia: a natural history. Univ. New South Wales Press, Sydney, 188 pp.
Lindenmayer, D. B., J. Dubach, and K. L. Viggers. 2002. Geographic dimorphism in the mountain brushtail possum *(Trichosurus caninus)*: the case for a new species. Aust. J. Zool. 50:369–93.
Logan, M. 2003. Effect of tooth wear on the rumination-like behavior, or merycism, of free-ranging koalas *(Phascolarctos cinereus)*. J. Mammal. 84:897–902.
Long, J. L. 2003. Introduced mammals of the world: their history, distribution and influence. CSIRO Publishing, Melbourne, 589 pp.
Long, J., M. Archer, T. Flannery, and S. Hand. 2002. Prehistoric mammals of Australia and New Guinea: one hundred million years of evolution. Univ. New South Wales Press, Sydney, 244 pp.
Lorini, M. L., J. A. De Oliveira, and V. G. Persson. 1994. Annual age structure and reproductive patterns in *Marmosa incana* (Lund, 1841) (Didelphidae, Marsupialia). Z. Säugetierk. 59:65–73.
Luckett, W. P. 1993. An ontogenetic assessment of dental homologies in therian mammals. *In* Szalay, F. S., M. J. Novacek, and M. C. McKenna, eds., Mammal phylogeny. Vol. 1: Mesozoic differentiation, multituberculates, monotremes, early therians, and marsupials, 182–204. Springer-Verlag, New York.
Lunde, D. P., and W. A. Schutt. 1999. The peculiar carpal tubercles of male *Marmosops parvidens* and *Marmosa robinsoni* (Didelphidae: Didelphinae). Mammalia 63:495–504.
Lunney, D. 2001. Causes of the extinction of native mammals of the Western Division of New South Wales: an ecological interpretation of the nineteenth century historical record. Rangel. J. 23:44–70.
Lunney, D., and C. R. Dickman, eds. 2002. A zoological revolution: using native fauna to assist in its own survival. Royal Zoological Society of New South Wales and Australian Museum, Sydney, 174 pp.
Luo, Z.-X., R. L. Cifelli, and Z. Kielan-Jaworowska. 2001. Dual origin of tribosphenic mammals. Nature 409:53–57.
Luo, Z.-X., Z. Kielan-Jaworowska, and R. L. Cifelli. 2002. In quest for a phylogeny of Mesozoic mammals. Acta Palaeontol. Pol. 47:1–78.
Luo, Z.-X., Q. Ji, J. R. Wible, and C.-X. Yuan. 2003. An early Cretaceous tribosphenic mammal and metatherian evolution. Science 302:1934–40.
Lyne, A. G. 1964. Observations on the breeding and growth of the marsupial *Perameles nasuta* Geoffroy, with notes on other bandicoots. Aust. J. Zool. 12:322–39.

———. 1976. Observations on oestrus and the oestrous cycle in the marsupials *Isoodon macrourus* and *Perameles nasuta*. Aust. J. Zool. 24:513–21.

Macdonald, A. A., J. E. Hill, Boeadi, and R. Cox. 1993. The mammals of Seram, with notes on their biology and local usage. *In* Edwards, I. D., A. A. Macdonald, and J. Proctor, eds., Natural history of Seram, Maluku, Indonesia, 161–90. Intercept, Andover, UK.

Majnep, I. S., and R. Bulmer. 1990. Kalam hunting traditions. Working Papers in Anthropology, Archaeology, Linguistics and Maori Studies, Univ. Auckland, Auckland, 47 pp.

Mansergh, I., and L. Broome. 1994. The mountain pygmy-possum of the Australian Alps. New South Wales Univ. Press, Sydney, 114 pp.

Mares, M. A., and J. K. Braun. 2000. Systematics and natural history of marsupials from Argentina. *In* Choate, J. H., ed., Reflections of a naturalist: papers honoring Professor Eugene D. Fleharty, 23–45. Stenberg Museum of Natural History, Fort Hays State Univ., Hays.

Marshall, A. J. 1966. On the disadvantages of wearing fur. *In* Marshall, A. J., ed., The great extermination: a guide to Anglo-Australian cupidity, wickedness and waste, 9–42. William Heinemann, Melbourne.

Marshall, L. G. 1978*a*. Evolution of the Borhyaenidae, extinct South American predaceous marsupials. Univ. Calif. Publs Geol. Sci. 117:1–89.

———. 1978*b*. *Chironectes minimus*. Mammal. Species 109:1–6.

———. 1981. Review of the Hathlyacyninae, an extinct subfamily of South American "dog-like" marsupials. Fieldiana: Geol., New Series 7:1–120.

———. 1982. Systematics of the extinct South American marsupial family Polydolopidae. Fieldiana: Geol., New Series 12:1–109.

———. 1987. Systematics of the Itaboraian (middle Paleocene) age "opossum-like" marsupials from the limestone quarry at São José de Itaboraí, Brazil. *In* Archer, M., ed., Possums and opossums: studies in evolution, 1:91–160. Surrey Beatty & Sons and Royal Zoological Society of New South Wales, Sydney.

Marshall, L. G., and C. de Muizon. 1995. Part II: the skull. *In* Muizon, C. de, ed., *Pucadelphys andinus* (Marsupialia, Mammalia) from the early Paleocene of Bolivia. Mém. Mus. Natl. d'Hist. Natur., Paris 165:21–90.

Marshall, L. G., and D. Sigogneau-Russell. 1995. Part III: the postcranial skeleton. *In* Muizon, C. de, ed., *Pucadelphys andinus* (Marsupialia, Mammalia) from the early Paleocene of Bolivia. Mém. Mus. Natl. d'Hist. Natur., Paris 165:91–164.

Marshall, L. G., J. A. Case, and M. O. Woodburne. 1990. Phylogenetic relationships of the families of marsupials. *In* Genoways, H. H., ed., Current mammalogy, 2:433–505. Plenum Press, New York.

Martin, H. A. 1998. Tertiary climatic evolution and the development of aridity in Australia. Proc. Linn. Soc. New South Wales 119:115–36.

Martin, R. W. 1981. Age-specific fertility in three populations of the koala, *Phascolarctos cinereus* Goldfuss, in Victoria. Aust. Wildl. Res. 8:275–83.

Martin, R. W., and K. A. Handasyde. 1991. Population dynamics of the koala *(Phascolarctos cinereus)* in south-eastern Australia. *In* Lee, A. K., K. A. Handasyde, and G. D Sanson, eds., Biology of the koala, 75–84. Surrey Beatty & Sons, Sydney.

Masters, P. 1993. The effects of fire-driven succession and rainfall on small mammals in spinifex grassland at Uluru National Park, Northern Territory. Wildl. Res. 20:803–13.

Masters, P., C. R. Dickman, and M. Crowther. 2003. Effects of cover reduction on mulgara *Dasycercus cristicauda* (Marsupialia: Dasyuridae), rodent and invertebrate populations in central Australia: implications for land management. Austral Ecol. 28:658–65.

Maxwell, S., A. A. Burbidge, and K. Morris, eds. 1996. The 1996 Action Plan for Australian marsupials and monotremes. Wildlife Australia, Canberra, 234 pp.

McAllan, B. 2003. Timing of reproduction in carnivorous marsupials. *In* Jones, M. E., C. R. Dickman, and M. Archer, eds., Predators with pouches: the biology of carnivorous marsupials, 147–68. CSIRO Publishing, Melbourne.

McAllan, B. M., and C. R. Dickman. 1986. The role of photoperiod in the timing of reproduction in the dasyurid marsupial *Antechinus stuartii*. Oecologia 68:259–64.

McAllan, B. M., C. R. Dickman, and M. S. Crowther. 2005. Photoperiod as a reproductive cue in the marsupial genus *Antechinus:* ecological and evolutionary consequences. Biol. J. Linn. Soc. In press.

McGowran, B., and Q. Li. 1994. The Miocene oscillation in southern Australia. Rec. South Aust. Mus. 27:197–212.

McIlwee, A. P., and C. N. Johnson. 1998. The contribution of fungus to the diets of three mycophagous marsupials in *Eucalyptus* forests, revealed by stable isotope analysis. Funct. Ecol. 12:223–31.

McKenna, M. C., and S. K. Bell. 1997. Classification of mammals above the species level. Columbia Univ. Press, New York, 631 pp.

McLean, I. G., N. T. Schmitt, P. J. Jarman, C. Duncan, and C. D. L. Wynne. 2000. Learning for life: training marsupials to recognise introduced predators. Behaviour 137:1361–76.

McNamara, J. A. 1997. Some smaller macropod fossils of South Australia. Proc. Linn. Soc. New South Wales 117:97–105.

Medellín, R. A. 1994. Mammal diversity and conservation in the Selva Lacandona, Chiapas, Mexico. Cons. Biol. 8:780–99.

Meltzer, D. J. 2004*a*. Peopling of North America. *In* Gillespie, A., S. C. Porter, and B. Atwater, eds., The Quaternary period in the United States, 539–63. Elsevier Science, New York.

———. 2004*b*. On possibilities, prospecting and patterns: thinking about a pre-LGM human presence in the Americas. *In* Madsen, D., ed., Entering America: Northeast Asia and Beringia before the Last Glacial Maximum, 269–87. Univ. Utah Press, Salt Lake City.

Melzer, A., F. Carrick, P. Menkhorst, D. Lunney, and B. St. John. 2000. Overview, critical assessment, and conservation implications of koala distribution and abundance. Cons. Biol. 14:619–28.

Menzies, J. I. 1991. A handbook of New Guinea marsupials and monotremes. Kristen Press, Madang, 138 pp.

Merchant, J. C. 1976. Breeding biology of the agile wallaby, *Macropus agilis* (Gould) (Marsupialia: Macropodidae), in captivity. Aust. Wildl. Res. 3:93–103.

Merchant, J. C., and J. H. Calaby. 1981. Reproductive biology of the red-necked wallaby *(Macropus rufogriseus banksianus)* and Bennett's wallaby *(M. r. rufogriseus)* in captivity. J. Zool. Lond. 194:203–17.

Meserve, P. L., B. K. Lang, and B. D. Patterson. 1988. Trophic relationships of small mammals in a Chilean temperate rainforest. J. Mammal. 69:721–30.

Miles, M. A., A. A. De Souza, and M. M. Póvoa. 1981. Mammal tracking and nest location in Brazilian forest with an improved spool-and-line device. J. Zool. Lond. 195:331–47.

Moir, R. J., M. Somers, and H. Waring. 1956. Studies on marsupial nutrition. I. Ruminant-like digestion in a herbivorous marsupial *(Setonix brachyurus* Quoy and Gaimard). Aust. J. Biol. Sci. 9:293–304.

Montague, T. L., ed. 2000. The brushtail possum: biology, impact and management of an introduced marsupial. Manaaki Whenua Press, Lincoln, 292 pp.

Monteiro-Filho, E. L. A., and V. S. Dias. 1990. Observações sobre a biologia de *Lutreolina crassicaudata* (Mammalia: Marsupialia). Rev. Brasil. Biologia 50:393–99.

Mooney, N. 2004. The devil's new hell. Nature Aust. 28:34–41.

Morton, S. R. 1990. The impact of European settlement on the vertebrate animals of arid Australia: a conceptual model. Proc. Ecol. Soc. Aust. 16:201–13.

Muizon, C. de. 1998. *Mayulestes ferox*, a borhyaenoid (Metatheria, Mammalia) from the early Palaeocene of Bolivia: phylogenetic and palaeobiologic implications. Geodiv. 20:19–142.

Muizon, C. de, R. L. Cifelli, and R. Céspedes Paz. 1997. The origin of the dog-like borhyaenoid marsupials of South America. Nature 389:486–89.

Murray, P. F. 1984. Extinctions Downunder: a bestiary of extinct Australian late Pleistocene monotremes and marsupials. *In* Martin, P. S., and R. G. Klein, eds., Quaternary extinctions: a prehistoric revolution, 600–628. Univ. Arizona Press, Tucson.

———. 1998. Palaeontology and palaeobiology of wombats. *In* Wells, R. T., and P. A. Pridmore, eds., Wombats, 1–33. Surrey Beatty & Sons, Sydney.

Musser, G. G. 1987. The mammals of Sulawesi. *In* Whitmore, T. C., ed., Biogeographical evolution of the Malay Archipelago, 73–93. Clarendon Press, Oxford.

Nagorsen, D. W. 1996. Opossums, shrews and moles of British Columbia. UBC Press, Univ. British Columbia, and Royal British Columbia Museum, Vancouver, 169 pp.

Nelson, J., R. M. Knight, and C. Kingham. 2003. Perinatal sensory and motor development in marsupials with special reference to the northern quoll, *Dasyurus hallucatus. In* Jones, M. E., C. R. Dickman, and M. Archer, eds., Predators with pouches: the biology of carnivorous marsupials, 205–17. CSIRO Publishing, Melbourne.

Newell, G. R. 1999. Home range and habitat use by Lumholtz's tree-kangaroo *(Dendrolagus lumholtzi)* within a rainforest fragment in north Queensland. Wildl. Res. 26:129–45.

Newsome, A. E. 1965*a*. The distribution of red kangaroos, *Megaleia rufa* (Desmarest), about sources of persistent food and water in central Australia. Aust. J. Zool. 13:289–99.

———. 1965*b*. Reproduction in natural populations of the red kangaroo, *Megaleia rufa* (Desmarest), in central Australia. Aust. J. Zool. 13:735–59.

———. 1966. The influence of food on breeding in the red kangaroo in central Australia. CSIRO Wildl. Res. 11:187–96.

———. 1997. Reproductive anomalies in the red kangaroo in central Australia explained by Aboriginal traditional knowledge and ecology. *In* Saunders, N. R., and L. A. Hinds, eds., Marsupial biology: recent research, new perspectives, 227–36. Univ. New South Wales Press, Sydney.

Nilsson, M., A. Gullberg, A. E. Spotorno, U. Arnasson, and A. Janke. 2003. Radiation of extant marsupials after the K/T boundary: evidence from complete mitochondrial genomes. J. Mol. Evol. 57:1–10.

Noble, J. C. 1993. Relict surface-soil features in semi-arid mulga *(Acacia aneura)* woodlands. Rangel. J. 15:48–70.

———. 1999. Fossil features of mulga *Acacia aneura* landscapes: possible imprinting by extinct Pleistocene fauna. Aust. Zool. 31:396–402.

Nogueira, J. C., P. M. Martinelli, S. F. Costa, G. A. Carvalho, and B. G. O. Câmara. 1999. The penis morphology of *Didelphis, Lutreolina, Metachirus* and *Caluromys* (Marsupialia, Didelphidae). Mammalia 63:79–92.

Norris, C. A., and G. G. Musser. 2001. Systematic revision within the *Phalanger orientalis* complex (Diprotodontia, Phalangeridae): a third species of lowland gray cuscus from New Guinea and Australia. Am. Mus. Nov. 356:1–20.

Novacek, M. J., G. W. Rougier, J. R. Wible, M. C. McKenna, D. Dashzeveg, and I. Horovitz. 1997. Epipubic bones in eutherian mammals from the late Cretaceous of Mongolia. Nature 389:483–86.

Oakwood, M. 1997. The ecology of the northern quoll, *Dasyurus hallucatus.* Ph.D. diss., Australian National Univ., Canberra.

Oakwood, M., A. J. Bradley, and A. Cockburn. 2001. Semelparity in a large marsupial. Proc. Roy. Soc. Lond. B 268:407–11.

O'Connell, M. 1979. Ecology of didelphid marsupials from northern Venezuela. *In* Eisenberg, J. F., ed., Vertebrate ecology in the northern Neotropics, 73–87. Smithsonian Institution Press, Washington, D.C.

O'Connell, M. A. 1983. *Marmosa robinsoni.* Mammal. Species 203:1–6.

———. 1989. Population dynamics of neotropical small mammals in seasonal habitats. J. Mammal. 70:532–48.

———. 2001. American opossums. *In* Macdonald, D. W., ed., The new encyclopedia of mammals, 808–13. Oxford Univ. Press, Oxford.

Osgood, W. H. 1921. A monographic study of the American marsupial *Caenolestes.* Field Mus. Nat. Hist., Zool. Series 14:1–162.

Owen, D. 2003. Thylacine: the tragic tale of the Tasmanian tiger. Allen & Unwin, Sydney, 228 pp.

Paddle, R. 2000. The last Tasmanian tiger: the history and extinction of the thylacine. Cambridge Univ. Press, Cambridge, 273 pp.

Palma, R. E. 1995. Range expansion of two South American mouse opossums (*Thylamys,* Didelphidae) and their biogeographic implications. Rev. Chilena Hist. Nat. 68:515–22.

———. 2003. Evolution of American marsupials and their phylogenetic relationships with Australian metatherians. *In* Jones, M. E., C. R. Dickman, and M. Archer, eds., Predators with pouches: the biology of carnivorous marsupials, 21–29. CSIRO Publishing, Melbourne.

Palma, R. E., and A. E. Spotorno. 1999. Molecular systematics of marsupials based on the rRNA 12S mitochondrial gene: the phylogeny of Didelphimorphia and of the living fossil microbiotheriid *Dromiciops gliroides* Thomas. Mol. Phyl. Evol. 13:525–35.

Pascual, R., and A. A. Carlini. 1987. A new superfamily in the extensive radiation of South American Paleogene marsupials. *In* Patterson, B. D., and R. M. Timm, eds., Studies in Neotropical mammalogy: essays in honor of Philip Hershkovitz. Fieldiana: Zool., New Series 39:99–110.

Patterson, B. D. 2000. Patterns and trends in the discovery of new Neotropical mammals. Div. Dist. 6:145–51.

Patton, J. L., and L. P. Costa. 2003. Molecular phylogeography and species limits in rainforest didelphid marsupials of South America. *In* Jones, M. E., C. R. Dickman, and M. Archer, eds., Predators with pouches: the biology of carnivorous marsupials, 63–81. CSIRO Publishing, Melbourne.

Patton, J. L., B. Berlin, and E. A. Berlin. 1982. Aboriginal perspectives of a mammal community in Amazonian Perú: knowledge and utilization patterns among the Aguaruna Jívaro. *In* Mares, M. A., and H. H. Genoways, eds., Mammalian biology in South America. Pymatuning Symposia in Ecology, 6:111–28. Pymatuning Laboratory of Ecology, Univ. Pittsburgh, Linesville.

Patton, J. L., M. N. F. da Silva, and J. R. Malcolm. 2000. Mammals of the Rio Juruá and the evolutionary and ecological diversification of Amazonia. Bull. Am. Mus. Nat. Hist. 244:1–306.

Payton, I. 2000. Damage to native forests. *In* Montague, T. L., ed., The brushtail possum: biology, impact and management of an introduced marsupial, 111–25. Manaaki Whenua Press, Lincoln.

Peres, C. A. 1999. The structure of nonvolant mammal communities in different Amazonian forest types. *In* Eisenberg, J. F., and K. H. Redford, eds., Mammals of the Neotropics. Vol. 3: The central Neotropics: Ecuador, Peru, Bolivia, Brazil, 564–81. Univ. Chicago Press, Chicago.

Pérez-Hernández, R. 1989. Distribution of the Family Didelphidae (Mammalia-Marsupialia) in Venezuela. *In* Redford, K. H., and J. F. Eisenberg, eds., Advances in Neotropical mammalogy, 363–409. Sandhill Crane Press, Gainesville.

Pflieger, J. F., and T. Cabana. 1996. The vestibular primary afferents and the vestibulospinal projections in the developing and adult opossum, *Monodelphis domestica*. Anat. Embryol. 194:75–88.

Phillips, M. J., Y.-H. Lin, G. L. Harrison, and D. Penny. 2001. Mitochondrial genomes of a bandicoot and a brushtail possum confirm the monophyly of australidelphian marsupials. Proc. Roy. Soc. Lond. B 268:1533–38.

Pine, R. H., P. L. Dalby, and J. O. Matson. 1985. Ecology, postnatal development, morphometrics, and taxonomic status of the short-tailed opossum, *Monodelphis dimidiata*, an apparently semelparous annual marsupial. Ann. Carnegie Mus. 54:195–231.

Pocock, R. I. 1921. The external characteristics of the koala *(Phascolarctos)* and some related marsupials. Proc. Zool. Soc. Lond. 1921:591–607.

———. 1926. The external characters of *Thylacinus, Sarcophilus* and some related marsupials. Proc. Zool. Soc. Lond. 1926:1037–84.

Procter-Gray, E. 1984. Dietary ecology of the coppery brushtail possum, green ringtail possum and Lumholtz's tree-kangaroo in north Queensland. *In* Smith, A. P., and I. D. Hume, eds., Possums and gliders, 129–35. Surrey Beatty & Sons and Australian Mammal Society, Sydney.

Ramírez, C., R. Godoy, W. Eldridge, and N. Pacheco. 1981. Impacto ecológico del ciervo rojo sobre el bosque de Olivillo en Osorno, Chile. Ann. Mus. Hist. Nat. (Valparaíso) 14:197–215.

Randall, A. 1991. The value of biodiversity. Ambio 20:64–67.

Redford, K. H., and J. F. Eisenberg. 1992. Mammals of the Neotropics. Vol. 2: The southern cone: Chile, Argentina, Uruguay, Paraguay. Univ. of Chicago Press, Chicago, 430 pp.

Reid, F. A. 1997. A field guide to the mammals of Central America and southeast Mexico. Oxford Univ. Press, New York, 334 pp.

Reilly, S. M., and T. D. White. 2003. Hypaxial motor patterns and the function of epipubic bones in primitive mammals. Science 299:400–402.

Renfree, M. B. 1981. Embryonic diapause in marsupials. J. Reprod. Fert. Suppl. 29:67–78.

Renfree, M. B., E. M. Russell, and R. D. Wooller. 1984. Reproduction and life history of the honey possum, *Tarsipes rostratus*. *In* Smith, A. P., and I. D. Hume, eds., Possums and gliders, 427–37. Surrey Beatty & Sons and Australian Mammal Society, Sydney.

Retief, J. D., C. Krajewski, M. Westerman, R. J. Winkfein, and G. H. Dixon. 1995. Molecular phylogeny and evolution of the marsupial protamine P1 genes. Proc. R. Soc. Lond. B 259:7–14.

Reynolds, H. C. 1952. Studies on reproduction in the opossum *(Didelphis virginiana)*. Univ. Calif. Publs. Zool. 52:223–84.

Rich, T. H. 1991. Monotremes, placentals, and marsupials: their record in Australia and its biases. *In* Vickers-Rich, P., J. M. Monaghan, R. F. Baird, and T. H. Rich, eds., Vertebrate palaeontology of Australasia, 893–1004. Pioneer Design Studio and Monash Univ. Publications Committee, Melbourne.

Richardson, K. C., R. D. Wooller, and B. G. Collins. 1986. Adaptations to a diet of nectar and pollen in the marsupial *Tarsipes rostratus* (Marsupialia: Tarsipedidae). J. Zool. Lond. 208:285–97.

Ride, W. D. L. 1957. *Protemnodon parma* (Waterhouse) and the classification of related wallabies *(Protemnodon, Thylogale,* and *Setonix)*. Proc. Zool. Soc. Lond. 128:327–46.

Ride, W. D. L., P. A. Pridmore, R. E. Barwick, R. T. Wells, and R. D. Heady. 1997. Towards a biology of *Propleopus oscillans* (Marsupialia: Propleopinae, Hypsiprymnodontidae). Proc. Linn. Soc. New South Wales 117:243–328.

Roberts, R. G., R. Jones, N. A. Spooner, M. J. Head, A. S. Murray, and M. A. Smith. 1994. The human colonization of Australia: optical dates of 53,000 and 60,000 years bracket human arrival at Deaf Adder Gorge, Northern Territory. Quat. Sci. Rev. 13:575–83.

Roberts, R. G., T. F. Flannery, L. K. Ayliffe, H. Yoshida, J. M. Olley, G. J. Prideaux, G. M. Laslett, A. Baynes, M. A. Smith, R. Jones, and B. L. Smith. 2001. New ages for the last Australian megafauna: continent-wide extinction about 46,000 years ago. Science 292:1888–92.

Rolls, E. C. 1969. They all ran wild. Angus & Robertson, Sydney, 444 pp.

Rougier, G. W., J. R. Wible, and M. J. Novacek. 1998. Implications of *Deltatheridium* specimens for early marsupial history. Nature 396:459–63.

Rounsevell, D. E., R. J. Taylor, and G. J. Hocking. 1991. Distribution records of native terrestrial mammals in Tasmania. Wildl. Res. 18:699–717.

Russell, E. M. 1982. Patterns of parental care and parental investment in marsupials. Biol. Rev. 57:423–86.

———. 1985. The metatherians: Order Marsupialia. *In* Brown, R. E., and D. W. Macdonald, eds., Social odours in mammals, 45–104. Clarendon Press, Oxford.

Russell, E. M., and M. B. Renfree. 1989. Tarsipedidae. *In* Walton, D. W., and B. J. Richardson, eds., Fauna of Australia. Vol. 1B: Mammalia, 769–82. Australian Government Publishing Service, Canberra.

Russell, R. 1980. Spotlight on possums. Univ. Queensland Press, St. Lucia, 101 pp.

Sadleir, R. 2000. Evidence of possums as predators of native animals. *In* Montague, T. L., ed., The brushtail possum: biology, impact and management of an introduced marsupial, 126–31. Manaaki Whenua Press, Lincoln.

Salas, L. A. 2002. Comparative ecology and behavior of mountain cuscuses *(Phalanger carmelitae)*, silky cuscus *(Phalanger sericeus)* and coppery ringtail *(Pseudochirops cupreus)* at Mt. Stolle, Papua New Guinea. PhD. diss., Univ. of Massachusetts, Amherst.

Sánchez-Villagra, M. R. 2001. Ontogenetic and phylogenetic transformations of the vomeronasal complex and nasal floor elements in marsupial mammals. Zool. J. Linn. Soc. 131:459–79.

———. 2002. Comparative patterns of postcranial ontogeny in therian mammals: an analysis of relative timing of ossification events. J. Exp. Zool. 294:264–73.

Sánchez-Villagra, M. R., and K. K. Smith. 1997. Diversity and evolution of the marsupial mandibular angle. J. Mammal. Evol. 4:119–44.

Sanson, G. D. 1978. The evolution and significance of mastication in the Macropodidae. Aust. Mammal. 2:23–28.

———. 1980. The morphology and occlusion of the molariform cheek teeth in some Macropodinae (Marsupialia: Macropodidae). Aust. J. Zool. 28:341–65.

Santori, R. T., D. Astúa de Moraes, and R. Cerqueira. 1995. Diet composition of *Metachirus nudicaudatus* and *Didelphis aurita* (Didelphimorphia, Didelphidae). Mammalia 59:511–16.

Santori, R. T., D. Astúa de Moraes, C. E. V. Grelle, and R. Cerqueira. 1997. Natural diet at a Restinga forest and laboratory food preferences of the opossum *Philander frenata* in Brazil. Studies Neotrop. Fauna Env. 32:12–16.

Saulei, S. M. 1990. Forest research and development in Papua New Guinea. Ambio 19:397.

Saunders, G., B. Coman, J. Kinnear, and M. Braysher. 1995. Managing vertebrate pests: foxes. Australian Government Publishing Service, Canberra, 141 pp.

Schoener, T. W. 1974. Resource partitioning in ecological communities. Science 185:27–39.

Scott, L. K., I. D. Hume, and C. R. Dickman. 1999. Ecology and population biology of long-nosed bandicoots *(Perameles nasuta)* at North Head, Sydney Harbour National Park. Wildl. Res. 26:805–21.

Seebeck, J. H., and R. W. Rose. 1989. Potoroidae. *In* Walton, D. W., and B. J. Richardson, eds., Fauna of Australia. Vol. 1B: Mammalia, 716–39. Australian Government Publishing Service, Canberra.

Seebeck, J. H., A. F. Bennett, and D. J. Scotts. 1989. Ecology of the Potoroidae—a review. *In* Grigg, G. C., P. Jarman, and I. D. Hume, eds., Kangaroos, wallabies and rat-kangaroos, 1:67–88. Surrey Beatty & Sons, Sydney.

Sharman, G. B. 1955. Studies on marsupial reproduction. III. Normal and delayed pregnancy in *Setonix brachyurus*. Aust. J. Zool. 3:56–70.

———. 1965. Marsupials and the evolution of viviparity. Viewpoints Biol. 4:1–28.

———. 1970. Reproductive physiology of marsupials. Science 167:1221–28.

Sharman, G. B., G. M. Maynes, M. D. B. Eldridge, and R. L. Close. 1995. Proserpine rock-wallaby *Petrogale persephone* Maynes, 1982. *In* Strahan, R., ed., The mammals of Australia, 386–87. Reed Books, Sydney.

Shaw, G. 2005. Reproduction. *In* Armati, P. J., C. R. Dickman, and I. D. Hume, eds., Marsupials. Cambridge Univ. Press, Cambridge, in press.

Shield, J. 1968. Reproduction of the quokka, *Setonix brachyurus*, in captivity. J. Zool. Lond. 155:427–44.

Shimmin, G. A., D. A. Taggart, and P. D. Temple-Smith. 2002. Mating behaviour in the agile antechinus *Antechinus agilis* (Marsupialia: Dasyuridae). J. Zool. Lond. 258:39–48.

Slater, G. 1993. Husbandry strategies for the Tasmanian devil and brush-tailed phascogale at the Healesville Sanctuary. *In* Roberts, M., J. Carnio, G. Crawshaw, and M. Hutchins, eds., The biology and management of Australasian carnivorous marsupials, 73–78. Metropolitan Toronto Zoo, Ontario, and American Association of Zoological Parks and Aquariums, Washington, D.C.

Slaughter, B. H. 1968. Earliest known marsupials. Science 162:254–55.

Smith, A. 1982. Is the striped possum *(Dactylopsila trivirgata;* Marsupialia, Petauridae) an arboreal anteater? Aust. Mammal. 5:229–34.

Smith, A. P., and D. G. Quin. 1996. Patterns and causes of extinction and decline in Australian conilurine rodents. Biol. Cons. 77:243–67.

Smith, K. K. 1997. Comparative patterns of craniofacial development in eutherian and metatherian mammals. Evolution 51:1663–78.

Smythe, N. 1986. Competition and resource partitioning in the guild of neotropical terrestrial frugivorous mammals. Ann. Rev. Ecol. Syst. 17:169–88.

Solari, S. 2003. Diversity and distribution of *Thylamys* (Didelphidae) in South America, with emphasis on species from the western side of the Andes. *In* Jones, M. E., C. R. Dickman, and M. Archer, eds., Predators with pouches: the biology of carnivorous marsupials, 82–101. CSIRO Publishing, Melbourne.

Spencer, P. B. S., S. G. Rhind, and M. D. B. Eldridge. 2001. Phylogeographic structure within *Phascogale* (Marsupialia: Dasyuridae) based on partial cytochrome *b* sequence. Aust. J. Zool. 49:369–77.

Springer, M. S., M. Westerman, and J. A. W. Kirsch. 1994. Relationships among orders and families of marsupials based on 12S ribosomal DNA sequences and the timing of the marsupial radiation. J. Mammal. Evol. 2:85–115.

Springer, M. S., J. A. W. Kirsch, and J. A. Case. 1997. The chronicle of marsupial evolution. *In* Givnish, T. J., and K. J. Sytsma, eds., Molecular evolution and adaptive radiation, 129–61. Cambridge Univ. Press, Cambridge.

Springer, M. S., M. Westerman, J. R. Kavanagh, A. Burk, M. O. Woodburne, D. J. Kao, and C. Krajewski. 1998. The origin of the Australasian marsupial fauna and the affinities of the enigmatic monito del monte and marsupial mole. Proc. Roy. Soc. Lond. B 265:2381–86.

Steiner, K. E. 1981. Nectarivory and potential pollination by a neotropical marsupial. Ann. Missouri Bot. Garden 68:505–13.

Stoddart, D. M., and R. W. Braithwaite. 1979. A strategy for utilization of regenerating heathland habitat by the brown bandicoot *(Isoodon obesulus;* Marsupialia, Peramelidae). J. Anim. Ecol. 48:165–79.

Strahan, R., ed. 1995. The mammals of Australia. Reed Books, Sydney, 756 pp.

Streilein, K. E. 1982. Behavior, ecology, and distribution of South American marsupials. *In* Mares, M. A., and H. H. Genoways, eds., Mammalian biology in South America. Pymatuning Symposia in Ecology, 6:231–50. Pymatuning Laboratory of Ecology, Univ. Pittsburgh, Linesville.

Sunquist, M. E., and J. F. Eisenberg. 1993. Reproductive strategies of female *Didelphis.* Bull. Florida Mus. Nat. Hist. Biol. Sci. 36:109–40.

Szalay, F. S. 1982. A new appraisal of marsupial phylogeny and classification. *In* Archer, M., ed., Carnivorous marsupials, 2:621–40. Royal Zoological Society of New South Wales, Sydney.

———. 1994. Evolutionary history of the marsupials and an analysis of osteological characters. Cambridge Univ. Press, New York, 481 pp.

Taggart, D. A., W. G. Breed, P. D. Temple-Smith, A. Purvis, and G. Shimmin. 1998. Reproduction, mating strategies and sperm competition in marsupials and monotremes. *In* Birkhead, T. R., and A. P. Møller, eds., Sperm competition and sexual selection, 623–66. Academic Press, London.

Taylor, J. M., J. H. Calaby, and T. D. Redhead. 1982. Breeding in wild populations of the marsupial-mouse *Planigale maculata sinualis* (Dasyuridae, Marsupialia). *In* Archer, M., ed., Carnivorous marsupials, 1:83–87. Royal Zoological Society of New South Wales, Sydney.

Thomas, O. 1888. Catalogue of the Marsupialia and Monotremata in the collection of the British Museum (Natural History) London. Trustees of the British Museum (Natural History), London, 401 pp.

Thorne, A., R. Grün, G. Mortimer, N. A. Spooner, J. J. Simpson, M. T. McCulloch, L. Taylor, and D. Curnoe. 1999. Australia's oldest human remains: age of the Lake Mungo 3 skeleton. J. Human Evol. 36:591–612.

Toftegaard, C. L., and A. J. Bradley. 2003. Chemical communication in dasyurid marsupials. *In* Jones, M. E., C. R. Dickman, and M. Archer, eds., Predators with pouches: the biology of carnivorous marsupials, 347–57. CSIRO Publishing, Melbourne.

Triggs, B. 1988. The wombat: common wombats in Australia. Univ. of New South Wales Press, Sydney, 148 pp.

Troughton, E. Le G. 1932. Australian furred animals, their past, present, and future. Aust. Zool. 7:173–93.

———. 1944. The imperative need for federal control of post-war protection of nature. Proc. Linn. Soc. New South Wales 69:iv–xv.

———. 1959. The marsupial fauna: its origin and radiation. *In* Keast, A., R. L. Crocker, and

C. S. Christian, eds., Biogeography and ecology in Australia, 69–88. Uitgeverij Dr. W. Junk, The Hague.

Turner, V. 1982. Marsupials as pollinators in Australia. *In* Armstrong, J. A., J. M. Powell, and A. J. Richards, eds., Pollination and evolution, 55–66. Royal Botanic Gardens, Sydney.

Turner, V., and G. M. McKay. 1989. Burramyidae. *In* Walton, D. W., and B. J. Richardson, eds., Fauna of Australia. Vol. 1B: Mammalia, 652–64. Australian Government Publishing Service, Canberra.

Tyndale-Biscoe, C. H. 1965. The female urogenital system and reproduction of the marsupial *Lagostrophus fasciatus*. Aust. J. Zool. 13:255–67.

———. 1966. The marsupial birth canal. Symp. Zool. Soc. Lond. 15:233–50.

———. 1973. Life of marsupials. Edward Arnold, London, 254 pp.

———. 1979. Ecology of small marsupials. *In* Stoddart, D. M., ed., Ecology of small mammals, 343–79. Chapman and Hall, London.

———. 1984. Mammals-marsupials. *In* Lamming, G. E., ed., Marshall's physiology of reproduction, 4th ed., 1:386–454. Churchill Livingstone, Edinburgh.

———. 1989. The adaptiveness of reproductive processes. *In* Grigg, G. C., P. Jarman, and I. D. Hume, eds., Kangaroos, wallabies and rat-kangaroos, 1:277–85. Surrey Beatty & Sons, Sydney.

Tyndale-Biscoe, C. H., and M. B. Renfree. 1987. Reproductive physiology of marsupials. Cambridge Univ. Press, Cambridge, 476 pp.

VandeBerg, J. L., and E. S. Robinson. 1997. The laboratory opossum *(Monodelphis domestica)* in biomedical research. *In* Saunders, N. R., and L. A. Hinds, eds., Marsupial biology: recent research, new perspectives, 238–53. Univ. of New South Wales Press, Sydney.

Van Dyck, S. 1982. The relationships of *Antechinus stuartii* and *A. flavipes* (Dasyuridae, Marsupialia) with special reference to Queensland. *In* Archer, M., ed., Carnivorous marsupials, 2:723–66. Royal Zoological Society of New South Wales, Sydney.

———. 2002. Morphology-based revision of *Murexia* and *Antechinus* (Marsupialia: Dasyuridae). Mem. Qld Mus. 48:239–330.

Van Dyck, S., and M. S. Crowther. 2000. Reassessment of northern representatives of the *Antechinus stuartii* complex (Marsupialia: Dasyuridae): *A. subtropicus* sp. nov., and *A. adustus* new status. Mem. Qld Mus. 45:611–35.

van Tets, I. G. 1998. Can flower-feeding marsupials meet their nitrogen requirements on pollen in the field? Aust. Mammal. 20:383–90.

Van Valkenburgh, B. 1999. Major patterns in the history of carnivorous mammals. Ann. Rev. Earth Pl. Sci. 27:463–93.

Vaughan, T. A., J. M. Ryan, and N. J. Czaplewski. 2000. Mammalogy, 4th ed. Saunders College Publishing, Orlando, 565 pp.

Vieira, E. M., and D. Astúa de Moraes. 2003. Carnivory and insectivory in Neotropical marsupials. *In* Jones, M. E., C. R. Dickman, and M. Archer, eds., Predators with pouches: the biology of carnivorous marsupials, 271–84. CSIRO Publishing, Melbourne.

Vieira, E. M., and E. L. A. Monteiro-Filho. 2003. Vertical stratification of small mammals in the Atlantic rain forest of south-eastern Brazil. J. Tropical Ecol. 19:501–507.

Vieira, E. M., and A. R. T. Palma. 1996. Natural history of *Thylamys velutinus* (Marsupialia, Didelphidae) in central Brazil. Mammalia 60:481–84.

Voss, R. S., and L. H. Emmons. 1996. Mammalian diversity of Neotropical lowland rainforests: a preliminary assessment. Bull. Am. Mus. Nat. Hist. 230:1–115.

Voss, R. S., and S. A. Jansa. 2003. Phylogenetic studies on didelphid marsupials. II. Nonmolecular data and new IRBP sequences: separate and combined analyses of didelphine relationships with denser taxonomic sampling. Bull. Am. Mus. Nat. Hist. 276:1–82.

Voss, R. S., D. P. Lunde, and N. B. Simmons. 2001. The mammals of Paracou, French Guiana: a Neotropical lowland rainforest fauna. Part 2: nonvolant species. Bull. Am. Mus. Nat. Hist. 263:1–236.

Wainwright, M. 2002. The natural history of Costa Rican mammals. Zona Tropical, Miami, 384 pp.

Warburton, B., G. Tocher, and N. Allan. 2000. Possums as a resource. *In* Montague, T. L., ed., The brushtail possum: biology, impact and management of an introduced marsupial, 251–61. Manaaki Whenua Press, Lincoln.

Warburton, N., C. Wood, C. Lloyd, S. Song, and P. Withers. 2003. The 3-dimensional anatomy of the north-western marsupial mole (*Notoryctes caurinus* Thomas 1920) using computed tomography, X-ray and magnetic resonance imaging. Rec. West. Aust. Mus. 22:1–7.

Ward, S. J. 1998. Numbers of teats and pre- and post-natal litter sizes in small diprotodont marsupials. J. Mammal. 79:999–1008.

———. 2003. Biased sex ratios in litters of carnivorous marsupials: why, when and how? *In* Jones, M. E., C. R. Dickman, and M. Archer, eds., Predators with pouches: the biology of carnivorous marsupials, 376–82. CSIRO Publishing, Melbourne.

Westerman, M., M. S. Springer, J. Dixon, and C. Krajewski. 1999. Molecular relationships of the extinct pig-footed bandicoot *Chaeropus ecaudatus* (Marsupialia: Perameloidea) using 12S rRNA sequences. J. Mammal. Evol. 6:271–88.

Williams, P. A. 2003. Are possums important dispersers of large-seeded fruit? New Zealand J. Ecol. 27:221–23.

Williams, S. E., R. G. Pearson, and P. J. Walsh. 1996. Distributions and biodiversity of the terrestrial vertebrates of Australia's Wet Tropics: a review of current knowledge. Pac. Cons. Biol. 2:327–62.

Wilson, B. A., and G. R. Friend. 1999. Responses of Australian mammals to disturbance: a review. Aust. Mammal. 21:87–105.

Wilson, B. A., C. R. Dickman, and T. P. Fletcher. 2003. Dasyurid dilemmas: problems and solutions for conserving Australia's small carnivorous marsupials. *In* Jones, M. E., C. R. Dickman, and M. Archer, eds., Predators with pouches: the biology of carnivorous marsupials, 407–21. CSIRO Publishing, Melbourne.

Wilson, D. E., and D. M. Reeder, eds. 1993. Mammal species of the world: a taxonomic and geographic reference, 2nd ed. Smithsonian Institution Press, Washington, D.C., 1206 pp.

Winter, J. W. 1988. Ecological specialization of mammals in Australian tropical and subtropical rainforest: refugial or ecological determinism? Proc. Ecol. Soc. Aust. 15:127–38.

———. 1997. Responses of non-volant mammals to late Quaternary climatic changes in the Wet Tropics region of north-eastern Australia. Wildl. Res. 24:493–511.

Winter, J. W., and L. K.-P. Leung. 1995*a*. Common spotted cuscus *Spilocuscus maculatus* (Desmarest, 1818). *In* Strahan, R., ed., The mammals of Australia, 266–68. Reed Books, Sydney.

———. 1995*b*. Southern common cuscus *Phalanger intercastellanus* Thomas, 1895. *In* Strahan, R., ed., The mammals of Australia, 268–70. Reed Books, Sydney.

Wood, D. H. 1970. An ecological study of *Antechinus stuartii* (Marsupialia) in a south-east Queensland rainforest. Aust. J. Zool. 18:185–207.

Woodburne, M. O. 1984. Families of marsupials: relationships, evolution, and biogeography. *In* Broadhead, T. W., ed., Mammals: notes for a short course. Univ. Tennessee Dept. Geol. Sci. Studies Geol. 8:48–71.

Woodburne, M. O., and J. A. Case. 1996. Dispersal, vicariance, and the late Cretaceous to early Tertiary land mammal biogeography from South America to Australia. J. Mammal. Evol. 3:121–61.

Woodburne, M. O., and W. J. Zinsmeister. 1982. Fossil land mammal from Antarctica. Science 218:284–86.

———. 1984. The first land mammal from Antarctica and its biogeographic implications. J. Paleontol. 58:913–48.

Woodburne, M. O., T. H. Rich, and M. S. Springer. 2003. The evolution of tribospheny and the antiquity of mammalian clades. Mol. Phyl. Evol. 28:360–85.

Wood Jones, F. 1923–24. The mammals of South Australia. Part I. The monotremes and the carnivorous marsupials. Part II. The bandicoots and the herbivorous marsupials. Government Printer, Adelaide, 270 pp.

———. 1949. The study of a generalized marsupial (*Dasycercus cristicauda* Krefft). Trans. Zool. Soc. Lond. 26:409–501.

Woolley, P. 1974. The pouch of *Planigale subtilissima* and other dasyurid marsupials. J. Roy. Soc. West. Aust. 57:11–15.

Woolley, P. A. 1982. Phallic morphology of the Australian species of *Antechinus* (Dasyuridae, Marsupialia): a new taxonomic tool? *In* Archer, M., ed., Carnivorous marsupials, 2:767–81. Royal Zoological Society of New South Wales, Sydney.

———. 1989. Nest location by spool-and-line tracking of dasyurid marsupials in New Guinea. J. Zool. Lond. 218:689–700.

———. 2003. Reproductive biology of some dasyurid marsupials of New Guinea. *In* Jones, M. E., C. R. Dickman, and M. Archer, eds., Predators with pouches: the biology of carnivorous marsupials, 169–82. CSIRO Publishing, Melbourne.

Wroe, S. 2003. Australian marsupial carnivores: recent advances in palaeontology. *In* Jones, M. E., C. R. Dickman, and M. Archer, eds., Predators with pouches: the biology of carnivorous marsupials, 102–23. CSIRO Publishing, Melbourne.

Wroe, S., and M. Archer. 2004. Origins and early radiations of marsupials. *In* Merrick, J. R., M. Archer, G. M. Hickey, and M. S. Y. Lee, eds., Evolution and zoogeography of Australasian vertebrates, 517–40. Auscipub, Sydney.

Wroe, S., M. Ebach, S. Ahyong, C. de Muizon, and J. Muirhead. 2000. Cladistic analysis of dasyuromorphian (Marsupialia) phylogeny using cranial and dental characters. J. Mammal. 81:1008–24.

Wroe, S., T. Myers, F. Seebacher, B. Kear, A. Gillespie, M. Crowther, and S. Salisbury. 2003. An alternative method for predicting body mass: the case of the Pleistocene marsupial lion. Paleobiology 29:403–11.

Wroe, S., M. Crowther, J. Dortch, and J. Chong. 2004*a*. The size of the largest marsupial and why it matters. Proc. Roy. Soc. Lond. B (Suppl.) 271:S34–S36.

Wroe, S., C. Argot, and C. R. Dickman. 2004*b*. On the rarity of big fierce carnivores and the primacy of isolation and area: tracking large mammalian carnivore diversity on two isolated continents. Proc. Roy. Soc. Lond. B 271:1203–11.

Wroe, S., J. Field, R. Fullagar, and L. S. Jermiin. 2004*c*. Megafaunal extinction in the late Quaternary and the global overkill hypothesis. Alcheringa 28:291–331.

Yalden, D. W. 1988. Feral wallabies in the Peak District, 1971–1985. J. Zool. Lond. 215:369–74.

Order Didelphimorphia

American Opossums

This marsupial order of 4 Recent families, 15 genera, and 66 species is found naturally from southeastern Canada, through the eastern United States and Mexico, into South America to about 47° S in Argentina. Opossums also occur on some of the islands in the Lesser Antilles. The vernacular name "opossum" generally applies to members of this order, which now is found only in the New World. The term "possum" is used mainly for members of certain marsupial families in the Australasian region.

This order has essentially the same living content as the family Didelphidae, as the latter was accepted by nearly all authorities, for example, Archer (1984) and Kirsch (1977*b*), until about a decade ago. Subsequent taxonomists, including Aplin and Archer (1987), Marshall, Case, and Woodburne (1990), and Wilson and Reeder (1993), have tended to give ordinal rank to the group. Hershkovitz (1992*a*, 1992*b*), who is followed here with respect to names and sequence of families and genera, regarded the group as an order but applied to it the respective terms Didelphoidea and Didelphida and used the name Didelphimorphia for a cohort. Szalay (1993, 1994) also used the name Didelphida for the order but included therein what is here regarded as the order Paucituberculata, as well as the Didelphimorphia, which he reduced to subordinal rank.

The members of this order are small to medium in size. For *Gracilinanus*, which includes the smallest species, adult head and body length is about 70–135 mm and tail length is 100–155 mm. In *Didelphis*, the largest genus, adult head and body length is 325–500 mm and tail length is 255–535 mm. The tail is usually long, scaly, very scantily haired, and prehensile, but in some forms it is short and/or rather hairy. Some genera have long, projecting guard hairs in the pelage. The muzzle is elongate, and the ears have well-developed conches. The limbs are short in many species, with the hind limbs slightly longer. All four feet have five separate digits. The great toe is large, clawless, and opposable. A distinct marsupial pouch is present in *Didelphis*, *Chironectes*, *Philander*, and *Lutreolina*. In the other genera it is absent, or it may consist of two longitudinal folds of skin, separate at both ends, near the median line of the body. Females have 5–27 mammae.

The dental formula for the order is: (i 5/4, c 1/1, pm 3/3, m 4/4) $\times$ 2 = 50. The teeth are rooted and sharp. The upper incisors are conical, small, and unequal; the first is larger than, and separated from, the others. The canines are large. The last premolar is preceded by a multicuspidate, molariform, deciduous tooth, and the molars are tricuspidate. The facial part of the skull is long and pointed, but the cranial part is small. Reig, Kirsch, and Marshall (1987) provided extensive further details on the morphological characters of this group of marsupials.

Opossums are active mainly in the evening and at night. Most forms are arboreal or terrestrial, and one genus *(Chironectes)* is semiaquatic. Opossums are insectivorous, carnivorous, or, more commonly, omnivorous.

As in other marsupials, the gestation period is short and the developmental period is long. The gestation period of *Didelphis*, about 12.5–13 days, is among the shortest found in mammals. The front legs of a newborn opossum are well developed, and the fingers are supplied with sharp, deciduous claws (at least in *Didelphis*); these claws drop off some time after the young reaches the pouch. The hind legs of opossums at birth are much smaller than the forelegs and practically useless. In young opossums the passage from the nasal chamber to the larynx is so separated from the passage to the esophagus that the baby can swallow and breathe at the same time. Contrary to an earlier opinion, the mother does not pump milk into the young; they suckle normally, though continuously, under their own power (Banfield 1974; Lowery 1974). The part of the brain that regulates body temperature is not functional in a young opossum; the baby is kept warm solely by the mother's body heat.

The pouch young of opossums and probably other pouch marsupials breathe and rebreathe air that contains 8–20 times the normal content of carbon dioxide. This may serve some unknown, useful role, or it may merely show unusual tolerance. After leaving the pouch the young of some species travel with the mother for awhile, usually riding by clinging to the fur on her back.

A number of carnivores prey on opossums, and people eat some of the larger species. Didelphid pelts are sometimes used for inexpensive trimmings and garments. Opossums, particularly members of the genus *Didelphis*, are widely used in laboratory research. South American didelphids are susceptible to yellow fever virus.

The statement that opossums copulate through the nose and that sometime later the tiny young are blown into the pouch is entirely false. This idea may have originated with the observation that the pouch is often filled with newborn young soon after the female has investigated the pouch with her nose. Or possibly the tale stems from the possession of a forked penis by the male.

The known geological range of this order is early Cretaceous to early Miocene and Pleistocene to Recent in North America, late Cretaceous to Recent in South America, Eocene to early Miocene in Europe, and Oligocene in North Africa (Hershkovitz 1995; Kirsch 1977*a*, 1977*b*; Simons and Bown 1984).

A. Eight baby murine opossums *(Marmosa murina)* attached to the mother's nipples, photo from New York Zoological Society.
B. A female North American opossum *(Didelphis virginiana)* with a well-developed baby trying to get into her pouch, photo by Ernest P. Walker.

DIDELPHIMORPHIA; **Family MARMOSIDAE**

Pouchless Murine, or Mouse, Opossums

This family of 8 Recent genera and 53 species is found from northeastern Mexico, throughout Central and South America, to southern Argentina. Although traditionally considered a part of the Didelphidae, and maintained as such by Gardner (*in* Wilson and Reeder 1993), the Marmosidae were elevated to familial rank by Hershkovitz (1992*a*, 1992*b*). The latter authority, who is followed here with respect to sequence of genera, recognized five subfamilies: the Marmosinae, with *Gracilinanus, Marmosops, Marmosa,* and *Micoureus;* the Thylamyinae, with *Thylamys;* the Lestodelphyinae, with *Lestodelphys;* the Metachirinae, with *Metachirus;* and the Monodelphinae, with *Monodelphis.*

According to Hershkovitz (1992*b*), head and body length is 70–310 mm and tail length is 45–390 mm. There is no pouch, there are 9–27 mammae, the fifth digit on the forefoot has a sharp claw, and the hallux is opposable. In the subfamily Marmosinae the tail is long and prehensile, the claws are short, and the anklebones are specialized. The Thylamyinae have a prehensile but incrassate (seasonally thickened) tail, long and stout claws, parallel-sided nasal bones, and unspecialized anklebones. In the Lestodelphyinae the tail is short, incrassate, and nonprehensile, the claws are stout, the skull has a sagittal crest, and the foot bones are specialized. The relatively large and terrestrial Metachirinae have a long, nonprehensile tail and unspecialized foot bones. The shrewlike Monodelphinae have a short and nonprehensile tail, long claws, a sagittally crested cranium, and unspecialized anklebones.

Recently Hershkovitz (1995) described the genus *Adinolon,* apparently a component of the family Marmosidae from the latter part of the early Cretaceous of Texas. It may be the oldest known member of the Didelphimorphia and is among the earliest of marsupials.

DIDELPHIMORPHIA; MARMOSIDAE; **Genus GRACILINANUS**
Gardner and Creighton, 1989

Gracile Mouse Opossums

There are nine described species (Hershkovitz 1992*b;* Pacheco et al. 1995; Tate 1933):

G. agilis, Colombia, Peru, eastern and south-central Brazil, Bolivia, Paraguay, Uruguay, northeastern Argentina;
G. microtarsus, southeastern Brazil;
G. marica, Colombia, northern Venezuela;
G. emiliae, northern Brazil;
G. perijae, northern Colombia;
G. aceramarcae, Peru, western Bolivia;
G. dryas, Colombia, northwestern Venezuela;
G. longicaudus, central Colombia;
G. kalinowskii, southern Peru.

Hershkovitz (1992*b*) reported the presence of an undescribed species, perhaps with some affinity to *G. kalinowskii,* in northwestern Ecuador. He also indicated that *G. microtarsus* may be only a subspecies of *G. agilis.* The species of *Gracilinanus* long were placed in *Marmosa,* but Creighton (1985) suggested their distinction and most subsequent authorities have agreed. Except as noted, the information for the remainder of this account was taken

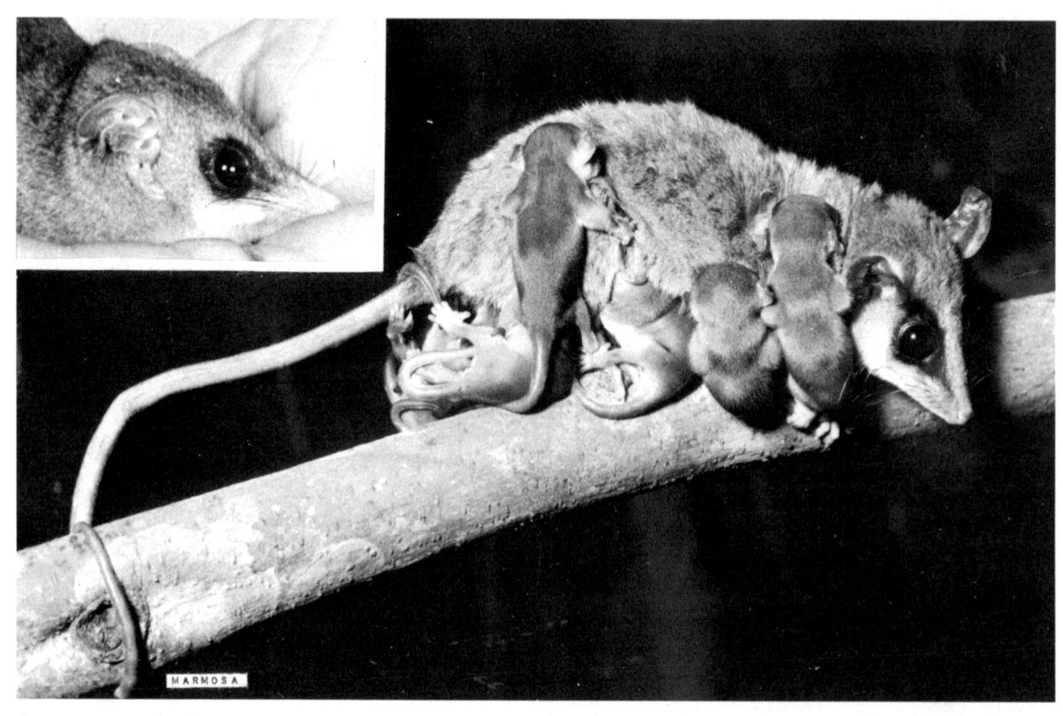

South American mouse opossums (*Marmosa* sp.), photo by Harald Schultz. Inset: murine opossum *(Micoureus demerarae)* showing partially coiled ear, photo by Ernest P. Walker.

from Gardner and Creighton (1989) and Hershkovitz (1992*b*).

Head and body length is 70–135 mm and tail length is 90–155 mm. Weight in *G. agilis* is 23–34 grams (Redford and Eisenberg 1992). The upper parts range from dull brownish gray to bright reddish brown; the underparts are paler, varying from almost white through cream to orange. There is a dark brown or blackish eye ring, the upper surface of the muzzle is pale, and the tail is scaled and weakly bicolored or unicolored fuscous. There is no pouch. Mammae number about 11–15 and are mostly abdominal, but usually a few are pectoral. The closely related genera *Marmosops, Marmosa,* and *Micoureus* lack pectoral mammae.

The skull of *Gracilinanus,* unlike those of *Marmosa* and *Micoureus,* lacks postorbital processes, the rostrum is slender, and the hard palate is highly fenestrated. The auditory bullae are tripartite and are relatively large compared with those of *Marmosa, Marmosops,* and *Micoureus.* The second upper premolar is always larger than the third, and the lower canine teeth are relatively shorter than those of *Marmosa, Micoureus,* and *Thylamys* but not as short or premolariform as in *Marmosops.* The caudal scales of *Gracilinanus* (except *G. kalinowskii*) and *Thylamys* are arranged in an annular pattern, whereas those of *Marmosa, Marmosops,* and *Micoureus* have a spiral pattern. Unlike that of *Thylamys,* the tail of *Gracilinanus* does not become seasonally thickened (incrassate) by fat deposition.

These tiny opossums inhabit forests and woodlands from coastal areas to elevations of about 4,500 meters in the Andes. They dwell in trees or shrubs but frequently forage on the ground. Locomotion is by a short, swift quadrupedal gait interchanged with overhand climbing and headfirst descents, assisted by use of the prehensile tail. *G. agilis* has been found in nests made of vegetation 1.6 meters above the ground (Redford and Eisenberg 1992). Individuals have been observed to enter torpor in response to cool weather. The diet consists of fruit, plant exudates, insects, and other small invertebrates. Breeding may continue throughout the year, at least when food is abundant. Litter size in *G. marica* has been reported to be six young (Hayssen, Van Tienhoven, and Van Tienhoven 1993). *G. aceramarcae,* known only from two restricted areas of declining habitat, is classified as critically endangered by the IUCN. *G. emiliae* and *G. dryas,* also losing their limited habitat, are classified as vulnerable, while *G. agilis, G. microtarsus,* and *G. marica* are designated as near threatened.

DIDELPHIMORPHIA; MARMOSIDAE; **Genus MARMOSOPS**
Matschie, 1916

There are nine species (Gardner *in* Wilson and Reeder 1993; Gardner and Creighton 1989; E. R. Hall 1981; Handley and Gordon 1979; Pine 1981):

M. noctivagus, Amazonian Brazil, Ecuador, Peru, Bolivia;
M. dorothea, Bolivia;
M. incanus, eastern and southeastern Brazil;
M. invictus, Panama;
M. parvidens, Colombia, Venezuela, Guyana, Surinam, Peru, Brazil;
M. handleyi, known only from the type locality in central Colombia;
M. fuscatus, eastern Colombia, western Venezuela, Trinidad;
M. cracens, northwestern Venezuela;
M. impavidus, western Panama to Peru and southern Venezuela.

Marmosops long was considered a synonym or subgenus of *Marmosa* but now is generally recognized as a full genus (Corbet and Hill 1991; Gardner *in* Wilson and Reeder 1993; Gardner and Creighton 1989; Hershkovitz 1992*a,* 1992*b*). The information for the remainder of this account was taken from Eisenberg (1989) and Hershkovitz (1992*b*).

Head and body length is 90–160 mm, tail length is 105–220 mm, and weight is about 24–85 grams. The upper parts are pale to dark brown or gray, the underparts are pale gray to white, and there is a dark eye ring. The tail is prehensile and, unlike that of *Thylamys,* is never incrassate

South American mouse opossum *(Marmosops incanus),* photo by Luiz Claudio Marigo.

(seasonally thickened by fat deposition). The caudal scales are arranged in a spiral pattern, not in an annular pattern as in *Gracilinanus.* Unlike those of *Marmosa* and *Micoureus,* the skull lacks strong postorbital processes and a lambdoidal crest. There is no pouch and there are only nine mammae, none of which are pectoral. Pregnant female *M. parvidens* have been found to contain six to seven embryos. *Marmosops* is nocturnal and arboreal, generally inhabits moist forests in either lowlands or mountains, and feeds on insects and fruit. Most species of *Marmosops* are losing their habitat to human activity. *M. handleyi,* known only from a single locality, is classified as critically endangered by the IUCN. *M. cracens,* also known only from one site, is classed as endangered, while *M. dorothea* is classed as vulnerable and *M. incanus, M. invictus, M. parvidens, M. fuscatus,* and *M. impavidus* are designated as near threatened.

DIDELPHIMORPHIA; MARMOSIDAE; **Genus MARMOSA**
Gray, 1821

Murine, or Mouse, Opossums

There are two subgenera and nine species (Cabrera 1957; Collins 1973; Gardner *in* Wilson and Reeder 1993; E. R. Hall 1981; Handley and Gordon 1979; Husson 1978; Kirsch 1977*a;* Kirsch and Calaby 1977; Pine 1972; Reig, Kirsch, and Marshall 1987; Streilen 1982*b*):

subgenus *Marmosa* Gray, 1821

M. lepida, Surinam to Ecuador and Bolivia;
M. murina, northern and central South America, Tobago;
M. rubra, eastern Ecuador, northeastern Peru;
M. tyleriana, southern Venezuela;
M. robinsoni, Belize, Ruatan Island (Honduras), Panama to Ecuador, Venezuela, Trinidad and Tobago, Grenada;
M. xerophila, northeastern Colombia, northwestern Venezuela;
M. mexicana, northeastern Mexico to western Panama;
M. canescens, western Mexico, Yucatan, Tres Marías Islands;

subgenus *Stegomarmosa* Pine, 1972

M. andersoni, southern Peru.

Until recently *Marmosa* commonly was considered to include the species now assigned to the genera *Gracilinanus, Marmosops, Micoureus,* and *Thylamys* (see accounts thereof).

Head and body length is 85–190 mm and tail length is 125–230 mm. Weight in *M. robinsoni,* one of the larger, more common species, is 60–130 grams for males and 40–70 grams for females. *M. murina,* a smaller species, weighs only 13–44 grams (Eisenberg 1989). In *M. mexicana* weight is 29–92 grams, but averages are 63.7 grams for males and only 35.2 grams for females (Alonso-Mejía and Medellín 1992). The upper parts range from gray through dark brown to bright reddish brown; the underparts are paler, varying from almost white through buff to yellowish. Almost all forms have dusky brown or black facial markings around the eyes. The fur is short, fine, and velvety in most species. The tail is strongly prehensile, has only a short basal brush of fur, and either is uniformly colored or has the ventral surface slightly paler than the dorsal. There is no pouch. Mammae number 9–19 depending on the species.

According to Hershkovitz (1992*b*), *Marmosa* differs from *Gracilinanus* in lacking pectoral mammae, in having a pronounced superior postorbital process on the skull, and in having its caudal scales arranged in a spiral, rather than

Mouse opossum *(Marmosa mexicana)*, photo by Barbara L. Clauson.

annular, pattern. *Marmosa* differs from *Marmosops* in having a throat gland present in the mature males and a lambdoidal crest on the skull. *Marmosa* differs from *Micoureus* in having a throat gland, shorter fur, and a ridged and beaded superior border of the frontal bone of the skull. *Marmosa* differs from *Thylamys* in having the tail never incrassate, weaker claws on the forefoot, the nasal bones of the skull flared at the frontomaxillary suture, and the second premolar tooth usually larger than the third (rather than the opposite).

Most species of *Marmosa* are forest dwellers. The genus ranges vertically from sea level to about 3,400 meters. Most specimens of *Marmosa* collected by Handley (1976) in Venezuela were taken near streams or in other moist areas, but *M. xerophila* was almost always found in dry situations. Mouse opossums are nocturnal and usually arboreal, though some species are terrestrial. They are often found on banana plantations and among small trees and vine tangles. They build nests of leaves and twigs in trees or shelter in abandoned birds' nests. They have also been located in ground nests, in holes in hollow logs, and under rocks. *M. mexicana* sometimes digs its own burrow but more commonly nests in trees (Alonso-Mejía and Medellín 1992). Most specimens of *M. canescens* taken in Sinaloa, Mexico, were found by day in their nests in hollows of dead cacti (Armstrong and Jones 1971). The species *M. robinsoni* does not seem to have a fixed abode but apparently spends the day in any suitable shelter when daylight overtakes it. This species reportedly is nomadic in Panama, occupying home ranges of about 0.22 ha. for two or three months and occurring at population densities of 0.31–2.25/ha. (Fleming 1972; Hunsaker 1977). In northern Venezuela *M. robinsoni* was estimated to occur at densities of 0.25–4.25 adults per hectare, and males were found to be more nomadic than females (O'Connell 1983). Mouse opossums are generally solitary, usually hunting and nesting alone. Fleming (1972) obtained data suggesting that while there was overlap between the home ranges of male *M. robinsoni* and between those of opposite sexes, the females had nonoverlapping home ranges and appeared intolerant of one another.

Murine opossums are courageous fighters for their size. Like some of the other marsupials, they can lower their ears by crinkling them down much as a sail is furled. Their eyes reflect light as brilliant ruby red. When moving along tree limbs and vines, *M. robinsoni* curves its tail loosely around the branch, except when climbing vertically, and sometimes leaps across gaps. None of its motions, however, are particularly rapid. The diet of *Marmosa* consists mainly of insects and fruits but also includes small rodents, lizards, and birds' eggs. Large grasshoppers are killed by a number of bites about the head and thorax; only the harder parts and lower legs are discarded. One mouse opossum in Mexico was noted hanging by its tail and eating a wild fig, which it held in its forefeet. Raw sugar is a particular favorite of *M. mexicana*. Banana and mango crops sometimes are damaged by these marsupials. Occasionally they are found in bunches of bananas in warehouses and stores in the United States, having remained in the bunch when it was shipped from the tropics. Occurrences in the New Orleans area, apparently resulting from such journeys, were documented by Lowery (1974).

Some species of *Marmosa* may breed throughout the year, while others are seasonal. According to Hunsaker (1977), *M. robinsoni* has breeding peaks in February, June, and July through December, an estrous cycle averaging 23 days, and a gestation period of 14 days. Litter size is seven to nine in the wild and slightly smaller in captivity. Weaning occurs when the young are 60–70 days old, they leave the mother a few days later, and young females have their first estrus when 265–75 days old. Life expectancy probably is under one year in the wild but is about three years in captivity. Reproductive ability declines in the second year of life.

Some slightly different information compiled by Collins (1973) is that gestation is just under 14 days, average size of 70 laboratory litters was about seven, females reach sexual maturity at approximately six months, and average life span in captivity is one to two years. One male *M. robinsoni* obtained at an unknown age lived for almost five more years, an apparent record longevity.

Godfrey (1975) found that litter size in a laboratory colony of *M. robinsoni* ranged from 1 to 13 and usually was 7–9. Litter sizes of up to 15 have been reported in Venezuela (O'Connell 1983). A study in the Panama Canal Zone found *M. robinsoni* to be a seasonal breeder, with reproduction occurring mainly from late March to September and litter size averaging 10 (Fleming 1973). According to Hunsaker and Shupe (1977), females of *M. murina* produce three litters annually and breed throughout the year. Eisentraut (1970) reported that *M. murina* has a gestation period of 13 days and a maximum litter size of 13. A female *M. canescens* taken on 5 September contained 13 embryos, one found on 4 September was accompanied by 8 nursing young, and females taken in October were lactating (Armstrong and Jones 1971).

As noted by Grimwood (1969), because of their small size and nocturnal habits populations of mouse opossums are not directly threatened by people; however, some of the high-altitude species might be jeopardized by widespread clearing of brush in the Andes region. *M. andersoni*, known only from a single montane locality, is now classified as critically endangered by the IUCN. *M. xerophila*, also restricted to a small area, is classed as endangered, while *M. lepida* is designated as near threatened.

DIDELPHIMORPHIA; MARMOSIDAE; **Genus MICOUREUS**
Lesson, 1842

There are four species (Cabrera 1957; Gardner *in* Wilson and Reeder 1993; Kirsch 1977*a;* Kirsch and Calaby 1977):

M. demerarae, Venezuela to Paraguay;
M. alstoni, Belize to Colombia;
M. constantiae, Mato Grosso of Brazil, Bolivia, northern Argentina;
M. regina, Colombia, Ecuador, Peru, Bolivia.

Micoureus long was considered a synonym or subgenus of *Marmosa* but now is generally recognized as a full genus (Corbet and Hill 1991; Gardner *in* Wilson and Reeder 1993; Gardner and Creighton 1989; Hershkovitz 1992*a*, 1992*b*). *M. demerarae* formerly was known as *Marmosa cinerea.* The information for the remainder of this account was taken from Eisenberg (1989), Hershkovitz (1992*b*), Redford and Eisenberg (1992), and Reig, Kirsch, and Marshall (1987).

Head and body length is 120–215 mm, tail length is 170–270 mm, and weight in *M. demerarae* is 50–230 grams. The upper parts are various shades of gray or brown, the underparts are cream or buff to yellow, and there is usually a dark eye ring. The pelage is thick, lax, and crinkly. The prehensile tail may be uniformly colored or bicolored, with the long terminal portion distinctly paler than the proximal portion. The fur extends at least 20 mm onto the tail, whereas in *Marmosa* it usually extends less than 20 mm. *Micoureus* also differs from *Marmosa* in lacking a throat gland and in having the superior border of the frontal bone of the skull projected as a ledge. There are 9–15 mammae, none of which are pectoral.

The only well-known species is *M. demerarae,* which lives mainly in tropical forest, forages in trees and on the ground, is nocturnal, and eats mostly insects and some fruit. The female builds an open nest by transporting leaves with her mouth or prehensile tail. Breeding in Venezuela is tied to rainfall and does not take place during the winter dry season. *M. alstoni* and *M. constantiae* are now designated as near threatened by the IUCN.

DIDELPHIMORPHIA; MARMOSIDAE; **Genus THYLAMYS**
Gray, 1843

Southern Mouse Opossums

There are five species (Cabrera 1957; Collins 1973; Gardner *in* Wilson and Reeder 1993; Gardner and Creighton 1989):

T. elegans, central Peru, Bolivia, Chile, northwestern Argentina;
T. macrura, Paraguay, southern Brazil;
T. pallidior, Bolivia, Argentina;
T. pusilla, central and southern Brazil, Argentina, Paraguay, southern Bolivia;
T. velutinus, southeastern Brazil.

Thylamys long was considered a synonym or subgenus of *Marmosa* but now is generally recognized as a full genus (Corbet and Hill 1991; Gardner *in* Wilson and Reeder 1993; Gardner and Creighton 1989; Hershkovitz 1992*a,* 1992*b*). *T. macrura* formerly was known as *Marmosa grisea.* Except as noted, the information for the remainder of this account was taken from Eisenberg (1989), Hershkovitz (1992*b*), and Redford and Eisenberg (1992).

Head and body length is 68–150 mm, tail length is 90–160 mm, and weight is about 18–55 grams. The upper parts are various shades of gray or brown, the underparts are yellowish or white, and there is a dark eye ring. The fur is very dense and soft. The tail is naked or only finely haired, except at the base, and is seasonally incrassate, that is, becoming thickened through storage of fat. In this last regard *Thylamys* differs from *Gracilinanus, Marmosops, Marmosa,* and *Micoureus.* It also differs in having stout and well-projecting claws on the forefoot, parallel-sided nasal bones, and the third upper premolar larger than the second. There are 15 mammae, including two pectoral pairs, and there is little or no pouch development.

These mouse opossums occur in a variety of habitats from sea level to the mountains and are both arboreal and terrestrial. *T. elegans* has been found in wet forest, brush, and scrub and may nest in trees, rocky embankments, or holes in the ground made by other animals. It stores fat and hibernates during the winter. *T. pusilla* frequently occurs in much drier areas, such as thorn forest, and has been found active even when there is snow on the ground. The diet of *T. elegans* consists primarily of insects but also includes fruit and small vertebrates. *T. elegans* seems to occur at low densities in all habitats; two individuals had an average home range of 289 sq meters. In Chile it reproduces from September to March, during which time a female can have two litters. It has as many as 15 young, usually 8–12. The IUCN now designates *T. macrura* as near threatened.

DIDELPHIMORPHIA; MARMOSIDAE; **Genus LESTODELPHYS**
Tate, 1934

Patagonian Opossum

The single species, *L. halli,* is known only from nine specimens taken at three localities in Patagonia, Argentina (L. G. Marshall 1977), and from a locality in the mountains of west-central Argentina (Redford and Eisenberg 1992). It occurs farther south than any other living marsupial.

Head and body length is 132–44 mm and tail length is 81–99 mm. One specimen weighed 76 grams (Redford and Eisenberg 1992). The fur is not particularly long, but it is dense, fine, and soft. The back is dark gray; the face is somewhat darker with no markings; the sides of the body are

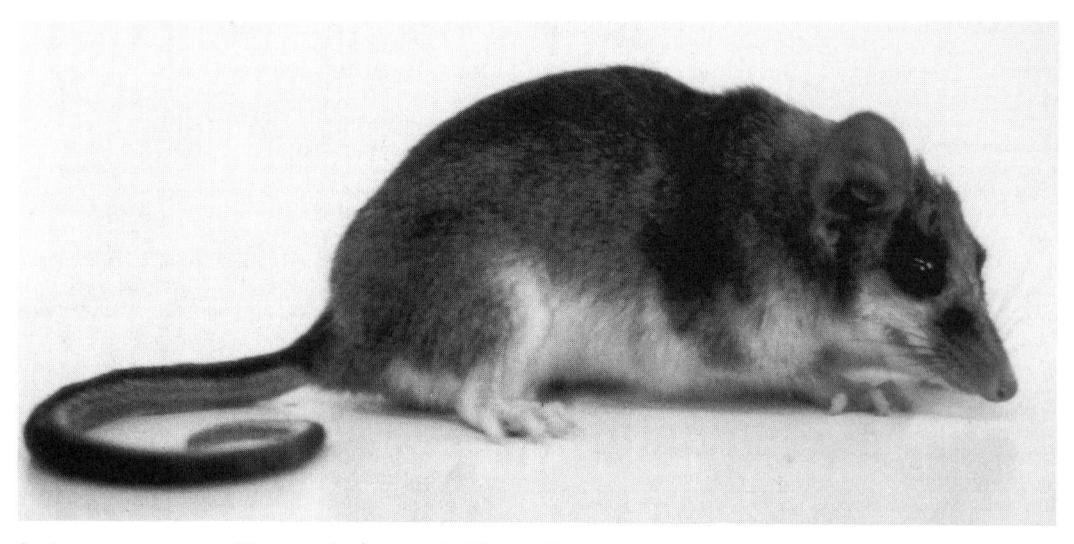

Southern mouse opposum *(Thylamys elegans)*, photo by Milton H. Gallardo.

Patagonian opossum *(Lestodelphys halli)*, photo by Oliver P. Pearson.

clear gray; and the forearms, hands, ankles, feet, and underparts are white. The cheeks, a patch over the eyes, and a patch at the posterior base of the ear are also white. There are dark shoulder and hip patches. The tail, which is furred like the body for about 20 mm at the base, then thickly covered with short, fine hairs, is dark grayish brown above and whitish below and at the tip.

In general appearance individuals of this genus resemble certain species of the subfamily Marmosinae, but they differ in characters of the feet and in the distinctive skull and dental features. The feet are stronger than those of the Marmosinae; the Patagonian opossum probably is more terrestrial than the murine opossums. In *Lestodelphys* the claw of the thumb and other digits extends considerably beyond the soft terminal pad; in the Marmosinae it is markedly shorter than the others and does not extend beyond the pad. The Patagonian opossum has a short, broad skull, small incisors, long, sharp, straight canines, and large molars. In *Lestodelphys* the bullar floor of the skull is complete and does not have a gap between the petrous and alisphenoid components such as are found in all other genera of the Marmosidae (Hershkovitz 1992*b*). As in the Thylamyinae, the tail occasionally becomes thickened near the base due to a seasonal accumulation of fat. The ears are short, rounded, and flesh-colored.

This opossum is said to inhabit pampas areas. One specimen was caught in a steel trap baited with a dead bird. Thomas (1921:138) commented: "This interesting little opossum . . . appears, from the structure of its skull, to be of a more carnivorous and predaceous nature than any of the other small members of the family. Ordinarily *Marmosa* feeds mainly on insects and fruit, and as insects are rare and fruit almost non-existent in its far-southern habitat, this opossum has had to acquire peculiar habits, and no doubt lives largely on mice and small birds." Redford and Eisenberg (1992) cited Oliver Pearson as reporting that captive *Lestodelphys* "killed live mice at lightning speed, eating everything—bones, teeth, and fur. A 70 gram animal will eat an entire 35 gram mouse in one night." Because it apparently occurs in a very restricted area of declining habitat, *L. halli* now is classified as vulnerable by the IUCN.

DIDELPHIMORPHIA; MARMOSIDAE; **Genus METACHIRUS**
Burmeister, 1854

Brown "Four-eyed" Opossum

The single species, *M. nudicaudatus*, is found from extreme southern Mexico to northeastern Brazil and northeastern Argentina (Husband et al. 1992; Medellín et al. 1992; Redford and Eisenberg 1992). Because of technical problems of nomenclature there has been recent argument regarding which generic name properly applies to this species. Hershkovitz (1976, 1981) supported *Metachirus*, which has been in more general use, while Pine (1973) favored *Philander* Tiedemann, 1808, at least for the time being. E. R. Hall (1981) followed Pine in applying *Philander* to the brown "four-eyed" opossum, but Corbet and Hill (1991), Gardner (1981 and *in* Wilson and Reeder 1993), and Kirsch and Calaby (1977) agreed with Hershkovitz. The generic designation *Philander* is actually applicable to the gray and black "four-eyed" opossums and is so employed here, again in accordance with Hershkovitz (1976, 1981) and his backers,

Brown "four-eyed" opossum *(Metachirus nudicaudatus)*, photo by Cory T. de Carvalho.

rather than the name *Metachirops*, used by Pine (1973) and E. R. Hall (1981).

Head and body length is 190–310 mm and tail length is 195–390 mm (Hershkovitz 1992*b*). Recorded weights in Argentina and Paraguay have ranged from 91 to 480 grams (Redford and Eisenberg 1992). A male and a female from Barro Colorado Island, Panama Canal Zone, each weighed 800 grams. The back and sides are brown, often dark cinnamon brown, and the rump may be washed with black. The face is dusky, almost black in some individuals, with a creamy white spot over each eye. These spots, suggesting eyes, are usually smaller and more widely separated than those of *Philander*. The underparts are buff to gray. The tail is furred for a short distance basally. The pelage is short, dense, and silky.

Although the common names and general appearance of *Metachirus* and *Philander* are similar, the two are not closely related. *Metachirus* may be distinguished externally by its brown coloration and longer tail. Unlike the gray and black "four-eyed" opossums, the females of *Metachirus* lack a pouch, having instead simple lateral folds of skin on the lower abdomen, in which are located the mammae. Females with five, seven, and nine mammae have been recorded (Collins 1973). Kirsch (1977*b*) observed that whereas *Philander* is probably the most aggressive of didelphids, *Metachirus* is almost quiet when held in the hand.

The brown "four-eyed" opossum lives in dense forests or in thickets in open, brushy country. It builds round nests of leaves and twigs in tree branches but occasionally makes its shelter under logs or rocks. Handley (1976) reported that all 18 specimens taken in Venezuela were caught on the ground, mostly near streams or in other moist areas. Members of this genus are completely nocturnal, rarely moving from the nest until dark. The diet includes fruits, insects, mollusks, amphibians, reptiles, birds, eggs, and small mammals. *Metachirus* has been accused of damaging fruit crops in some areas.

Limited data indicate that this opossum is seasonally polyestrous (Fleming 1973). It reportedly breeds in November in Central America, has litters of one to nine young, and probably has a maximum life span of three to four years (Hunsaker 1977). The single, 51-mm young of a female obtained on 18 December was then already able to stand alone. It later rode on its mother's back or hips and was fully independent by early February (Collins 1973).

DIDELPHIMORPHIA; MARMOSIDAE; **Genus MONODELPHIS**
Burnett, 1830

Short-tailed Opossums

There are 2 subgenera and 15 species (Anderson 1982; Cabrera 1957; Collins 1973; Gardner *in* Wilson and Reeder 1993; Kirsch and Calaby 1977; Massoia 1980; Pine 1975, 1976, 1977, 1979; Pine and Abravaya 1978; Pine, Dalby, and Matson 1985; Pine and Handley 1984; Soriano 1987):

subgenus *Monodelphis* Burnett, 1830

M. brevicaudata, Venezuela and Guianas to northern Argentina;
M. adusta, eastern Panama to western Venezuela and northern Peru;
M. osgoodi, southern Peru, western Bolivia;
M. kunsi, Bolivia, Brazil;
M. domestica, eastern and central Brazil, Bolivia, Paraguay;
M. maraxina, Isla Marajo in northeastern Brazil;
M. americana, northern and central Brazil;
M. sorex, southern Brazil, southeastern Paraguay, northeastern Argentina;
M. emiliae, Amazonian Brazil, northeastern Peru;
M. iheringi, southern Brazil;
M. theresa, eastern Brazil, Andes of Peru;
M. unistriata, southeastern Brazil;

subgenus *Minuania* Cabrera 1919

M. rubida, eastern Brazil;
M. scalops, southeastern Brazil;
M. dimidiata, southeastern Brazil, Uruguay, eastern Argentina.

Head and body length is 110–200 mm and tail length is 45–85 mm. Fadem and Rayve (1985) listed the weight of *M. domestica* as 90–150 grams in males and 80–100 grams in females. Other recorded weights are 24–78 grams for *M. brevicaudata*, 40–84 grams for *M. dimidiata*, 58–95 grams for *M. domestica*, and 741 grams for *M. scalops* (Eisenberg 1989; Redford and Eisenberg 1992). The tail is about half as long as the head and body, always shorter than the body alone. In most forms the tail is sparsely haired, with only a few millimeters at the base well furred. Coloration varies greatly depending upon the species. One group has a bright red-brown back and face, three faint, dusky brown stripes on the back, and buffy to buffy gray underparts. A second

Short-tailed opossum *(Monodelphis brevicaudata)*, photo by Robert S. Voss.

group has chestnut brown back and sides and paler underparts, with a wash of buffy tips to the hairs. A third group has a well-marked gray dorsal stripe from nose to rump, bright red-brown sides, ashy to buffy gray underparts, and a black tail. A fourth group, which includes *M. adusta* of Panama, has a dark brown back and sides and grayish underparts. The fur is short, dense, and rather stiff. The pouch is not developed. There are 11 to 17 or more mammae, depending on the species; most are arranged in a circle on the abdomen, but there also are some pectoral mammae (Hershkovitz 1992*b*). The tail is only slightly prehensile but is not incrassate.

The habits of these opossums are not well known. They apparently are among the least well adapted members of the Didelphidae for arboreal life and are usually found on the ground, though they can climb fairly well. Most specimens of *M. brevicaudata* collected in Venezuela by Handley (1976) were taken on the ground in moist areas. About half were caught in evergreen forest and about half in open areas. Nests of *Monodelphis* usually are built in hollow logs, in fallen tree trunks that bridge streams, or among rocks. Most species of the subgenus *Monodelphis* are thought to be nocturnal. Their food consists of small rodents, insects, carrion, seeds, and fruits. The species *M. domestica* received its name because of its habit in Brazil of living in dwellings, where it destroys rodents and insects. It is welcomed by the householders. One opossum was seen running through the double cane walls of an Indian hut carrying a piece of paper by curling its tail downward around the paper. Streilen (1982*a*, 1982*c*) found this species to be an efficient predator and particularly adept at capturing scorpions. Individuals were highly intolerant of one another, though conflicts rarely resulted in serious injury.

In a study in Argentina, Pine, Dalby, and Matson (1985) found *M. dimidiata* to be diurnal, with most activity concentrated in the late afternoon. This species was considered to be predominantly insectivorous and not an efficient predator of rodents. Breeding occurred in the summer months of December and January, and the young dispersed from March to May. The species evidently is semelparous; that is, individuals breed but once in their life, and very few, if any, survive past their second summer. One captive female produced a litter of 16 young.

In the tropical part of their range short-tailed opossums apparently breed throughout the year. The number of young in the subgenus *Monodelphis* varies from 5 to 14 depending on the species, and the newborn cling to the nipples of the mother. Later the young ride on the back and flanks of the female. In *M. domestica,* according to Fadem and Rayve (1985), females have as many as four litters annually, the gestation period is 14–15 days, there are usually 5–12 offspring, postpartum dependence lasts about 50 days, sexual maturity is attained at 4–5 months, and breeding has occurred at up to 39 months of age in males and 28 months in females. The period of estrus in that species was found to last 3–12 days, and the estrous cycle showed a bimodal distribution, lasting about 2 weeks in one group of captive females and about 1 month in another group. One specimen of *M. domestica* was captured in the wild and subsequently lived four years and one month (Marvin L. Jones, Zoological Society of San Diego, pers. comm., 1995).

Most species of *Monodelphis* evidently are declining because of habitat destruction. *M. kunsi,* which is thought to have been reduced by at least 50 percent in the last decade, is classified as endangered by the IUCN. *M. osgoodi, M. maraxina, M. sorex, M. emiliae, M. theresa, M. unistriata, M. rubida,* and *M. scalops* are classified as vulnerable, and *M. americana, M. iheringi,* and *M. dimidiata* are designated as near threatened.

DIDELPHIMORPHIA; **Family CALUROMYIDAE**

This family of two living genera and four species is found from southern Mexico, throughout Central and South America, to northern Argentina. This family usually has been considered a subfamily of the Didelphidae that includes *Glironia* and was maintained as such by Gardner (*in* Wilson and Reeder 1993). However, Hershkovitz (1992*a*, 1992*b*) regarded the Caluromyidae as a full family that excludes *Glironia*, which he placed in the separate family Glironiidae.

According to Hershkovitz (1992*a*), a critical distinction expressing the separate evolution of the Caluromyidae and the Didelphidae is found in the anklebones—the astragalus and calcaneus. In the Didelphidae this joint has a primitive pattern in which two separate facets on the calcaneal surface articulate with a corresponding pair of separate facets on the astragalar plantar surface. The Caluromyidae, however, have a partially derived pattern in which there has been coalescence of the once dual facets of the calcaneus into a single continuous facet. The Caluromyidae also can be distinguished by karyotype and retention of a cloaca.

Hershkovitz (1992*b*) pointed out that a major distinction between the Caluromyidae and Glironiidae is found in the auditory or tympanic bullae, the dome-shaped housing for the inner ear at the posterior base of the skull. In the Caluromyidae the bullae are bipartite, with the floor formed by the close junction (but not fusion) of the alisphenoid and petrous bones, thereby enclosing the ectotympanic bone. In the Glironiidae, and also the Didelphidae and Marmosidae (except *Thylamys*), the bullae are tripartite, with the ectotympanic forming a portion of the floor and there being a gap between the alisphenoid and petrous bones. Also, in the Caluromyidae the skull has a sagittal crest and the bony palate is nearly entirely ossified, whereas the Glironiidae lack this crest and have midpalatal and posterolateral vacuities.

DIDELPHIMORPHIA; CALUROMYIDAE; **Genus CALUROMYS**
J. A. Allen, 1900

Woolly Opossums

There are three species (Cabrera 1957; E. R. Hall 1981; Kirsch and Calaby 1977; Massoia and Foerster 1974):

C. derbianus, southern Mexico to Ecuador;
C. lanatus, Colombia to northern Argentina and southern Brazil;
C. philander, Venezuela to southern Brazil.

Head and body length is 180–290 mm, tail length is 270–490 mm, and weight (Bucher and Fritz 1977) is 200–500 grams. The pelage is long, fine, and woolly and extends along almost all of, or more than the basal half of, the tail. Some forms are pale gray or otherwise not well marked, but woolly opossums usually have an ornate color pattern, including the diagnostic dark median stripe on the face, which extends between the ears and the eyes almost to the nose. Some individuals show a pale stripe extending backward from the shoulder region. The ornate color pattern may also include indistinct dark patches around the eyes, a light creamy white to buffy white face, and a reddish and blackish body.

These opossums can generally be distinguished by the striped face, woolly pelage, and characteristics of the tail. The prehensile tail of *Caluromys* is longer than the head and body; its terminal part is not haired and varies from cream-colored to pinkish.

Woolly opossums generally inhabit forested country and are more arboreal than the other large opossums. Nearly all specimens of *C. lanatus* and *C. philander* collected by Handley (1976) in Venezuela were taken in trees, usually near streams or other moist areas. These animals live in tree hollows or limbs and are active mainly during the evening, night, or early morning. They are quite agile, presenting a sharp contrast to *Didelphis*. A nest of leaves made by *C. derbianus* was found in a vine tangle in a small tree. This species, like *Didelphis*, reportedly coils its tail to carry nesting materials (Hunsaker and Shupe 1977). Woolly opossums are ap-

Ecuadorean woolly opossum *(Caluromys lanatus)*, photo by Ernest P. Walker.

parently fairly common throughout their wide range, but their numbers never seem to reach the population levels of *Philander* and *Didelphis*. Perhaps their arboreal habits make them less conspicuous than terrestrial didelphids.

Woolly opossums are omnivorous. Their diet in the wild consists of a variety of fruits, seeds, leaves, soft vegetables, insects, and small vertebrates. They reportedly feed on carrion as well. One captive group had a decided preference for meat (Collins 1973), while another colony had a special liking for fruit (Bucher and Fritz 1977).

According to Eisenberg (1989), the nightly foraging area of *C. philander* varies from 0.3 to 1.0 ha. depending on food availability. Individuals are solitary and adult interaction tends to be agonistic, but there seems to be no strong territorial defense and home ranges overlap. Females of that species can produce three litters a year. Studies of captive *C. philander* by Atramentowicz (1992) indicate that gestation lasts 25 days, apparently the longest in the Didelphimorphia; litter size is 1–7, and the young leave the pouch at 3 months and are weaned at 4 months.

Captive females of both *C. derbianus* and *C. lanatus* have modal estrous cycle lengths of 27–29 days and are cyclic throughout the year (Bucher and Fritz 1977). In the wild in Nicaragua *C. derbianus* also seems to be reproductively active all year (Phillips and Jones 1968). Litter size there averaged 3.3 (2–4). In Panama the breeding period reportedly begins with the onset of the dry season in February (Enders 1966), and litter size averaged between 3 and 4 (1–6). Sexual maturity is said to be attained by *C. derbianus* at seven to nine months. A number of captive *Caluromys* have survived for more than three years (Collins 1973), and known record longevity is six years and four months for a specimen of *C. philander* (Jones 1982).

All three species are thought to be declining because of habitat destruction. *C. derbianus* is classified as vulnerable by the IUCN, and *C. lanatus* and *C. philander* are designated as near threatened. In the past these animals were trapped extensively for their pelts, but at present their fur is not popular. They are said to damage fruit crops occasionally in South America, but otherwise they are of little economic importance. Recently they have been shown to have potential use in laboratory research (Bucher and Fritz 1977).

DIDELPHIMORPHIA; CALUROMYIDAE; **Genus CALUROMYSIOPS**
Sanborn, 1951

Black-shouldered Opossum

According to De Vivo and Gomès (1989), Emmons (1990), and Izor and Pine (1987), the single species, *C. irrupta*, is known with certainty only from three localities in southern Amazonian Peru and one on the upper Jarú River in west-central Brazil; Simonetta's (1979) report of a specimen from extreme southern Colombia is doubtful. The generic distinction of *Caluromysiops* from *Caluromys* has been questioned (Gewalt *in* Grzimek 1990), but Hershkovitz (1992*a*) put the two in completely separate tribes based in part on differences in the molar teeth.

Head and body length is 250–330 mm and tail length is 310–40 mm. The upper parts are gray, with two separate black lines that begin on the forefeet and run onto the back, join on the shoulders, then separate again and run parallel to each other down the back and over the rump to the hind limbs. The face has faint, dusky lines running through the eyes. The underparts are gray with buffy tips to the hairs, and there is a faint dusky line along the middle of the belly. The upper side of the tail for the basal two-thirds to three-quarters of its length is slightly darker than the gray of the body; the remainder of the tail is creamy white. The tail is well furred, except the underside of the last three-fourths, which is naked. The fur is long, dense, and woolly.

Black-shouldered opossum *(Caluromysiops irrupta)*, photo from New York Zoological Society.

The skull is similar to that of *Caluromys*, but the molars are much larger and the rostrum is relatively shorter.

The black-shouldered opossum is thought to inhabit humid forests. According to Hunsaker (1977), it is nocturnal and arboreal in habit and probably has a diet like that of *Caluromys*. Emmons (1990) stated that it uses the upper levels of the forest and rarely descends even to the middle levels; it moves slowly and will spend hours in the same flowering tree, feeding periodically on nectar. Collins (1973) reported that the few specimens maintained in captivity thus far have been hardy in comparison with other didelphids. One survived for 7 years and 10 months. *Caluromysiops* is declining because of habitat destruction and is now classified as vulnerable by the IUCN.

DIDELPHIMORPHIA; **Family GLIRONIIDAE; Genus GLIRONIA**
Thomas, 1912

Bushy-tailed Opossum

The single genus and species, *Glironia venusta*, is known only from eight specimens collected in the Amazonian regions of Ecuador, Peru, Brazil, and Bolivia (Da Silva and Langguth 1989; Emmons 1990; Marshall 1978*c*). Gardner (*in* Wilson and Reeder 1993), along with most other recent authorities, placed the genus in the subfamily Caluromyinae of the family Didelphidae. However, Hershkovitz (1992*a*, 1992*b*) considered *Glironia* and the Caluromyinae to represent distinct families.

Head and body length is 160–205 mm and tail length is 195–225 mm. The upper parts are fawn-colored or cinnamon brown; a dark brown to black stripe extends through each eye and gives the appearance of a mask; the tail is tipped with white or has only a sprinkle of white hairs, and the underparts are gray or buffy white. The texture of the fur varies from soft and velvety to dense and woolly.

This genus is much like *Marmosa* in general appearance, but the tail is well furred and bushy to the tip. The extent of the naked area on the tail of *Glironia* varies, but there is at least a trace of a ventral naked area on the terminal few centimeters. The characters considered by Hershkovitz (1992*b*) to distinguish the Glironiidae as a family distinct from the Caluromyidae are given above in the account of the latter group.

Four of the known specimens of *Glironia* were collected by commercial animal dealers, and all were taken in heavy, humid tropical forests. *Glironia* is presumed to be arboreal because of its large, opposable hallux (Marshall 1978*c*). The diet is unknown but probably is comparable to that of *Marmosa*—insects, eggs, seeds, and fruits. Emmons (1990) reported seeing an individual at night in dense vegetation about 15 meters above the ground. It ran about the vines quickly, often jumping from one to another in a manner unlike that of other opossums, and seemed to be hunting insects. *Glironia* is declining because of habitat destruction and is now classified as vulnerable by the IUCN.

DIDELPHIMORPHIA; **Family DIDELPHIDAE**

Pouched Opossums

This family of four Recent genera and eight species is found from southeastern Canada, through the eastern United States and Mexico, to central Argentina. Corbet and Hill (1991), Gardner (*in* Wilson and Reeder 1993), and most other authorities have considered the Didelphidae to include the Marmosidae, the Caluromyidae, and the Glironiidae. However, the latter three groups were distinguished as full families (see accounts thereof) by Hersh-

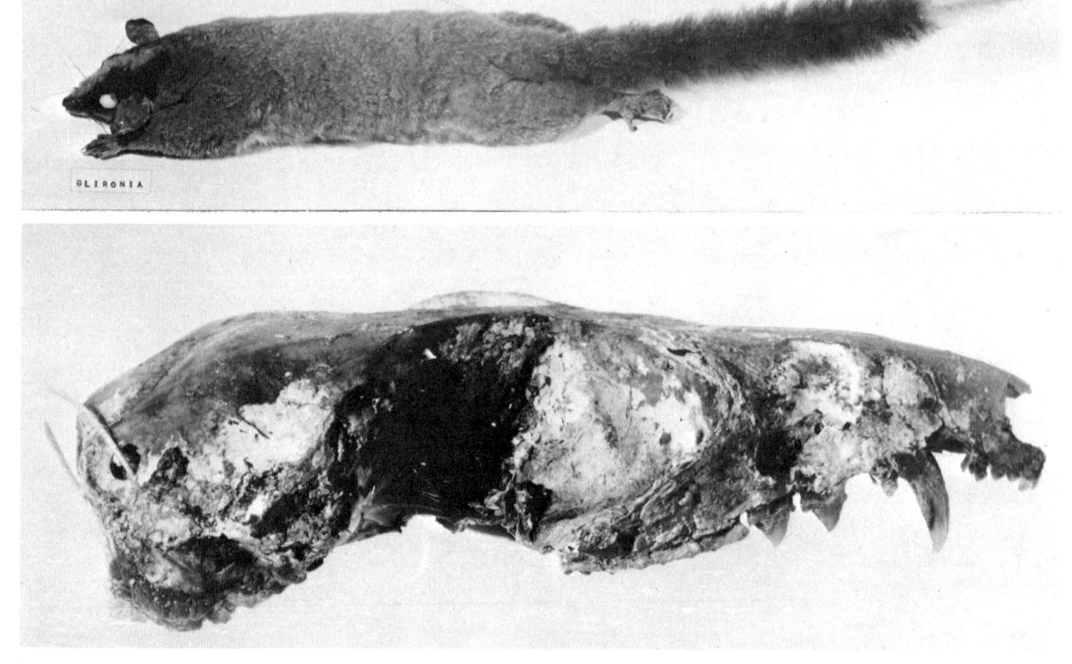

Bushy-tailed opossum *(Glironia venusta)*, photos from British Museum (Natural History).

A North American opossum *(Didelphis virginiana)* feigning death. The opposable great toe and naked, prehensile tail are both plainly shown. Photo from U.S. Fish and Wildlife Service.

kovitz (1992*a*, 1992*b*), who is followed here in that regard and with respect to sequence of genera.

Head and body length is 250–500 mm, tail length is 210–535 mm, and weight is 200–5,500 grams. There is much variation in pelage. The tail is thinly haired or naked for most of its length. Unlike those of the other families in the Didelphimorphia, female didelphids usually have a distinctive pouch, though there is some disagreement regarding its development in *Lutreolina*. According to Hershkovitz (1992*b*), the Didelphidae also generally are distinguished from the Marmosidae in having a more massive skull, with more prominent postorbital processes and sagittal crest, and stouter claws that are recurved and extend well beyond the digital tips (though not on the forefoot of *Lutreolina*). The skull of the Glironiidae lacks a sagittal crest. The skull of the Didelphidae differs from that of the Caluromyidae (see account thereof) in having large maxillary-palatine vacuities and tripartite tympanic bullae.

DIDELPHIMORPHIA; DIDELPHIDAE; **Genus PHILANDER**
Tiedemann, 1808

Gray and Black "Four-eyed" Opossums

There are two species (Cabrera 1957; Gardner *in* Wilson and Reeder 1993; Gardner and Patton 1972; E. R. Hall 1981):

P. opossum (gray "four-eyed" opossum), northeastern Mexico to northeastern Argentina;
P. andersoni (black "four-eyed" opossum), eastern Colombia, Ecuador, southern Venezuela, western Brazil, Peru.

The latter species formerly was known as *P. mcilhennyi*. Pine (1973) argued that the generic name *Metachirops* Matschie, 1916, is the correct designation for both species and that *Philander* actually applies to the brown "four-eyed" opossum (here called *Metachirus*). Both Husson (1978) and E. R. Hall (1981) agreed, but most others have not. For further comment see the account of *Metachirus*.

Head and body length is 200–350 mm and tail length is 253–329 mm. Emmons (1990) reported natural weight to be 200–660 grams, but Collins (1973) stated that healthy, well-fed captive males weighed 800–1,500 grams and females, 600–1,000 grams. The fur is rather straight and short in *P. opossum* but much longer in *P. mcilhennyi*. The upper parts are gray to black, with a white spot above each eye, and the underparts vary from yellowish to buffy white. The white spots above the eyes account for the vernacular name "four-eyed" opossum. The tail, which is furred for about 50–60 mm from the rump and naked toward the tip, is black or grayish black on its basal half and white toward the end, with a pink tip. The body is slim and usually lean, and the head is large, with an elongate, conical muzzle. The ears are naked. The tail is slender, tapering, and prehensile. Females have a distinct pouch. The number of mammae varies from five to nine.

These opossums inhabit forested areas and are often

Gray "four-eyed" opossum *(Philander opossum)*, photo from West Berlin Zoo (Gerhard Budich).

found near swamps and rivers. They are smaller and more agile than *Didelphis* and quick in their actions. Although good climbers and swimmers, they are mainly terrestrial. All 46 specimens taken in Venezuela by Handley (1976) were caught in moist areas, nearly always on the ground. They build globular nests, about 30 cm in diameter, in the lower branches of trees or in bushes, and they may also inhabit ground nests and burrows. They are thought to be mainly nocturnal, though Husson (1978) stated that in Surinam they are as active in the day as at night. Their diet includes small mammals, birds and their eggs, reptiles, amphibians, insects, freshwater crustaceans, snails, earthworms, fruits, and probably carrion. Occasional damage to fruit crops and cornfields has given them a bad reputation in certain areas. In Panama, Fleming (1972) found maximum population densities of 0.55–0.65/ha. Unlike *Didelphis,* these opossums are not known to feign death when danger threatens, but will open their mouths wide, hiss loudly, and fight savagely. When disturbed they may utter a long, chattering cry.

The species *P. opossum* reportedly breeds year-round in some areas, possibly including Veracruz, Mexico, but is seasonal in others (Hunsaker and Shupe 1977). Collins (1973) cited records of females with pouch young being taken in Nicaragua from February to October, in Panama from April to July, and in Colombia in June, September, and October. Jones, Genoways, and Smith (1974) caught a female with six nursing young in March on the Yucatan Peninsula. Phillips and Jones (1969) collected data indicating that the main reproductive season in Nicaragua extends from March through July. Litter size there averaged 6.05 (3–7). According to Fleming (1973), *P. opossum* is seasonally polyestrous in Panama, with two or more litters probably being produced by each female from January to November. Litter size averaged 4.6 (2–7). Husson (1978) found females with young in Surinam during January, March, and April. Litter size there ranged from 1 to 7 but averaged only 3.4. These records support the statement by Phillips and Jones (1969) that litter size tends to be larger in the north. Collins (1973) reported that the small pouch young of a female received on 7 May were weaned on 23 July, that first estrus in females occurred at 15 months of age, and that maximum known longevity for captives was only 2 years and 4 months.

DIDELPHIMORPHIA; DIDELPHIDAE; **Genus DIDELPHIS**
Linnaeus, 1758

Large American Opossums

There are four species (Cabrera 1957; Gardner 1973 and *in* Wilson and Reeder 1993; E. R. Hall 1981):

D. albiventris, Colombia and Venezuela to central Argentina;
D. marsupialis, southern Tamaulipas (eastern Mexico) to Brazil and Bolivia;
D. aurita, eastern Brazil, Paraguay, northeastern Argentina;
D. virginiana (Virginia opossum), naturally from New Hampshire to Colorado, and from southern Ontario to Costa Rica.

Cerqueira's (1985) separation of *D. aurita* from *D. marsupialis* was not accepted by Corbet and Hill (1991).

Head and body length is 325–500 mm and tail length is 255–535 mm. The animals weigh about 0.5–5.5 kg. The pelage, consisting of underfur and white-tipped guard hairs, is unique in the family Didelphidae. Guard hairs are lacking or few in number in the other opossums. Coloration is gray, black, reddish, or, rarely, white. Head markings are sometimes present in the form of three dark streaks, one running through each eye and another running along the midline of the crown of the head. The basal tenth of the tail is furred, and the remainder is almost naked. At the northern limits of their range many of the individuals that survive the winter lose part of the tail and the ears from frostbite.

As in the other didelphids, all four feet have five digits, and all digits have sharp claws, except that the first toe of the hind foot is clawless, thumblike, and opposable to the other digits in grasping. The female has a well-developed pouch in which are commonly located 13 mammae arranged in an open circle with 1 in the center. Number and arrangement of mammae, however, are variable.

Opossums usually inhabit forested or brushy areas but also have been found in open country near wooded watercourses. The species *D. virginiana* generally has been reported to favor moist woodlands or thick brush near streams or swamps (Banfield 1974; Jackson 1961; McManus 1974; Schwartz and Schwartz 1959). According to Cerqueira (1985), *D. albiventris* is found in open and deciduous forests and in mountainous areas, while *D. marsupialis* is restricted to humid, broad-leaved forests. In

Opossums *(Didelphis virginiana)*, photo by Leonard Lee Rue III.

Venezuela, Handley (1976) collected most specimens of *D. marsupialis* on the ground in moist areas but found most *D. albiventris* in dry situations. Opossums are largely nocturnal, spending the day in rocky crevices, in hollow tree trunks, under piles of dead brush, or in burrows (sometimes dug by themselves). They construct rough nests of leaves and grasses, using their curled-up tail and their mouth to transport dry vegetation. They are mainly terrestrial, moving with a slow, ambling gait, and are strong swimmers. They can climb well, with the help of the prehensile tail, and can even hang by the tail (Lowery 1974). In the northern part of its range, *D. virginiana* accumulates fat in the autumn and may become inactive during severe winter weather, remaining in its nest for several days, but it does not hibernate. Females have a greater tendency toward winter inactivity and also are more sedentary in their movements at other times. Opossums have an extremely varied diet that includes small vertebrates, carrion, invertebrates, and many kinds of vegetable matter.

Records compiled by Hunsaker (1977) showed that for *D. virginiana* in the United States, population density averaged 0.26/ha. (0.02–1.16/ha.), home range averaged about 20 ha. (4.7–254.0 ha.), and nightly foraging distance was 1.6–2.4 km; and that for *D. marsupialis* in Panama, density was 0.09–1.32/ha., with the animals being nomadic and remaining in an area for only two or three months. According to Hunsaker and Shupe (1977), *D. virginiana* also is nomadic and stays in a particular area for six months to a year; there is no territoriality, but individuals will defend the space occupied at a given time. In a study in Venezuela, Sunquist, Austad, and Sunquist (1987) found the home range size of *D. marsupialis* to average 11.3 ha. in the dry season and 13.2 ha. in the wet season. Male home ranges overlapped one another extensively, and each overlapped the ranges of several females. The latter occupied exclusive home ranges for at least part of the year.

Opossums usually are thought to be solitary and antisocial, either avoiding one another or acting aggressively. However, Holmes (1991) found that most interaction among unrelated individuals kept in large enclosures was neutral or affiliative, with females sometimes nesting together. The animals readily formed stable dominance hierarchies, with females usually dominant. There seems agreement that there almost always is extreme agonistic behavior when two males meet, but if opposite sexes meet during the breeding season, initial aggressive displays turn to courtship and the two animals may spend several days together. If a male is placed with a female that is not in estrus, she becomes aggressive, but the male does not return her attacks. The vocal repertoire of *D. virginiana* consists of a hiss, a growl, and a screech, which are used in agonistic and defensive situations, and a metallic lip clicking, which is heard under a variety of conditions, including mating (McManus 1970, 1974).

Death feigning, referred to popularly as "playing possum" and technically known as catatonia, is a passive defensive tactic of *D. virginiana* employed occasionally, but not always, in the face of danger. In this state the opossum becomes immobile, lies with the body and tail curled ventrally, usually opens the mouth, and apparently becomes largely insensitive to tactile stimuli. The condition may last less than a minute or as long as six hours (McManus 1974). Although catatonia seems partly under the conscious control of the animal, physiological changes suggest a state analogous to fainting in humans (Lowery 1974). Death feigning may cause a pursuing predator to lose the visual cue of motion or to become less cautious in its approach, thereby giving the opossum a better chance of escape. McManus (1970) thought that anal secretions, sometimes accompanying catatonia, might contribute to deterring a predator. He also suggested that catatonia may have evolved primarily as a means of reducing intraspecific aggression. If so, the condition may be vaguely comparable to submission in certain other animals. According to Hunsaker and Shupe (1977), death feigning has been reported for, but is rare in, *D. albiventris*.

Except as noted, the following life history data on *D. virginiana* were derived from Collins (1973), Hunsaker (1977), Lowery (1974), Schwartz and Schwartz (1959), and Tyndale-Biscoe and Mackenzie (1976). Females are polyestrous, have an estrous cycle of about 28 days, and are receptive for 1 or 2 days. First matings usually occur in January or February in most of the United States but can be as early as mid-December in Louisiana (Edmunds, Goertz, and Linscombe 1978) or as late as March in Wisconsin (Jackson 1961). There are two litters per year in most areas, though Jackson (1961) thought there usually was only one in Wisconsin, and the possibility of a third occurring in the South has been suggested. The second peak of mating takes place in the spring or early summer, and an average of 110 days separates the two litters. The gestation period is 12.5–13 days. The tiny young are remarkably undeveloped but have pronounced claws, which are used in the scramble from the birth canal to the mother's pouch. The young measure only 10 mm in length and weigh 0.13 grams. Twenty could fit in a teaspoon, and it would take 217 to make one ounce. There usually are about 21 young, but there is a report of 56 being born at once. Since there normally are only 13 mammae, and since the young must continually grasp the mammae to suckle, some newborn often cannot be accommodated and quickly perish. Additional mortality usually occurs later. The number of pouch young generally observed is 5–13 and reportedly averages about 8–9 in northern areas and 6–7 in the South. The young first release their grip on the mammae when they are about 50 days old, begin to leave the pouch temporarily at 70 days, and are completely weaned and independent at 3–4 months. After the pouch becomes too small to hold all the young, but before weaning, some of them ride on the mother's back. Sexual maturity is attained by about 6–8 months, and females apparently have only 2 years of reproductive activity. Very few opossums survive their third year of life, though some laboratories have maintained captives for three to five years.

For *D. marsupialis* in Panama, Fleming (1973) found that females were polyestrous, breeding began in January, and two or possibly three litters were born between then and October. Litter size averaged 6 (2–9). Tyndale-Biscoe and Mackenzie (1976) reported that reproduction in this species also began in January on the llanos of eastern Colombia. A second litter was produced there in April or May, and there was no breeding from September to December. Mean litter size was 6.5 (1–11). In western Colombia the breeding period was the same, but a third litter was known to be produced in August. Mean litter size there was 4.5 (1–7). No reproductive season could be determined for *D. albiventris* in Colombia, but litter size averaged only 4.2 (2–7). In Brazil *Didelphis* reportedly has its first litter in late August or September (Collins 1973).

The Virginia opossum has adapted better to the presence of people than have most mammals. When European colonists first arrived in North America *D. virginiana* apparently did not occur north of Pennsylvania. Subsequently it extended its range, the first being taken in Ontario in 1858 (Hunsaker 1977) and in New England during the early twentieth century (Godin 1977). It also moved westward on the Great Plains, its progress facilitated by human agricultural development (Armstrong 1972; Jones 1964). In 1890 captives were released in California, and boosted by other introductions, the Virginia opossum eventually spread all along the West Coast. At present it is well established from southwestern British Columbia to the vicinity of San Diego. Introduced populations also occupy parts of Arizona, western Colorado, and Idaho (E. R. Hall 1981).

Opossums are sometimes hunted or trapped by people for food and sport and as predators of poultry. Leopold (1959) wrote that in Mexico they are considered chicken and egg thieves, they are hunted for food and local use of their fur, and certain of their parts are believed to have medicinal value. In Peru both *D. marsupialis* and *D. albiventris* have a bad reputation for poultry killing, but populations have not been adversely affected by human settlement (Grimwood 1969). The value of opossum fur and the consequent commercial harvest have varied widely. During a single year in the 1920s, when prices were high, 518,295 opossum skins were sold in Louisiana alone (Lowery 1974) and 350,286 from Kansas were sold (Hall 1955). In the 1970–71 annual season in the United States 101,278 pelts of *D. virginiana* were reported sold at an average price of $0.85. The corresponding figures for the 1976–77 season were 1,069,725 and $2.50 (Deems and Pursley 1978). The take for 1983–84 in Canada and the United States was 515,832 skins, and the average price in Canada was $1.75 (Novak, Obbard, et al. 1987). The number of skins taken in the 1991–92 season in the United States was 145,290, the average price $1.26 (Linscombe 1994). In addition to its other uses, *Didelphis* is finding increasing employment in laboratory research.

DIDELPHIMORPHIA; DIDELPHIDAE; **Genus CHIRONECTES**
Illiger, 1811

Water Opossum, or Yapok

The single species, *C. minimus*, occurs from southern Mexico and Belize to northeastern Argentina (Marshall 1978*a;* McCarthy 1982). It also may occur on Trinidad.

Head and body length is 270–400 mm, tail length is 310–430 mm, and weight reportedly ranges from 604 to 790 grams (Marshall 1978*a*). Collins (1973), however, stated that the weight of a captive female stabilized at about 1,200 grams. The pelage is relatively short, fine, and dense. The back is marbled gray and black, the rounded black areas coming together along the midline. The muzzle, a band through the eye to below the ear, and the crown are blackish; a prominent grayish white, crescentlike band passes from the front of one ear to the other, just above the eyes. The chin, chest, and belly are white. The striking color pattern is unique among marsupials and may serve as camouflage while the yapok is swimming, the dorsal bands blending with ripples and presenting a disruptive appearance to aerial predators (Brosset 1989). The long, ratlike tail is well furred only at the base and is black near the body and yellowish or whitish toward the end. The facial bristles are stout, long, and placed in tufts. One of the wrist bones is enlarged and simulates, in some respects, a sixth digit on the forefoot. The ears are moderately large, naked, and rounded.

This opossum is the only marsupial well adapted for a semiaquatic life. It has dense, water-repellent pelage, webbed hind feet, a streamlined body, and, in females, a rear-opening, waterproof pouch. Both sexes actually have a pouch. In the male it cannot be fully closed, as in the female, but the scrotum is pulled up into it when the animal is swimming or moving swiftly. The female is able to swim with the young in her pouch. A well-developed sphincter muscle closes the pouch, creating a watertight compartment, and the young can tolerate low oxygen levels for many minutes (Marshall 1978*a;* Rosenthal 1975).

The yapok is confined mainly to tropical and subtropical habitats, where it frequents freshwater streams and lakes.

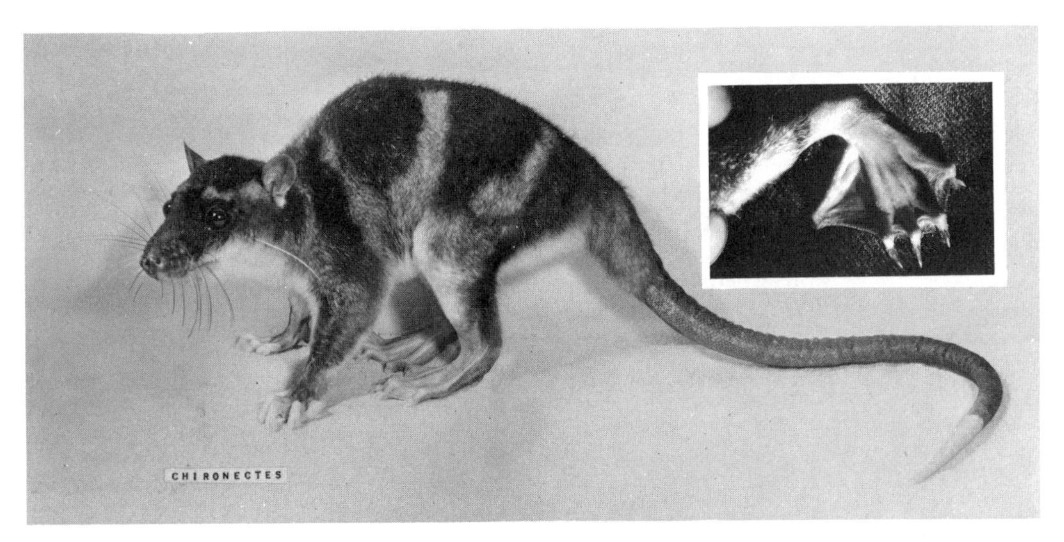

Water opossum, or yapok *(Chironectes minimus)*. Inset: hind foot. Photos by Richard Peacock.

In some areas it is found at considerable elevations along mountain rivers. Grimwood (1969) reported that one was killed in Peru at an altitude of 900 meters, and Handley (1976) collected specimens at up to 1,860 meters in the mountains of Venezuela. This opossum generally is considered rare throughout its range, but this view may stem from the nocturnal habits of the animal and the inaccessibility of its habitat. The main den, a subterranean cavity, usually is reached through a hole in the stream bank just above water level. A ground nest of leaves or grasses in a dimly lit area may be used as a place of rest during the day. Like *Didelphis*, the yapok collects nesting material with the forepaws, pushing it under the body into a bundle that is then held by the tail (Marshall 1978*a*).

The water opossum is an excellent swimmer and diver, the broadly webbed hind feet serving as effective paddles. Only the rear feet are used in swimming, thereby avoiding the reduced propulsion efficiency that might result from interference between the fore- and hind limbs. The pelage does not soak up water, so the nose can remain above the surface without assistance from the front feet (Fish 1993). *Chironectes* can climb but rarely does so; its tail, though prehensile, is too thick to be of effective use. It is largely carnivorous, feeding on such aquatic life as crayfish, shrimp, and fish. Prey is located in the water by contact with the forefeet, which are not used for propulsion but are held out in front while the animal is swimming (Marshall 1978*a*). The yapok also may eat some aquatic vegetation and fruits (Hunsaker 1977).

In Brazil the young are born in December and January; one female was seen in February with 5 young. Litters of 2–5 have been recorded, with the average reportedly being 3.5 (Collins 1973; Hunsaker 1977). Rosenthal (1975) stated that three captive-born litters at the Lincoln Park Zoo numbered 4, 5, and 5. Experience with these litters indicated that the young of *Chironectes* develop more rapidly than those of any other didelphid. According to Collins (1973), the greatest known longevity for a captive yapok is 2 years and 11 months.

Grimwood (1969) observed that yapok skins had been considered worthless but were beginning to command a price in Peru. He thought that intensive commercial hunting could readily reduce populations and that protective measures were necessary. Emmons (1990) noted that the genus seems rare in most regions but is common in some Central American rivers. McLean and Ubico (1993), who reported the first record for Guatemala, suggested that it was very rare in that country. The IUCN designates *Chironectes* as near threatened.

DIDELPHIMORPHIA; DIDELPHIDAE; **Genus LUTREOLINA**
Thomas, 1910

Thick-tailed Opossum

The single species, *L. crassicaudata*, is known to occur east of the Andes in Bolivia, southern Brazil, Paraguay, Uruguay, and northern Argentina. Another population, far to the north, long was known only from two specimens taken in Guyana (Marshall 1978*d*). Recently, however, Handley (1976) collected four more specimens from eastern Venezuela, and Lemke et al. (1982) reported that nine had been taken in eastern Colombia. It is possible that additional investigation will reveal that this species also occupies the intervening region of central South America.

Head and body length is 210–445 mm and tail length is 210–310 mm. Adults weigh 200–540 grams. The fur is short, dense, and soft but not water-repellent. The upper parts are a rich, soft yellow, buffy brown, or dark brown, and the underparts are reddish ochraceous or pale to dark brown. The pelage of some live individuals has a peculiar purplish tinge. There may be faint markings on the face, but there are no eye spots or other prominent markings.

The body form is long and low, almost weasel-like. The ears are short and rounded and barely project above the fur. The limbs and feet are short and stout, and the pads are small and narrow. The tail is characteristic in its extremely thick, heavily furred base; only about 5 cm of the undersurface of the tip is naked, though in some individuals all the terminal half is thinly haired, showing the scales. The tail is not as prehensile as in other didelphids. The thumbs and great toe are not fully opposable. Although some reports have indicated otherwise, specimens examined by Lemke et al. (1982) did have a well-developed pouch. Reig, Kirsch, and Marshall (1987) also confirmed the presence of

Thick-tailed opossums *(Lutreolina crassicaudata)*, female on left, male on right, photo from New York Zoological Park through Joseph Davis.

a pouch. The mammae, at least in one specimen, numbered nine.

The thick-tailed opossum is restricted mostly to grassland, savannahs, and gallery woodland, often near the shores of streams and lakes. It is considered to be the species of didelphid most adapted to life on the pampas. It climbs well, is agile on the ground, and seems suited for wetland habitat. Although one captive reportedly was clumsy in the water, most information indicates that *Lutreolina* is an excellent swimmer under natural conditions (Grzimek 1975; Hunsaker 1977; Marshall 1978*d*). In wooded areas it often shelters in tree holes, but in wetlands it constructs a snug round nest of grasses and rushes among the reeds, and on the pampas it often utilizes abandoned armadillo and viscacha burrows.

This opossum is nocturnal, emerging after dark to prey on small mammals, birds, reptiles, fishes, and insects. It occasionally raids chicken houses and pigeon lofts. It is said to be sometimes savage in temperament but can be tamed. *Lutreolina* actually appears somewhat more sociable than other didelphids, and a group of one male and two females was maintained together successfully (Collins 1973). An average core home range of 0.38 ha. was determined in Argentina (Eisenberg 1989). This opossum breeds in the spring and again later in the year after the young of the first litter have become independent. The young are raised in a nest of dry grass. The gestation period is thought to be about two weeks. One specimen lived for three years in captivity (Collins 1973). Roig (1991) reported that *Lutreolina* had disappeared from a large part of central and northern Argentina because of human habitat disruption.

Order Paucituberculata

PAUCITUBERCULATA; **Family CAENOLESTIDAE**

"Shrew" Opossums

This order, containing the single Recent family Caenolestidae, with two genera and seven species, is found in western South America (Bublitz 1987). The group sometimes is placed only at the level of a superfamily (Marshall 1987) or infraorder (Szalay 1994), but full ordinal status was supported by Aplin and Archer (1987), Kirsch (1977*b*), and Marshall, Case, and Woodburne (1990). Some authorities have suggested that there may be only a single valid genus and as few as three species (Marshall 1980). In contrast, Gardner (*in* Wilson and Reeder 1993) recognized three genera and five species. All species prefer densely vegetated, humid habitat, typically in scrub adjacent to meadows of high, moist Andean paramo (Kirsch 1977*a*).

These marsupials are somewhat shrewlike in appearance. Head and body length is 90–135 mm and tail length is 65–135 mm. The head is elongate and conical in shape. The eyes are small and vision is poor, but the sense of smell is well developed. Sensory vibrissae are present on the snout and cheeks. The tail is entirely covered with stiff, short hairs. The hind limbs are slightly longer than the forelimbs, and all have five separate, clawed digits. The humerus is large and heavy in comparison with the slender forearm; the hind foot is relatively long and narrow. Females of *Caenolestes* have four mammae, and females of *Rhyncholestes* have five. A marsupium is absent in adult females, though a rudimentary pouch may possibly be present in the young.

The teeth are rooted, sharp, and cutting. The dental formula is: (i 4/3–4, c 1/1, pm 3/3, m 4/4) × 2 = 46 or 48. According to Marshall (1984), there is one large, laterally compressed, procumbent incisor in each lower jaw, followed by six or seven tiny, spaced, vestigial teeth, among them a vestigial lower canine. The molar teeth are four-sided or almost triangular in outline, and there is a sharp reduction in size from the first to the fourth. Szalay (1994) stated that the Paucituberculata are diagnosed by emphasized vertical shearing between the third upper and the first lower molar and by a carpus in which the lunate and magnum are in contact but are indented by a slight lateral wedge of the scaphoid.

Comparatively few specimens have been collected, probably more because of the inhospitable nature of the habitat than because of the rarity of the animals (Kirsch and Waller 1979). Habits are not well known but may be like those of large forest shrews. Caenolestids use runways on the surface of the ground and are mainly terrestrial but climb well. They are active during the evening and night and prey on invertebrates and small vertebrates.

The living Paucituberculata are highly relictual; there are seven extinct families of the order, with a morphological and ecological diversity comparable to that of the Australian order Diprotodontia (Aplin and Archer 1987). The geological range of the family Caenolestidae is early Eocene to Recent in South America (Kirsch 1977*a*). Fossil specimens from the Tertiary were known before description of the living representatives of the family (Grzimek 1975).

Common "shrew" opossum *(Caenolestes obscurus)*, photo by John Kirsch through Larry Collins.

PAUCITUBERCULATA; CAENOLESTIDAE; **Genus CAENOLESTES**
Thomas, 1895

Common "Shrew" Opossums

There are five species (Barkley and Whitaker 1984; Bublitz 1987; Cabrera 1957; Kirsch and Calaby 1977):

- *C. fuliginosus,* Andes of northern and western Colombia, extreme western Venezuela, and Ecuador;
- *C. caniventer,* Andes of southwestern Ecuador and northern Peru;
- *C. convelatus,* Andes of western Colombia and north-central Ecuador;
- *C. inca,* Andes of south-central Peru;
- *C. gracilis,* Andes of southeastern Peru.

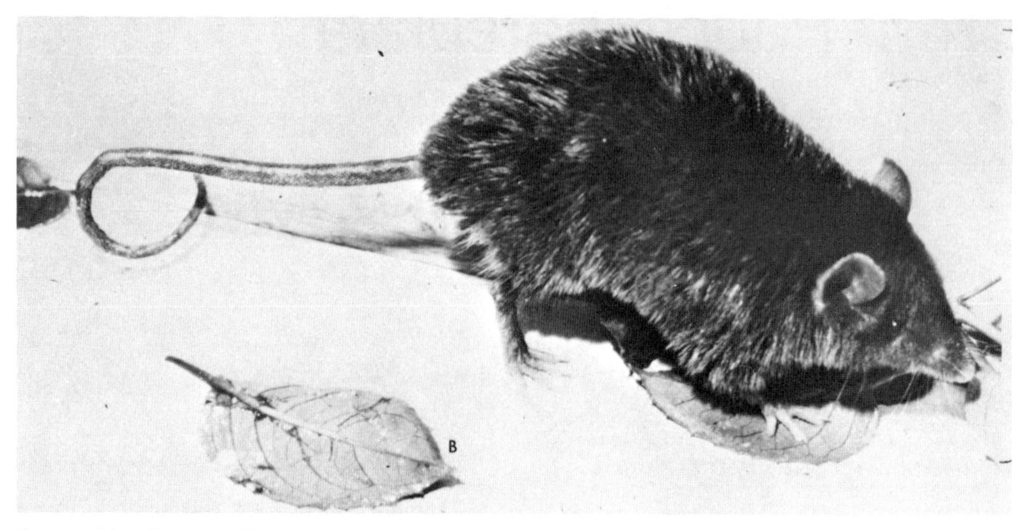

Common "shrew" opossum *(Caenolestes inca)*, photo by John Kirsch through Larry Collins.

The last two species are sometimes put in the genus *Lestoros* Oehser, 1934 (synonyms of which are *Orolestes* Thomas, 1917, and *Cryptolestes* Tate, 1934). Various authorities long argued that *Lestoros* would be shown to be invalid once a sufficient number of specimens had been examined (Marshall 1980). Bublitz (1987) did study a large collection and did place *Lestoros* in the synonymy of *Caenolestes*. Gardner (*in* Wilson and Reeder 1993) retained *Lestoros* as a full genus but treated *gracilis* as a synonym of *C. inca*.

Head and body length is 90–135 mm and tail length is 93–139 mm. Tail length is approximately the same as head and body length in all species. Kirsch and Waller (1979) recorded weights of 25.0–40.8 grams for adult males and 16.5–25.4 grams for females. The pelage is soft and thick over the entire body, but it appears loose and uneven because of the different textures of the hairs. Coloration varies somewhat with the species, but all have dark upper parts, generally deep sooty brown, blackish brown, fuscous black, or plumbeous black. Some species are uniformly colored, or nearly so, whereas some are markedly lighter below. The tail is scantily haired and about the same color as the back, except that in some specimens the tip is white.

These marsupials could with equal justification be called shrewlike or ratlike in general form. The head is elongate, the eyes are very small, and the ears are rounded and project above the pelage. The tail tapers gradually and is nonprehensile. In the best-known species, *C. fuliginosus*, the forefoot has five digits: the outer toes are small and bear blunt nails, and the other three digits have sharp, curved claws. On the hind foot the great toe is small and bears a small nail, and the other four digits have well-developed, curved claws. Females have four mammae but no pouch.

Common "shrew" opossums occur in the alpine forest and meadow zone of the Andes at altitudes of about 1,500–4,000 meters. They prefer cool, wet areas covered by thick vegetation. They are nocturnal and terrestrial and move from one favored area to another by means of runways through the surface vegetation. Kirsch and Waller (1979) indicated that the habitat of the southern Peruvian species is considerably drier than that of the more northerly species. These authorities also reported that *C. fuliginosus* and *C. inca* may bound at high speed but are not saltatorial. They are primarily terrestrial, but they are agile climbers and use their tail as a prop as they move up a vertical surface. *C. fuliginosus* reportedly has well-developed senses of hearing and smell but poor vision. It had long been considered to be mainly insectivorous, but Kirsch and Waller (1979) found that it was readily trapped by baits of meat and could efficiently kill newborn rats. Stomach contents of a series of *C. caniventer* from Peru suggest that the diet consists largely of invertebrate larvae but also includes small vertebrates, fruit, and other vegetation (Barkley and Whitaker 1984).

Of the six female *C. fuliginosus* collected by Kirsch and Waller (1979) in southern Colombia from 25 August to 4 September 1969, one was small, four were lactating, and one had enlarged but empty uteri. None had attached young, so apparently the breeding season must begin several weeks before August. Three of the lactating females had all four mammae enlarged, and one had three of the mammae enlarged, suggesting a litter size as great or greater than the mother's capacity to suckle. Barnett (1991) collected a pregnant female *C. caniventer* with two fetuses in southern Ecuador on 9 September.

PAUCITUBERCULATA; CAENOLESTIDAE; **Genus RHYNCHOLESTES**
Osgood, 1924

Chilean "Shrew" Opossums

There are two species (Bublitz 1987):

R. raphanurus, Chiloe Island in southern Chile;
R. continentalis, mainland Chile just north of Chiloe Island.

Gardner (*in* Wilson and Reeder 1993) considered *continentalis* a synonym of *R. raphanurus.*

Head and body length is 110–28 mm and tail length is 65–87 mm. A specimen taken by Pine, Miller, and Schamberger (1979) weighed 21 grams. The pelage is loose and soft. The coloration is dark brown above and below with no markings, and the tail is blackish. The external appearance is much like that of *Caenolestes* except that the tail is short-

Chilean "shrew" opossum *(Rhyncholestes raphanurus)*, photo by P. L. Meserve.

er and, at least periodically, thickened at the base. The skull is narrow and elongate, especially the facial part. The lateral upper incisors are unique among living marsupials in that they have two cusps. Females have seven mammae, the seventh being in a medial position; there is no pouch (Patterson and Gallardo 1987).

According to Patterson and Gallardo (1987), *Rhyncholestes* apparently is restricted to temperate rainforests at elevations from sea level to 1,135 meters. Most specimens have been taken on the ground, alongside logs, and in dense cover. Some have been collected near burrow entrances at the bases of trees or under fallen logs. The genus appears to be mainly terrestrial and nocturnal; it is insectivorous but also eats earthworms and plant matter. Patterson, Meserve, and Lang (1990) described it as semifossorial, feeding chiefly on soil-inhabiting invertebrates and fungi. Kelt and Martínez (1989) found that the tails of animals caught during the autumn were thicker than those of specimens taken in the summer, thus suggesting that fat was being deposited in preparation for winter torpor. However, some individuals were taken on packed snow during midwinter. Kelt and Martínez also found lactating females in February, March, May, October, November, and December.

These marsupials appear to be rare; only three specimens were known through the 1970s, but subsequently larger numbers have been collected. Miller et al. (1983) warned that the dense forest habitat is now shrinking because of logging. Such declines have led the IUCN to classify *R. raphanurus* as vulnerable. Patterson, Meserve, and Lang (1990) noted that *Rhyncholestes* tends to occupy tall, wet forest but exhibits broad elevational range, wide habitat tolerance, and local abundance.

Order Microbiotheria

MICROBIOTHERIA; **Family MICROBIOTHERIIDAE; Genus DROMICIOPS**
Thomas, 1894

Monito del Monte

This order contains one family, the Microbiotheriidae, with the single living genus and species *Dromiciops australis,* which occurs only from the vicinity of Concepción south to Chiloe Island in south-central Chile and east to slightly beyond the Argentine border in the mountains. Except perhaps for the extinct Bibymalagasia, the Microbiotheria have the most restricted Recent distribution of any mammalian order. Until about the 1970s *Dromiciops* generally was placed in the family Didelphidae. Subsequent investigation indicated that it is distinct from that group and more appropriately referred to the Microbiotheriidae, an otherwise extinct family of New World marsupials, which, in turn, was considered part of the group referred to here as the order Didelphimorphia (Archer 1984; Kirsch 1977*b;* Kirsch and Calaby 1977; Marshall 1982). Even more recently the characters of *Dromiciops* have suggested that it represents a full order distinct from all other living groups of marsupials (Aplin and Archer 1987; Gardner *in* Wilson and Reeder 1993; Hershkovitz 1992*a;* Szalay 1982).

The affinities of *Dromiciops* have become a central point of interest and contention with respect to marsupial phylogeny. On the basis of both morphological and biochemical evidence, Reig, Kirsch, and Marshall (1987) regarded the genus as very distinctive but clearly part of the Didelphimorphia and having no special similarity to Australian marsupials. Aplin and Archer (1987) divided the marsupials into two great cohorts, the Ameridelphia, which includes most living New World genera, and the Australidelphia, comprising all living Australasian genera. Considering anatomy, serology, and cytology, they, and also Marshall, Case, and Woodburne (1990), assigned *Dromiciops* and the

Monito del monte *(Dromiciops australis),* photo by Luis E. Peña

order Microbiotheria to the Australidelphia. Based to a considerable extent on the analysis of limb bones, Szalay (1994) recognized approximately the same cohort division but designated a new order, Gondwanadelphia, with the New World Microbiotheria and the Australasian Dasyuromorphia as suborders. It was suggested that this group arose in South America, spread to Antarctica and then Australia when the southern continents were joined during the late Cretaceous, and became the founding stock of the entire great Australasian marsupial radiation.

Hershkovitz (1992*a*, 1992*b*, 1995) disagreed with Szalay's assessment of limb bones and with the assignment of *Dromiciops* and the Microbiotheria to an otherwise Australian cohort. He argued that *Dromiciops* is the sole living member of an entirely separate cohort, the Microbiotheriomorphia, that diverged from the other marsupials before the development of the existing orders. All living marsupials besides *Dromiciops* can be united in the cohort Didelphimorphia; an early migration of the latter group from South America led to the Australasian radiation. The development of the Microbiotheriomorphia may have occurred as early as the Jurassic in either North or South America. The retention today by *Dromiciops* of a basicaudal cloaca such as found in monotremes but in no other marsupials is suggestive of a very primitive and independent origin. Moreover, *Dromiciops* has a number of key characters that are found regularly in placental mammals but have been lost in other marsupials, such as an entotympanic bone at the base of the auditory bullae and a nonstaggered third lower incisor tooth. All other marsupials with comparable dentition are seen to have undergone a reduction in muzzle length, with associated crowding of teeth, so that the third lower incisor is not in line with the other teeth and is usually supported by a bony buttress; this condition is characteristic of the cohort Didelphimorphia.

Head and body length of *Dromiciops* is 83–130 mm, tail length is 90–132 mm, and weight is 16–42 grams (Marshall 1978*b*; Redford and Eisenberg 1992). The fur is silky, short, and dense. The upper parts are brown, with several ashy white patches on the shoulders and rump arranged in vague whorls in a manner somewhat like the color pattern of *Chironectes*. The underparts are buffy to pale buffy. Other than pronounced black rings around the eyes there are no well-defined markings on the face.

Dromiciops may be recognized by its short, furry ears; thick, hairy tail; and features of the skull and dentition. The dental formula is the same as in the Didelphidae, but the canines, last molars, and first incisors are relatively smaller than those of typical didelphids. The tympanic bullae of *Dromiciops* are greatly inflated and occupy a relatively large area on the base of the skull. The face is quite short. The basal third of the moderately prehensile tail is furred like the body; the terminal two-thirds is also well furred, but the hairs are straight and slightly different in color, being nearly all dark brown. The only naked part of the tail is a narrow strip 25–30 mm long on the underside of the tip. Females have four mammae in a small but distinct pouch.

The monito del monte inhabits dense, humid forests, especially areas with thickets of Chilean bamboo (*Chusquea* sp.). It sometimes is said to be arboreal, but Pearson (1983) considered it scansorial, noting that while it is a good climber, it may take refuge underground when released from a trap. Patterson, Meserve, and Lang (1990) stated that the highly scansorial habits of this genus are underscored by its prehensile tail, opposable toes, and unqualified aversion to entering enclosed live traps. It makes small round nests, about 200 mm in diameter, of sticks and the water-repellent leaves of *Chusquea*, lined with grasses and mosses. The nests may be under rocks or fallen tree trunks, in hollow trees, on branches, or suspended in lianas or patches of *Chusquea* (Marshall 1978*b*; Rageot 1978). *Dromiciops* is almost entirely nocturnal. In some areas, at least, it hibernates for lengthy periods when cold temperatures prevail and food is scarce. Hibernation is preceded by an accumulation of fat in the basal part of the tail, which may more than double the weight of an individual in one week (Kelt and Martínez 1989; Rageot 1978). *Dromiciops* also has been observed to enter periods of daily torpor even when food was readily available (Grant and Temple-Smith 1987). The natural diet seems to consist mainly of insects and other invertebrates, though Rageot (1978) observed that captives could be maintained on vegetable matter, and Pearson (1983) reported that captives ate large quantities of apple, as well as grubs, flies, and small lizards. The voice has been recorded as a chirring trill with a slight coughing sound at the end.

This marsupial reportedly lives in pairs, at least during the mating season in spring (October–December). Females with young have been recorded from November to May, and litters of one to five have been observed (Collins 1973; Marshall 1978*b*). Four stages have been described in the development of the young: (1) in the pouch, attached to the mammae; (2) in the nest; (3) nocturnal trips on the mother's back; and (4) loose association with other members of the family. Both males and females become sexually mature during their second year of life. Rageot (1978) reported catching a female that then lived in captivity for two years and two months.

Dromiciops has declined in recent years through loss of its restricted habitat and now is classified as vulnerable by the IUCN, though apparently it once was common in some areas. It is called "colocolo" by the natives of the Lake Region of Chile, who have several superstitions about it. Some believe that it is very bad luck to see a colocolo or to have one living about the house. Some residents of the area have even been known to burn their houses to the ground after seeing one of these inoffensive little animals running about in the house.

Geologically, according to Marshall (1982) and as accepted by Aplin and Archer (1987), the family Microbiotheriidae is known only from the late Oligocene and early Miocene of southern Argentina. However, Reig, Kirsch, and Marshall (1987) considered the family to comprise additional extinct genera from the late Cretaceous of North and South America and from the middle Paleocene to the early Miocene of South America.

Order Dasyuromorphia

Australasian Carnivorous Marsupials

This order of 3 Recent families, 19 genera, and 64 species is found in Australia, Tasmania, and New Guinea and on some nearby islands. Essentially this order has the same living content as the family Dasyuridae as the latter was accepted by Simpson (1945) and most other authorities until about the 1960s. Subsequent investigation led to the distinction of the Myrmecobiidae and Thylacinidae as separate families but eventually to their union with the remaining Dasyuridae at the ordinal level (Aplin and Archer 1987; Groves *in* Wilson and Reeder 1993; Marshall, Case, and Woodburne 1990). Szalay (1994) regarded the Dasyuromorphia as a suborder of his new order Gondwanadelphia, within which he also placed the Microbiotheria as a suborder, but Hershkovitz (1992*a*, 1995) saw no close relationship between those groups.

The members of the Dasyuromorphia are small to medium in size and vary widely in appearance; head and body length is about 50–1,300 mm. They have a general superficial resemblance to the Didelphimorphia and indeed sometimes were combined with the latter group in an order designated the Marsupicarnivora or Polyprotodontia. However, the Dasyuromorphia have only four upper and three lower incisor teeth on each side of the jaw compared with five upper and four lower in the Didelphimorphia. In both orders the incisors are polyprotodont, being relatively numerous, small, and sharp. In this regard the two orders differ from the diprotodont marsupials, which have only one or two lower incisor teeth on each side of the jaw, including a very strongly developed pair. The two orders also share a didactylous condition–having all their digits separate–whereas in the Australasian orders Diprotodontia and Peramelemorphia the second and third digits of the hind foot are syndactylous, being joined together by integument. Unlike the Didelphimorphia, the Dasyuromorphia lack a caecum and never have a prehensile tail. According to Szalay (1994), the talonids of the molar teeth of the Dasyuromorphia are reduced relative to the size of the trigonids compared with the dentition of the Didelphimorphia, and there is a lack of magnum contact with the lunate in the carpus.

Clemens, Richardson, and Baverstock (1989) indicated that the Didelphimorphia are characterized by retention of many primitive features, including polyprotodonty and didactyly, but also reduction of the number of incisors, presence of a large neocortex, the loss of the calcaneal vibrissae, and a number of adaptations of tarsal morphology for terrestrial existence. The known geological range of the order is middle Miocene to Recent in Australasia. However, according to Morton, Dickman, and Fletcher (1989), the group appears to have differentiated from the syndactylous Australian marsupials during the Eocene, at about the time when the last connection between Antarctica and Australia was broken. Subsequently, in the Oligocene the Thylacinidae split off from the basic dasyurid stock, and then in the Miocene the Myrmecobiidae and Dasyuridae also diverged.

DASYUROMORPHIA; **Family DASYURIDAE**

Marsupial "Mice" and "Cats" and Tasmanian Devil

This family of 17 Recent genera and 62 species occurs in Australia, Tasmania, New Guinea, and some adjacent islands. The sequence of genera presented here follows that of Archer (1982*a*, 1984), who recognized the following subfamilies: Muricinae, for the genus *Murexia;* Phascolosoricinae, for *Phascolosorex* and *Neophascogale;* Phascogalinae, for *Phascogale* and *Antechinus;* Sminthopsinae, for *Planigale, Ningaui, Sminthopsis,* and *Antechinomys* (which Archer considered part of *Sminthopsis*); and Dasyurinae for the remaining genera. Based on an analysis of mitochondrial DNA, Krajewski et al. (1994) recommended abolishing the subfamilies Muricinae and Phascolosoricinae and transferring their constituent genera to, respectively, the Phascogalinae and Dasyurinae; a close relationship of *Murexia* and the New Guinean species of *Antechinus* was suggested.

Dasyurids are small to medium in size. The genus *Planigale* includes the world's smallest marsupials, some measuring only 95 mm in total length and weighing about 5 grams. The largest living member of the family is the Tasmanian devil *(Sarcophilus)*, which may weigh over 10 kg. The tail of a dasyurid is long, hairy, and nonprehensile. The limbs usually are about equal in length, and the digits are separate. The forefoot has five digits, the hind foot four or five. The longest front toe is the third; digits 2 through 5 of the hind foot are well developed, but the great toe, which lacks a claw, is small or absent. Morton, Dickman, and Fletcher (1989) noted that arboreal species have broad hind feet and a mobile hallux, while species that are both arboreal and terrestrial have longer hind feet and a reduced hallux. Some genera walk plantigrade, that is, on the soles of the feet, and others walk digitigrade, on the toes. The mar-

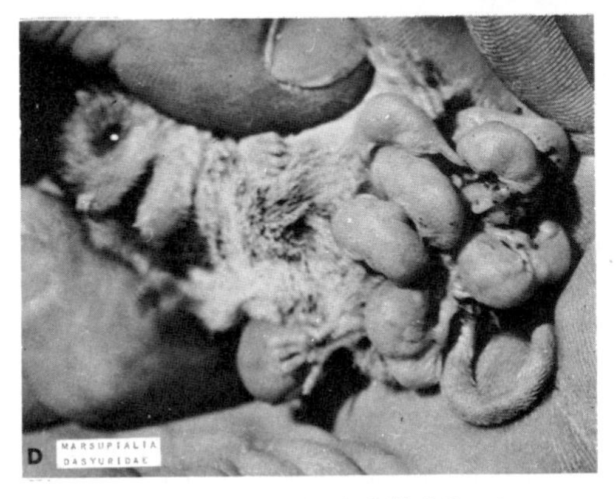

A–C. Narrow-footed marsupial "mice" *(Sminthopsis crassicaudata)*, photos by Shelly Barker. D. Seven baby narrow-footed marsupial "mice" (*Sminthopsis* sp.) attached to their mother's nipples, photo by Vincent Serventy.

supium is absent in some; when present it usually opens posteriorly and often is poorly developed. The pouches of some genera become conspicuous only during the breeding season. Females have 2–12 mammae, the usual number being 6 or 8.

The basic pattern of the dasyurid dentition resembles that of *Didelphis*. The incisors are small, pointed, or bladelike, and the canines are well developed and large with a sharp cutting edge. The molars have three sharp cusps. The teeth in this family are specialized for an insectivorous or carnivorous diet—rooted, sharp, and cutting. The dental formula is: (i 4/3, c 1/1, pm 2–3/2–3, m 4/4) × 2 = 42 or 46.

Dasyurids have acute senses and are considered alert and intelligent. They are active animals and move rapidly. Most of the members of this family are terrestrial, though some marsupial "mice" are primarily arboreal. The usual shelter is a hollow log, a hole in the ground, or a cave. The mouselike forms are usually silent, but the Tasmanian devil growls and screeches loudly. Dasyurids are active mainly at night and prey upon almost any living thing that they can overpower.

Morton, Dickman, and Fletcher (1989) provided an extensive summary of the physiology and natural history of the Dasyuridae. They identified six different life history strategies based on variation in age of sexual maturity, frequency of estrus, seasonality, and duration of male reproductive effort. As explained by Schmitt et al. (1989), male dasyurids tend to die in large numbers at the end of their very first breeding season, such mortality being nearly complete in certain genera, notably *Antechinus* and *Phascogale*. This phenomenon is thought to result from stress mediated through an increase in the concentration of plasma glucocorticoid and to be exacerbated by a reduction in the concentration of plasma-corticosteroid-binding globulin associated with a progressive rise in plasma androgen. Increased glucocorticoids, which inhibit most stages of the inflammatory and immune responses of animals, results in debilitating effects manifested by anemia, lymphocytopenia and neutrophilia, splenic hypertrophy, gastrointestinal hemorrhage and disease, immune suppression and disease, degeneration of major organs, and negative nitrogen balance.

The geological range of the family is middle Miocene to Recent in Australia and middle Pliocene to Recent in New Guinea (Archer 1982*a;* Marshall 1984).

DASYUROMORPHIA; DASYURIDAE; **Genus MUREXIA**
Tate and Archbold, 1937

Short-haired Marsupial "Mice"

There are two species (Flannery 1995; Kirsch and Calaby 1977; Ziegler 1977):

M. longicaudata, New Guinea, Japen Island (Yapen) to northwest, and Aru Islands to southwest;
M. rothschildi, southeastern Papua New Guinea.

A specimen from Normanby Island, southeast of New Guinea, originally was assigned to *M. longicaudata* but actually seems to represent an unknown species of *Antechinus* (Flannery 1995).

Head and body length is 105–285 mm and tail length is 145–283 mm. Flannery (1990*b*) listed weights of 32–40 grams for male *M. rothschildi,* 114–434 grams for male *M. longicaudata,* and 57–88 grams for the much smaller female *M. longicaudata.* The latter species is dull grayish

Short-haired marsupial "mouse" *(Murexia longicaudata)*, photo by P. A. Woolley and D. Walsh.

brown above with buffy white underparts and has a long, sparsely haired tail with a few longer hairs at its tip. In *M. rothschildi* the upper parts are similarly colored, but a broad black dorsal stripe is present; the underparts in this species are light brown. The fur is short and dense.

The skull is heavy and strong with deep zygomatic arches. The foot pads are striated, indicating that the animals probably are scansorial (Collins 1973). Two pairs of mammae are present.

The species *M. longicaudata* is found in all lowland and midmountain forests of New Guinea, from sea level to 2,200 meters. *M. rothschildi,* which is known from only 10 specimens, has an altitudinal range of 600–2,000 meters. These animals are presumed to be carnivorous and at least partly diurnal and arboreal (Flannery 1990*b;* Ziegler 1977). Woolley (1989) found *M. longicaudata* to utilize a spherical nest about 20 cm wide composed of leaves and located in a tree about 4 meters above the ground.

DASYUROMORPHIA; DASYURIDAE; **Genus PHASCOLOSOREX**
Matschie, 1916

There are two species (Flannery 1990*b*):

P. dorsalis, highlands of eastern and extreme western New Guinea;
P. doriae, western New Guinea.

There is some question whether *P. doriae* and *P. dorsalis* are really two separate species, for they are remarkably alike in basic characters and differ most markedly only in coloration and body size. *P. doriae* is generally the larger animal; it ranges in head and body length from 117 to 226 mm and in tail length from 116 to 191 mm. Its general body color is dark grizzled orange brown with bright rufous on its underside, whereas *P. dorsalis* is a grizzled gray brown with chestnut red underneath. The head and body length of *P. dorsalis* is 134–67 mm and the tail length is 110–60 mm.

One of the most distinctive characters of the genus *Phascolosorex* is the presence of a thin black stripe that runs from the head region all the way down the middle of the back to the base of the tail. *Murexia rothschildi,* with

Short-haired marsupial "mouse" *(Murexia rothschildi)*, photo by P. A. Woolley and D. Walsh.

which it may otherwise be confused, has a broad black stripe down its back. The tail is black and short-haired except at the base, where it is covered by thick, short, soft fur like that covering the body. The tail may or may not terminate in white. The ears are relatively short and more sparsely haired than the body. The feet, which are black in *P. doriae* and brown in *P. dorsalis,* possess striated pads and relatively short claws. There are four mammae in the pouch area.

These marsupials occur in mountain forests. *P. dorsalis* is more abundant at high altitudes but has been collected at 1,210–3,100 meters; *P. doriae,* the more geographically restricted species, has been recorded only in the lower to middle elevations, at 900–1,900 meters. The differences in locality of the two species when they both occur in the same general area are more strikingly vertical than horizontal. They usually are thought to be nocturnal and scansorial in habit (Collins 1973). However, extended laboratory observations by Woolley et al. (1991) demonstrate that *P. dorsalis* is predominantly diurnal; in this regard it is unique among the Dasyuridae. Lidicker and Ziegler (1968) referred to *P. dorsalis* as a "relatively rare high mountain species." Dwyer (1983) described it as "a rainforest species said to be terrestrial and often diurnal in activity. An immature male of 13.3 grams was hand caught in daylight in September while 'sunning' itself on a rock in a small forest clearing. A second immature male of 15 grams was collected in September from beneath the fronds of *Pandanus.*"

DASYUROMORPHIA; DASYURIDAE; **Genus NEOPHASCOGALE**
Stein, 1933

Long-clawed Marsupial "Mouse"

The single species, *N. lorentzii,* is found in the western and central mountains of New Guinea (Flannery 1990*b*).

This marsupial resembles a tree shrew in general appearance. Head and body length is 170–230 mm and tail length is 185–215 mm. The color ranges from deep rufous to dull, pale cinnamon on the upper parts, and the fur is profusely speckled with white hairs or white-tipped hairs. The underparts are rich rufous with whitish, subterminal bands to the hairs. The head is deep rufous brown, and the

Phascolosorex dorsalis, photo by P. A. Woolley and D. Walsh.

backs of the ears are white. The limbs are rusty in color and blend to brown on the feet. The tail is rufous, except for the last third, which is white. The fur is long, soft, and dense. A few specimens taken have been melanistic.

There are long claws on all the feet, and the pads are striated. Structural differences that support separation from *Phascogale* are mainly in the dentition: there is no gap between incisors 1 and 2, and the last premolar is considerably smaller than the first two. The long-clawed marsupial "mouse" inhabits humid moss forests at altitudes of 1,500–3,400 meters. It is partly diurnal and probably is largely arboreal (Ziegler 1977). Collins (1973) reported that two specimens that survived several months in captivity were fed strips of beef, cicadas, and cockchafer beetles. Live insects were consumed greedily.

DASYUROMORPHIA; DASYURIDAE; **Genus PHASCOGALE**
Temminck, 1824

Brush-tailed Marsupial "Mice," or Tuans

There are two species (Cuttle *in* Strahan 1983; Kitchener *in* Strahan 1983; Ride 1970; Smith and Medlin 1982):

P. tapoatafa, extreme northern and southwestern Western Australia, northern Northern Territory, northern and southeastern Queensland, eastern New South Wales, southern Victoria, southeastern South Australia;
P. calura, originally known from northeastern and southern Western Australia, southern Northern Territory, southeastern South Australia, southwestern New South Wales, and northwestern Victoria.

In *P. tapoatafa* head and body length is 160–230 mm, tail length is 170–220 mm, and weight is 110–235 grams (Cuttle *in* Strahan 1983). *P. calura* is considerably smaller, with a head and body length of 93–122 mm, a tail length of 119–45 mm, and a weight of 38–68 grams (Kitchener *in* Strahan 1983). Both species are grayish above and whitish below. The tail in *P. tapoatafa* is black except for the gray base; in *P. calura* it is black except for a reddish basal area.

Phascogale is distinguished from the other broad-footed marsupial "mice" by a tail whose terminal half is covered with a silky brush of long black hairs. In *P. tapoatafa* these hairs are capable of erection, producing a striking "bottle brush" effect. This is normal when the animal is active. Soderquist (1994) pointed out that this action is intended to distract predators and deflect their strike away from the body. When the animal is at rest the hairs are pressed along the tail and are not conspicuous. The ears are relatively large, thin, and almost naked. Females have eight mammae (Cuttle 1982). There is no true pouch, but there is a pouch area marked by light-tipped brown hairs coarser in texture than the body fur, which begins to develop protective folds of skin about two months before parturition. Strong, curved claws are present on all digits except the innermost digit of each hind foot, which is clawless.

These marsupial "mice" are usually arboreal, frequenting both the heavy, humid forest and the more sparsely timbered arid regions. Normally they build their nests of leaves and twigs in the forks or holes of trees, but in some localities they build them on the ground. Tuans are nocturnal and active and are as agile as squirrels. They capture and eat small mammals, birds, lizards, and insects. Apparently they do not eat carrion. They sometimes prey on poultry and can be destructive, but the benefits derived by their extermination of mice and insects probably outweigh any harm they may do. Ride (1970) wrote that tuans also feed on nectar. When disturbed, *P. tapoatafa* utters a low, rasping hiss, which apparently is an alarm note. When an-

Long-clawed marsupial "mouse" *(Neophascogale lorentzii)*, photo by P. A. Woolley and D. Walsh.

gered, tuans emit a series of staccato "chit-chit" sounds. Sometimes, when excited, tuans slap the pads of their forefeet down together while holding an alert, rigid pose, thus producing a sharp rapping sound. At times they also make a rapid drumming noise by quick vibrations of the tail.

Radio-tracking studies by Soderquist (1995) of *P. tapoatafa* in Victoria showed that female home ranges averaged 41 ha. and had little or no overlap with one another. Adult females generally were highly agonistic, but a mother sometimes relinquished a portion of her range to a newly independent daughter. Male ranges averaged 106 ha. and overlapped extensively with one another and with those of females. Observations in the wild (Soderquist and Ealey 1994) indicated that females are able to dominate males and to deter mating. Young animals disperse at an average age of 162 days, but females either stay within their mother's range or settle adjacent to it (Soderquist and Lill 1995).

The reproductive biology of *P. tapoatafa* is very much like that of *Antechinus stuartii* (Cuttle 1982 and *in* Strahan 1983; Soderquist 1993). The breeding season is restricted, females are monestrous, and ovulation is synchronized. Studies in Victoria indicate that mating occurs in June (winter), gestation lasts about 30 days, and births take place in late July and early August. Litters usually consist of eight young but sometimes contain as few as one. They remain attached to the nipples for about 40–50 days and are then left in the nest while the mother forages. They begin to emerge just before weaning at an age of around 5 months and reach adult size at about 8 months. Males disappear entirely from the wild population just after mating and before they are a year old, evidently succumbing to stress-related diseases (see also account of family Dasyuridae). Captive males have lived for more than three years but are not reproductively viable after their first breeding season. Females may survive to breed for a second year in the wild but are known to have survived to a third year only in captivity. The reproductive pattern of *P. calura* is similar (Kitchener 1981 and *in* Strahan 1983). Females give birth to eight young from mid-June to mid-August, and they are weaned before the end of October.

The IUCN designates *P. calura* as endangered. It has disappeared from more than 90 percent of its range and now survives only in isolated reserves of southwestern Western Australia (Kennedy 1992). Kitchener (*in* Strahan 1983) noted that this species depends on large areas of dense, undisturbed vegetation and that its decline was probably associated with destruction of such habitat through grazing by domestic cattle and sheep. Thornback and Jenkins (1982) noted, however, that much suitable habitat remains and that predation by introduced cats and foxes may be a factor. *P. tapoatafa* still is widespread, but it has lost considerable habitat (Kennedy 1992) and is designated as near threatened by the IUCN.

Brush-tailed marsupial "mouse" *(Phascogale calura)*, photo by M. Archer.

DASYUROMORPHIA; DASYURIDAE; **Genus ANTECHINUS**
Macleay, 1841

Broad-footed Marsupial "Mice"

There are 10 species (Baverstock et al. 1982; Kirsch and Calaby 1977; Ride 1970; Van Dyck 1980, 1982*a*, 1982*b*; Woolley 1982*b*; Ziegler 1977):

A. melanurus, New Guinea;
A. naso, New Guinea;
A. wilhelmina, central mountains of New Guinea;
A. godmani, northeastern Queensland;
A. stuartii, Victoria, eastern New South Wales, Queensland;
A. flavipes, eastern and southern Australia;
A. leo, Cape York Peninsula of northern Queensland;
A. bellus, Northern Territory;
A. swainsonii, southeastern Australia, Tasmania;
A. minimus, coastal southeastern Australia, Tasmania, islands of Bass Strait.

The genera *Parantechinus, Dasykaluta,* and *Pseudantechinus* (see accounts thereof) sometimes have been considered part of *Antechinus.* The species once known as *Antechinus maculatus* has been transferred to the genus *Planigale* (Archer 1975, 1976). On the basis of electrophoretic analysis, Dickman et al. (1988) and McNee and Cockburn (1992) reported that *A. stuartii* actually comprises two morphologically and behaviorally cryptic species, one found in Queensland and northeastern New South Wales, the other in southern New South Wales and Victoria.

Head and body length is about 75–175 mm and tail length is about 65–155 mm. Within any species males generally are larger than females. Green (1972) reported that in *A. swainsonii,* a species of about average size, a series of male specimens weighed 48–90 grams and a series of females weighed 31–55 grams. In *A. minimus,* a smaller species, males weighed 30–57 grams, and females 24–52 grams. Emison et al. (1978) found a sample of *A. stuartii* to average 35.2 grams for males and 26.4 grams for females. The fur of *Antechinus* is short, dense, and rather coarse. Coloration of the upper parts varies from pale pinkish fawn through gray to coppery brown; the underparts are buffy, creamy, or whitish. The tail, usually the same color as the back, is short-haired in most species.

The short, broad feet are distinctive. The great toe is present but is small and clawless. All species have transversely striated pads on the bottom of the feet. In the semiarboreal species, such as *A. flavipes,* the pads are prominent and strongly striated; in *A. swainsonii* and *A. minimus,* which are poor climbers, the pads are small and the striations faint (especially so in *A. minimus,* which occurs in treeless habitat). These two closely related species have long and strong foreclaws modified for digging. *A. flavipes* has short, hooked claws, is much more active, and is a great climber.

A pouch is known to develop during the breeding sea-

Broad-footed marsupial "mouse" *(Antechinus flavipes)*, photo by Howard Hughes through Australian Museum, Sydney.

Marsupial "mouse" *(Antechinus stuartii)*, photo by H. J. Aslin.

son in most species but may be absent in others or may consist only of prominent skin folds around the mammae. The number of mammae varies between and within species, having been reported as 10–12 in *A. flavipes*, 6–10 in *A. stuartii*, 6–8 in *A. swainsonii*, 6–8 in *A. minimus* (Collins 1973), 6 in *A. bellus* (Taylor and Horner 1970), 4 in *A. melanurus* (Dwyer 1977), and 10 in *A. leo* (Van Dyck 1980).

The species found in New Guinea, as well as *A. godmani* and *A. swainsonii*, occur mainly in dense, moist forests (although the latter species is a ground dweller); *A. stuartii* and *A. flavipes* can inhabit a wide variety of forest and brushland habitat as long as there is sufficient cover; *A. bellus* dwells on tropical savannahs; and *A. minimus* occurs in grassy areas on the mainland but in Tasmania is restricted to wet sedgelands (Emison et al. 1978; Green 1972; Ride 1970; Robinson et al. 1978; Woolley 1977; Ziegler 1977). Broad-footed marsupial "mice" are secretive, active, nocturnal animals characterized by rapid movements. The claws and ridged foot pads of some species enable them to run upside down on the ceilings of rock caverns. They construct nests in hollow trees, fallen logs, rock crevices, abandoned birds' nests, or pockets in cave ceilings (Collins 1973).

The diet consists mainly of invertebrates. Nagy et al. (1978) found *A. stuartii* in winter to consume about 60 percent of its weight in arthropods each day. Small vertebrates, including house mice that have been introduced in Australia, are also taken. Green (1972) reported that captive *A. swainsonii* accepted items ranging in size from mosquitoes to house mice and domestic sparrows.

Population density of *A. minimus* on Great Glennie Island in Bass Strait was estimated at 80/ha. (Wainer 1976). Density of the best-known species, *A. stuartii*, was determined as 3/ha. and 18/ha. in two sections of wet sclerophyll forest (Nagy et al. 1978). Home ranges of three females near Brisbane were 0.80, 0.98, and 1.15 ha., and males utilized larger but uncalculated areas (Wood 1970). Studies by Cockburn and Lazenby-Cohen (1992), Lazenby-Cohen (1991), and Lazenby-Cohen and Cockburn (1988) indicate that outside of the breeding season both male and female *A. stuartii* forage in clearly defined individual home ranges but that neither sex is territorial. As temperatures drop during the winter individuals may leave their nocturnal foraging range to spend the day in a communal nest that may hold 18 animals simultaneously. The total number of individuals visiting a single nest during the winter may be more than 50. When the mating season begins males abandon their foraging range and aggregate in a few of the nest trees, where they spend most of the night. Males always nest communally and always in groups that include females. Such mating aggregations are comparable to the "leks" that have been described for various large hoofed mammals. Prior to the mating season females nest alone or in small unisexual groups. Even during the breeding season females continue to use their foraging range and may nest alone, but they make excursions to the aggregations, thus determining the time and place of mating.

Each investigated Australian species (*A. flavipes, A. stuartii, A. bellus, A. swainsonii*, and *A. minimus*) has been shown to be monestrous, to have a breeding season restricted to about three months, and to produce only one litter per year. This reproductive period often occurs in winter (July–September) but may be in autumn or spring in some areas. Births usually are highly synchronized; in any one population of *A. stuartii* they all take place within two weeks (Collins 1973; Lee, Bradley, and Braithwaite 1977). Wood (1970) reported that in the population of *A. stuartii* near Brisbane mating occurred in late September, births occurred in October, and lactation lasted until February. Wainer (1976) found that in *A. minimus* on Great Glennie Island mating occurred in late May and most litters were born in July. The young remained in the pouch until late August or September and then suckled in the nest until November. Available data indicate that in New Guinea *A. melanurus* and probably *A. naso* breed throughout the year (Dwyer 1977). Gestation has been reported as 23–27 days in *A. flavipes* and 26–35 days in *A. stuartii*. Litter size is 3–12 (Collins 1973). The young of *A. stuartii* measure 4–5 mm in length at birth and weigh 0.016 gram. They are weaned and independent after about 90 days and attain sexual maturity at about 9–10 months (Collins 1973; Marlow 1961). Young males are forced by their mothers to disperse, and thus the subsequent winter aggregations consist largely of unrelated males, but young females remain in their natal vicinity (Cockburn and Lazenby-Cohen 1992).

A remarkable feature of the biology of *Antechinus* is the abrupt and total mortality of males following mating, when they are 11–12 months old (Arundel, Barker, and Beveridge 1977; Barker et al. 1978; Emison et al. 1978; Lazenby-Cohen and Cockburn 1988; Lee, Bradley, and Braithwaite 1977; Morton, Dickman, and Fletcher 1989; Nagy et al. 1978; M. P. Scott 1987; Wainer 1976; Wilson and Bourne 1984; Wood 1970). Although the period of mortality varies geographically, it occurs at the same time each year in any given population. This phenomenon is known to take place in five species—*A. flavipes, A. stuartii, A. bellus, A. swainsonii*, and *A. minimus*—but is best understood and is especially sudden in *A. stuartii*. In this species the males become more active and aggressive during the breeding season. As the breeding season approaches a climax they move about considerably, even in daylight and mainly from one male aggregation to another. They apparently are subject to intensive physiological stress resulting from continuous competition for the females that visit the male aggregations as well as from gluconeogenic mobilization of body protein, a process that sustains them temporarily. Increasing levels of testosterone and other androgens evidently depress plasma-corticosteroid-binding globulin and result in a rise in corticosteroid concentration. The trauma associated with stress and endocrine changes causes suppression of the immune system, major ulceration of the gastric mucosa, exposure to parasites and pathological conditions, and death. Even males captured during the mating season die in the laboratory at the same time as those remaining in the wild. Males taken before sexual maturity, however, have been maintained until at least two years and eight months (Rigby 1972), and captive females have survived for more than three years (Collins 1973). In the wild, many females also die after rearing their first litter, but some live for at least another year (Lee, Bradley, and Braithwaite 1977). Limited evidence suggests that a synchronized male die-off may not occur in *A. melanurus* of New Guinea (Dwyer 1977; Woolley 1971*b*).

Broad-footed marsupial "mice" generally are not under serious human pressure, though some populations have been reduced by predation from domestic cats. Green (1972) considered *A. minimus* to be "rare and endangered" in Tasmania because of its limited habitat and the threats posed by mineral development, grazing of domestic livestock, and flooding by hydroelectric dams. Aitken (1977*b*) also expressed concern for the habitat of the mainland populations of *A. minimus.* The subspecies *A. minimus maritimus* and also *A. swainsonii insulanus* of coastal Victoria are designated as near threatened by the IUCN. Laurance (1990) determined that deforestation had reduced the historical range of *A. godmani* by about 30 percent, resulting in extensive fragmentation of remaining populations. Both *A. godmani* and *A. leo* are now classed as near threatened by the IUCN. Cockburn and Lazenby-Cohen (1992) cautioned that the cutting of large trees with hollows might adversely affect the breeding requirements of *A. stuartii.*

DASYUROMORPHIA; DASYURIDAE; **Genus PLANIGALE**
Troughton, 1928

Planigales, or Flat-skulled Marsupial "Mice"

There are five species (Aitken 1971*b*, 1972; Archer 1976; Baverstock et al. 1982; Lumsden, Bennett, and Robertson 1988; Ziegler 1972):

P. maculata, northern and eastern Australia, mainly near the coast;
P. novaeguineae, southern New Guinea;
P. ingrami, northern parts of Western Australia, Northern Territory, and Queensland;
P. tenuirostris, southern Queensland, northern New South Wales, eastern South Australia;
P. gilesi, eastern South Australia, southwestern Queensland, northern and western New South Wales, northwestern Victoria.

Archer (1976) observed that *P. maculata* and *P. novaeguineae* are very similar and might prove conspecific; he observed as well that it is not clear whether *P. ingrami* and *P. tenuirostris* are separate species. The species *P. gilesi* is distinct from all the others in possessing only two, rather than three, premolars in both the upper and the lower jaw. In contrast, analysis of mitochondrial DNA (Painter, Krajewski, and Westerman 1995) indicates that *P. gilesi* is immediately related to *P. novaeguineae* and *P. ingrami* and that *P. maculata* is the most divergent member of the genus, possibly with closer affinity to *Sminthopsis.* Such analysis also suggests that two specimens from the Pilbara region of northwestern Western Australia represent undescribed species, one related to *P. maculata* and one possibly with affinity to *P. ingrami.*

Head and body length for the genus is about 50–100 mm and tail length is about 45–90 mm. The smallest species, *P. ingrami* and *P. tenuirostris,* weigh only about 5 grams. A series of specimens of *P. maculata,* one of the largest species, averaged 15.3 grams for males and 10.9 grams for females (Morton and Lee 1978). The upper parts are pale tawny olive, darker tawny, or brownish gray; the underparts are olive buff, fuscous, or light tan. The feet are light grayish olive or pale brown, and the tail is grayish or brownish. The fur is soft and dense on the body; the tail is short-haired and nontufted. The central areas of the foot pads are usually smooth but sometimes are striated, the latter condition perhaps being most common in *P. novaeguineae.* In appearance and behavior planigales resemble the true shrews, *Sorex.*

These marsupials are remarkable for their extremely flat skull, which has an almost straight upper profile and a depth of as little as 6 mm. The same condition, however, is closely approached in some species of *Antechinus* (Archer 1976; Ride 1970). The pouch becomes fairly well developed during the breeding season and opens to the rear. The known number of mammae is 5–10 (possibly up to 15) in *P. maculata,* 6–12 in *P. ingrami* and *P. tenuirostris,* and 12 in *P. gilesi* (Archer 1976).

Planigales occur mainly in savannah woodland and grassland, though *P. maculata* also has been reported from rainforest. They may shelter in rocky areas, clumps of grass, the bases of trees, or hollow logs, and in captivity they have been observed to build saucer-shaped nests of dry grass (Archer 1976). Although primarily terrestrial, they climb fairly well (Collins 1973). Most seem to be nocturnal, but Aitken (1972) reported that *P. gilesi* was active for short periods throughout the day. Planigales are avid predators, feeding on insects, spiders, small lizards, and small mammals such as *Leggadina* (Collins 1973). They are capable of catching and eating grasshoppers almost as large as themselves. Captive females of *P. ingrami* ate six to eight grasshoppers 50 mm long every day and appeared to thrive on this diet.

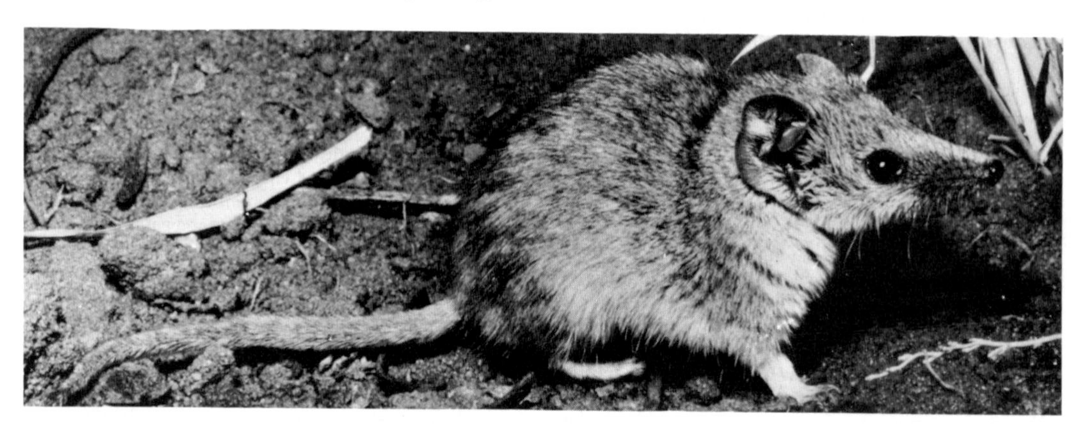

Flat-skulled marsupial "mouse" *(Planigale tenuirostris)*, photo by Dick Whitford.

In captivity, females of *P. maculata* are polyestrous, can breed throughout the year, and can produce several litters annually. Gestation is 19–20 days and litter size is 5–11 young. Males are capable of breeding until they are at least 24 months old (Aslin 1975). In a study of wild populations of this species in the Northern Territory, Taylor, Calaby, and Redhead (1982) found breeding to occur in all months; litter size ranged from 4 to 12 young and averaged 8. Read (1984) reported that *P. gilesi* has an estrus of 3 days and an estrous cycle of 21 days, while *P. tenuirostris* has an estrus of 1 day, an estrous cycle of 33 days, and a gestation period of 19 days; the breeding season of each species extends from July–August to mid-January, and some females produce two litters per season. Three reports on *P. ingrami* are that it breeds from February to April and has litters of 4–6, that in northeastern Queensland it breeds during the wet season (December–March) and has litters of 4–12, and that females collected in the northern part of Western Australia in December–January had pouch young (Archer 1976). The young of this species are capable of an independent existence at approximately three months of age and are mature within their first year (Collins 1973).

The subspecies *P. ingrami subtilissima,* found in the Kimberley Division in the northern part of Western Australia, and the species *P. tenuirostris* are classified as endangered by the USDI. These designations seem to have been applied mainly because relatively few specimens of either taxon had been collected. Recently both forms have been found to occur over a more extensive region than previously thought (Thornback and Jenkins 1982). The IUCN does not assign a classification to either form but does now designate the species *P. novaeguineae* as vulnerable. *P. novaeguineae* is restricted to a small area of suitable habitat that could be quickly eliminated by human encroachment.

DASYUROMORPHIA; DASYURIDAE; **Genus NINGAUI**
Archer, 1975

Ningauis

There are three species (Archer 1975; Baverstock et al. 1983; Kitchener, Stoddart, and Henry 1983):

N. timealeyi, northwestern Western Australia;
N. yvonneae, southern Western Australia, southern South Australia, central New South Wales, western Victoria;
N. ridei, interior Western Australia, southwestern Northern Territory, western South Australia.

Head and body length is 46–57 mm, tail length is 59–79 mm, and weight is 2–13 grams (Dunlop *in* Strahan 1983; McKenzie *in* Strahan 1983). The upper parts are dark brown to black, the underparts are usually yellowish, and the sides of the face are salmon to buffy brown. The tail is thin, without a brush or crest. *N. ridei* and *N. yvonneae* have seven mammae, and *N. timealeyi* has six (Kitchener, Stoddart, and Henry 1983). This genus is considered to be most similar to *Sminthopsis* but to differ in smaller size, longer hair, broader hind feet with enlarged apical granules, and various cranial and dental characters. Archer (1981) stated that *Ningaui* resembles *Sminthopsis* in having wide molars and narrow nasals and in lacking posterior cingula on the upper molars but differs from *Sminthopsis* in lacking squamosal-frontal contact in the skull.

Specimens have been collected in dry grassland and savannah, and the various species appear adapted to life under arid conditions. They are nocturnal, sheltering by day in dense hummocks, small burrows, or hollow logs. They prey on insects, other invertebrates, and possibly small vertebrates. Captives have climbed along spinifex leaves and used their tails in a semiprehensile manner (McKenzie *in* Strahan 1983). Captive *N. yvonneae* were often found to enter daily torpor, especially in response to low temperature and withdrawal of food (Geiser and Baudinette 1988). According to Joan M. Dixon (National Museum of Victoria, pers. comm., c. 1981), a female ningaui captured in Victoria on 28 December 1977 lived until 6 April 1980. It was fed on a wide variety of insects as well as oranges and apples. The animal was vocal throughout its period in captivity and used a high-pitched rasping noise to demand food.

Dunlop (*in* Strahan 1983) wrote that male *N. timealeyi* become aggressive toward each other during the breeding season and that females with pouch young drive other adults away. In years with good rainfall the reproductive period of this species may extend from September to March (spring and summer); in other years breeding is restricted to November–January. Usually five to six young are carried in the simple pouch to the stage of weaning. By March in most years the population of *N. timealeyi* consists predominantly of the now independent young. They attain sexual maturity in late winter.

In a study of a group of *N. ridei* captured in the Northern Territory, Fanning (1982) found females to be polyestrous and the breeding season to extend from early Sep-

Ningaui *(Ningaui ridei)*, photo by N. L. McKenzie.

Narrow-footed marsupial "mouse" *(Sminthopsis longicaudata)*, photo by P. A. Woolley and D. Walsh.

tember to late February (spring and summer). Males uttered a distinctive mating call, "tsitt," and females responded similarly. Gestation lasted from about 13 to 21 days, and most litters contained seven young. They remained attached to the nipples until they reached about 42–44 days, then were left in a nest constructed by the mother, and became independent at 76–81 days. Fleming and Cockburn (1979) reported that a female *N. yvonneae* with five young was taken in western Victoria in November 1978 and that immature animals were captured there in June, August, and December 1977.

DASYUROMORPHIA; DASYURIDAE; **Genus SMINTHOPSIS**
Thomas, 1887

Dunnarts, or Narrow-footed Marsupial "Mice"

There are 20 species (Archer 1979*b*, 1981; Aslin 1977; Baverstock, Adams, and Archer 1984; Cole and Gibson 1991; Collins 1973; Hart and Kitchener 1986; Kirsch and Calaby 1977; Kitchener, Stoddart, and Henry 1984; Mahoney and Ride *in* Bannister et al. 1988; McKenzie and Archer 1982; Morton 1978*a;* Morton, Wainer, and Thwaites 1980; Pearson and Robinson 1990; Ride 1970; Van Dyck 1985, 1986; Van Dyck, Woinarski, and Press 1994; Waithman 1979):

S. murina, northeastern Queensland, New South Wales, Victoria, southeastern South Australia;
S. dolichura, southern Western Australia, southern South Australia;
S. fuliginosus, southwestern Western Australia;
S. gilberti, southwestern Western Australia;
S. griseoventer, southern Western Australia;
S. aitkeni, Kangaroo Island off southeastern South Australia;
S. leucopus, northeastern Queensland, eastern New South Wales, southern and southeastern Victoria, Tasmania;
S. longicaudata, northwestern Western Australia;
S. ooldea, southeastern Western Australia, southern Northern Territory, western South Australia;
S. psammophila, southwestern Northern Territory, south-central WesternAustralia, southern South Australia;
S. granulipes, inland parts of southwestern Western Australia;
S. macroura, central Western Australia to western Queensland and northern New South Wales;
S. bindi, northern Northern Territory;
S. virginiae, southern New Guinea, Aru Islands, northeastern Western Australia to northeastern Queensland;
S. douglasi, northwestern Queensland;
S. butleri, northeastern Western Australia;
S. archeri, southwestern New Guinea, Cape York Peninsula of northern Queensland;
S. youngsoni, northern Western Australia, Northern Territory;

Narrow-footed marsupial "mouse" *(Sminthopsis crassicaudata)*, photo by Stanley Breeden.

S. hirtipes, west-central Western Australia, southern Northern Territory;
S. crassicaudata, mostly inland parts of the southern half of Australia.

Groves (*in* Wilson and Reeder 1993) noted that *S. butleri* also occurs in New Guinea, but Flannery (1990*b*) indicated that the pertinent records are referable to *S. archeri.*

Head and body length is about 70–120 mm and tail length is about 55–130 mm except in *S. longicaudata.* In this species the length of the tail is approximately 200 mm, or about twice the length of the head and body. Adult *S. crassicaudata* (average head and body length about 83 mm) usually weigh 10–15 grams. An adult male *S. leucopus* (head and body length 112 mm) weighed 30 grams (Green 1972). The average weight of *S. macroura* is 19 grams in males and 16 grams in females (Godfrey 1969). The fur is soft, fine, and dense. The back and sides are buffy to grayish, the underparts are white or grayish white, the feet are usually white, and the tail is brownish or grayish. Some species have a median facial stripe.

This genus is differentiated from other marsupial "mice" largely by features of the skull and dentition. The feet are slender, and the pads are striated or granulated. The hind part of the soles lacks pads. Most species have 8–10 mammae, some have 6, and one species apparently has only 2. The pouch is relatively better developed than in most other marsupial "mice."

In some species the tail accumulates fat and becomes carrot-shaped during times of abundant food. This fat reserve may be utilized when food is scarce, and the tail will then become thin. In *S. murina, S. leucopus, S. virginiae, S. longicaudata,* and *S. psammophila* the tail never becomes fat, not even under the best of conditions (Ride 1970).

Most of the permanently thin-tailed species live in moist forest or savannah, but *S. longicaudata* and *S. psammophila* occupy arid grassland and desert. As would be expected, those species capable of storing food in the tail are mainly inhabitants of dry country (Ride 1970), though *S. crassicaudata* is sometimes found in moist areas (Morton 1978*a*). Dunnarts dig burrows or construct nests of grasses and leaves, which are placed in hollow logs or under bushes or stumps. During the summer *S. crassicaudata* reportedly shelters among rocks. This species proceeds by means of bipedal bounds when traveling at top speed, but over short distances it has a peculiar quadrupedal ramble during which it holds its tail above the ground in a stiff upward curve. It and the other species are mainly terrestrial, but some are agile climbers. Most species are known to be strictly nocturnal. Ewer (1968) found that *S. crassicaudata* sheltered by day and then had alternating periods of activity and rest during the night. Individuals of that species, along with *S. macroura* and *S. murina,* also have been reported occasionally to enter a state of torpor during periods of low food supplies (Geiser et al. 1984; Morton 1978*d;* Ride 1970). Dunnarts are mainly insectivorous, but small vertebrates, such as lizards and mice, also are eaten. Most insects are caught on the ground, but *S. murina* sometimes leaps high into the air to catch moths in flight.

For *S. crassicaudata* in two parts of Victoria, Morton (1978*b*) reported population densities of 0.5/ha. and 1.3/ha. Individuals occupied home ranges, but precise size could not be determined as the areas utilized shifted over a period of months. Home ranges overlapped among the same sex and between sexes even during the breeding season, and no territorial behavior by males was observed. Breeding females, however, tended to be sedentary and possibly defended a small territory around the nest. Both sexes usually nested alone during the breeding season, but at other times up to 70 percent of the population shared nests in groups of 2–8 that included members of either or both sexes. Such groups apparently were nonpermanent, random aggregations. If nest sharing did occur during the breeding season, it generally involved pairs of a male and an estrous female. Males tolerated one another to some extent during the breeding season but may have attempted to monopolize estrous females. Collins (1973) reported that captive dunnarts could be maintained in pairs or small groups if sufficient space and multiple nesting facilities were provided but that an adult male might be attacked by a female with a litter. Aslin (1983) found captive adult *S. ooldea* to be generally intolerant of one another, even of the opposite sex, and to fight to the death when caged together. This species attains sexual maturity at 10 months and gives birth to five to eight young from September to January.

The reproductive traits of *S. crassicaudata* are fairly well known (Collins 1973; Ewer 1968; Godfrey and Crowcroft 1971; Morton 1978*c*). Females are polyestrous and are able to produce litters continuously in captivity. Estrus occurs in repeated cycles of 25–37 days, extending through a season of at least 6 months and perhaps the entire year. In the wild, breeding appears to be restricted to a period of 6–8 months, starting in June or July, during which each female is thought to give birth to two litters. The gestation period is about 13 days. Litter size is 3–10, usually 7–8 in the wild and slightly smaller in captivity. The young are carried in the mother's pouch until they are about 42 days old and then are left in the nest until they are about 63 days old. They are then practically self-sufficient, and the family soon breaks up. Minimum age of sexual maturity is 115 days in females and 159 days in males.

The breeding season of *S. macroura* was found to last from July to February in a captive colony in South Australia (Godfrey 1969). Females are polyestrous, with an average cycle length of 26.2 days. Gestation is 12.5 days. Litter size is one to eight, but litters of only one or two are not reared. The young are carried in the pouch for 40 days, then suckled in the nest for 30 more days. For *S. murina* gestation reportedly is 13–16 days and there are up to eight young (Collins 1973).

Morton (1978*c*) found that in the wild few female *S. crassicaudata* lived past 18 months and few males lived past 16 months. In captivity dunnarts seem to live and reproduce over a longer period (Collins 1973). The longevity record is held by a specimen of *S. macroura* that survived for 4 years and 11 months (Jones 1982).

The IUCN classifies the species *S. psammophila, S. douglasi,* and *S. aitkeni* as endangered and *S. butleri* as vulnerable. All are highly restricted in distribution and/or population size, and further declines are expected. *S. longicaudata* and *S. psammophila* are listed as endangered by the USDI and are on appendix 1 of the CITES. *S. longicaudata* had been known by only five specimens, the last found in 1975 (Thornback and Jenkins 1982), but was recently rediscovered alive in the Northern Territory (*Oryx* 28 [1994]: 97). Aitken (1971*c*) reported that *S. psammophila* had been known from only a single specimen collected in 1894 in the Northern Territory but that a colony was discovered on the Eyre Peninsula of South Australia in 1969. This colony is near a wildlife reserve with comparable habitat, so the species may have a good chance for survival in the area. Additional specimens have since been found in both South and Western Australia (Pearson and Robinson 1990). The habitat of *S. douglasi* has been almost entirely cleared for grazing (Thornback and Jenkins 1982). It was known by only four specimens, the last collected in 1972, but was rediscovered alive by Woolley (1992). Archer (1979*a*) stated that two other species, *S. granulipes* and *S. hirtipes,* should be considered vulnerable or of uncertain status. Both have restricted ranges and may be subject to adverse habitat modification. In addition to the above full species, the IUCN now classifies the subspecies *S. murina tatei,* of northeast-

Julia Creek dunnarts *(Sminthopsis douglasi)*, female with 60-day-old young, photo by P. A. Woolley and D. Walsh.

ern Queensland, as near threatened and an unnamed subspecies of *S. griseoventer* on Boullanger Island as critically endangered. The latter comprises a single population of fewer than 250 individuals and is continuing to decline.

DASYUROMORPHIA; DASYURIDAE; **Genus ANTECHINOMYS**
Krefft, 1866

Kultarr

The single species, *A. laniger*, is distributed over much of the inland region from Western Australia to Queensland and New South Wales. Archer (1977) recognized only this one species but observed that there were two distinctive allopatric forms. Lidicker and Marlow (1970) considered these two forms to be full species: *A. laniger* in New South Wales and Queensland and *A. spenceri* in more westerly areas. On the basis of skeletal and dental characters, Archer (1981) treated *Antechinomys* only as a subgenus of *Sminthopsis*, but Woolley (1984) pointed out that biochemical analysis and phallic morphology indicate that the two genera are distinct. The former arrangement was followed by Groves (*in* Wilson and Reeder 1993), the latter by Mahoney and Ride (*in* Bannister et al. 1988).

Head and body length is 80–110 mm and tail length is about 100–145 mm. Males are usually larger than females. The upper parts are grayish; the underparts usually are whitish, with gray hairs at the base. There is a dark ring around the eye and a dark patch in the middle of the forehead. The ears are relatively large. The fur is long, soft, and fine with few guard hairs except on the back and rump. More than the basal half of the very long tail is fawn-colored, then the hairs increase in length, and the terminal third is covered with long brown or black hairs. The upper third of the limbs is furred like the body; the remaining two-thirds is covered by short, fine, white fur. The face is well provided with vibrissae, and usually there are long vibrissae arising from the carpal pads on the wrists.

The body form is well adapted for a bounding locomotion. The feet are narrow and have granular pads. Each hind foot is elongated and has one large, well-haired, cushion-like pad on the sole and only four toes, the first being absent. The forelimbs also are graceful and elongated. It was once thought that the long hind feet betokened a bipedal, hopping means of progression comparable to that of such rodents as *Notomys, Dipodomys,* and *Jaculus;* however, recent studies have demonstrated that the kultarr actually moves in a graceful gallop, springing rapidly from its hind feet and landing on its forefeet (Ride 1970).

During the breeding season the pouch becomes fairly well developed on females. It consists of folds of skin that enlarge from the sides to partially cover the nipples. The fold is least developed on the posterior side, but unlike in most dasyurids, the pouch does not open to the rear (Lidicker and Marlow 1970). The number of mammae has been recorded at 4, 6, 8, and 10, there usually being 8 in the form once called *A. laniger* and 6 in the form once called *A. spenceri* (Archer 1977).

The kultarr occurs in a wide variety of mostly dry habitats—savannah, grassland, and desert associations. It apparently does not dig its own burrow, but nests in logs, stumps, or vegetation (Collins 1973). Perhaps it also uses the burrows of other animals or nests in deep cracks in the ground. It appears to be strictly terrestrial and nocturnal. The natural diet consists mainly of insects and other small invertebrates. It is not definitely known whether vertebrates, such as lizards and mice, are taken under normal conditions (Archer 1977). Torpor has been induced experimentally through the withholding of food and may be an adaptive mechanism to deal with deteriorating environmental conditions in the wild (Geiser 1986).

Field and laboratory investigations by Woolley (1984) show that there is a long breeding season. Animals in southwestern Queensland are in reproductive condition from mid-winter to mid-summer (July–January). Females

Kultarr *(Antechinomys laniger)*, photo from Australian Museum, Sydney, through Basil Marlow. Inset photo by W. D. L. Ride.

Dibbler *(Parantechinus apicalis)*, photo by P. A. Woolley and D. Walsh.

are polyestrous, being able to enter estrus up to six times during the season. The estrous cycle lasts about 35 days, gestation 12 days or less. A female may rear up to six young per litter. Weaning occurs after about 3 months and sexual maturity at 11.5 months. Both sexes may live to breed in more than one season. Collins (1973) referred to two adults each of which had lived for nearly three years in captivity.

Although the kultarr has a wide range, it once seemed to be rare. The USDI designates *A. laniger* as endangered, but this classification is intended to apply only to the population of the species that inhabits New South Wales and Queensland. Thornback and Jenkins (1982) noted that the species as a whole is relatively common and widespread.

DASYUROMORPHIA; DASYURIDAE; **Genus PARANTECHINUS**
Tate, 1947

There are two species (Archer 1982*a;* Begg *in* Strahan 1983; Woolley *in* Strahan 1983):

P. bilarni, Arnhem Land in northern Northern Territory;
P. apicalis (dibbler), southwestern Western Australia.

Neither Kirsch and Calaby (1977) nor Ride (1970) recognized *Parantechinus* as generically distinct from *Antechinus,* but recent systematic work involving cranial and dental features, biochemical analysis, and phallic morphology suggests that the two genera are not closely related (Archer 1982*a;* Baverstock et al. 1982; Woolley 1982*b*).

In *P. bilarni* head and body length is 57–100 mm, tail length is 82–115 mm, and weight is 12–44 grams. The upper parts are grizzled brown, the underparts are pale gray, and there are cinnamon patches behind the large ears. The tail is long and never fat (Begg *in* Strahan 1983). In *P. apicalis* head and body length is about 140–45 mm, tail length is 95–115 mm, and weight is 30–100 grams. The upper parts are brownish gray speckled with white, and the underparts are grayish white tinged with yellow. This species is distinguished from the species of *Antechinus* by its tapering and hairy tail, a white ring around the eye, and the freckled appearance of its rather coarse fur (Woolley 1991*c* and *in* Strahan 1983). The pouch of females consists merely of folds of skin on the lower abdomen enclosing the mammae, which number six in *P. bilarni* and eight in *P. apicalis.*

P. bilarni occurs in rugged, rocky country commonly covered with open eucalyptus forest and perennial grasses. Its diet consists mainly of insects. Mating occurs from late June to early July, litters contain four to five young, weaning occurs late in the year, and sexual maturity is attained by the following June (Begg *in* Strahan 1983).

P. apicalis was found in dense heathland and apparently made nests in dead logs or stumps. Captives tend to be nocturnal but may emerge from cover during the day to bask in the warmth of the sun. They burrow through leaf litter, which suggests that they do the same in the wild to search for insects. They also eat chopped meat, honey, and nectar. The latter food, plus the climbing ability of captives, indicates that the nectar of flowers is sought in the wild (Woolley *in* Strahan 1983). Mating in *P. apicalis* occurs early in the year, in March or April, females are monestrous, and gestation lasts 44–53 days (Woolley 1971*b,* 1991*c*). There are up to eight young per litter; they remain dependent for 3–4 months and reach sexual maturity at 10–11 months. Recently, Dickman and Braithwaite (1992) showed that some populations of *P. apicalis* experience a synchronized male die-off following the mating season, as occurs in most species of *Antechinus* (see account thereof). However, observations of other populations in the wild and captivity indicate that such is not inevitable (Woolley 1991*c*).

P. bilarni was not discovered until 1948, and *P. apicalis* is one of the rarest of mammalian species. Geologically Recent fossil remains indicate that it once occurred as far north as Shark Bay on the west-central coast of Western Australia, but it had not been collected for 83 years when, in 1967, two specimens were taken alive in the extreme southwestern corner of that state. Despite intensive efforts, only seven more individuals were captured until 1984, and two were found dead. Clearing of habitat for agricultural purposes has been responsible for the decline within historical time, but a 75-ha. reserve now protects the place where the dibbler was rediscovered (Morcombe 1967; Woolley 1977, 1980; Woolley and Valente 1982). In 1984 another dead specimen was found in Fitzgerald River National Park on the south coast of Western Australia, and subsequently another 17 individuals were trapped in and near the park. Meanwhile, in 1985 the species was discovered in abundance on Boullanger and Whitlock islands off the west coast (Fuller and Burbidge 1987; Muir 1985). *P. apicalis* is classified as endangered by both the IUCN and USDI.

DASYUROMORPHIA; DASYURIDAE; **Genus DASYKALUTA**
Archer, 1982

The single species, *D. rosamondae,* is found in the Pilbara region of northwestern Western Australia (Woolley *in* Strahan 1983). Since its description in 1964 this species had been placed in the genus *Antechinus,* but Archer (1982*a*) considered it to represent a separate and not closely related genus. The latter arrangement was accepted by Groves (*in* Wilson and Reeder 1993), though Mahoney and Ride (*in* Bannister et al. 1988) referred the species to *Parantechinus.*

According to Woolley (*in* Strahan 1983), head and body length is 90–110 mm, tail length is 55–70 mm, and weight is 20–40 grams. The overall coloration is russet brown to coppery. The fur is rather rough, the head and ears are short, and the forepaws are strong and well haired on the back. *Dasykaluta* has a general form similar to that of *Dasycercus* but is distinguished by its small size, coloration, and lack of black hair on the tail. Females have eight mammae.

Dasykaluta dwells in spinifex grassland and feeds on insects and small vertebrates. Observations on both wild-caught and laboratory-maintained animals (Woolley 1991*a*) show that females are monestrous and there is a short annual breeding season. Mating occurs in September and the young are born in November. The total period of pregnancy averages 50 days but ranges from 38 to 62 days, the great variation perhaps being associated with arrested development in some cases. Normally there are eight young, but there may be as few as one. They open their eyes after 58–60 days and are weaned at 90–120 days. Both sexes attain sexual maturity at about 10 months and thus are able to participate in the first breeding season following their birth. However, as in *Antechinus* (see account thereof), all mature males perish shortly thereafter. Females produce only one litter annually, but some live to breed for a second and possibly a third season.

DASYUROMORPHIA; DASYURIDAE; **Genus PSEUDANTECHINUS**
Tate, 1947

There are three species (Kitchener 1988; Kitchener and Caputi 1988; Woolley *in* Strahan 1983):

P. macdonnellensis, Western Australia, Northern Territory;
P. ningbing, northeastern Western Australia;
P. woolleyae, northwestern Western Australia.

Neither Kirsch and Calaby (1977) nor Ride (1970) recognized *Pseudantechinus* as being generically distinct from

Dasykaluta rosamondae, photo by P. A. Woolley and D. Walsh.

Pseudantechinus macdonnellensis, photo by C. W. Turner through Basil Marlow.

Antechinus, but recent systematic work involving cranial and dental features, biochemical analysis, and phallic morphology suggests that the two genera are not closely related (Archer 1982*a;* Baverstock et al. 1982; Woolley 1982*b*). The generic distinction of *Pseudantechinus* was accepted by Groves (*in* Wilson and Reeder 1993) and Morton, Dickman, and Fletcher (1989), but Corbet and Hill (1991) and Mahoney and Ride (*in* Bannister et al. 1988) considered it a synonym of *Parantechinus.* Electrophoretic studies by Cooper and Woolley (1983) suggested that the closest relative of *P. ningbing* is actually *Dasycercus cristicauda.* The IUCN recognizes *P. mimulus,* restricted to a small area of northeastern Northern Territory, as a species separate from *P. macdonnellensis* and classifies it as vulnerable.

According to Woolley (*in* Strahan 1983), head and body length in *P. macdonnellensis* is 95–105 mm, tail length is 75–85 mm, and weight is 20–45 grams. The upper parts are grayish brown, there are chestnut patches behind the ears, and the underparts are grayish white. The tail tapers and becomes very thick at the base for purposes of fat storage. *P. ningbing* is similar in general coloration and form but is smaller, and its tail is longer and has long hairs covering the base. Females have six mammae.

P. macdonnellensis is found mainly on rocky hills and breakaways but also lives in termite mounds in some areas. It is predominantly nocturnal but may emerge from shelter among the rocks to sunbathe. The diet consists mainly of insects. Observations by Woolley (1991*b*) show that there is a short annual breeding season during the austral winter. Mating occurs in June and early July in the eastern part of the range and in August and early September farther west. The gestation period is 45–55 days. There commonly are five or six young, they open their eyes after 60–65 days, they are weaned at about 14 weeks, and they are able to breed in the first season following their birth. Unlike the usual situation in *Antechinus,* there is no mass die-off of males after reproduction; indeed, males are potentially capable of breeding for at least three years. Females are monestrous and may breed in at least four seasons. *P. ningbing* also apparently lives into at least a second breeding season. It mates in June and gives birth during a period of just two to three weeks in late July and early August after a gestation of 45–52 days. The young are weaned at about 16 months and reach sexual maturity at 10–11 months (Woolley 1988).

DASYUROMORPHIA; DASYURIDAE; **Genus MYOICTIS**
Gray, 1858

Three-striped Marsupial "Mouse"

The single species, *M. melas,* is found in New Guinea and on Japen, Waigeo, and Salawatti islands to the northwest and the Aru Islands to the southwest (Flannery 1990*b*, 1995).

Head and body length is about 170–250 mm and tail length is approximately 150–230 mm. A single adult weighed 200 grams (Flannery 1990*b*). This animal was named *melas,* meaning "black," because the first specimen known to science was a melanistic individual. Normal individuals are among the most colorful of all marsupials: the upper parts are a richly variegated chestnut mixed with black and yellow. Except in occasional melanistic individuals there are three dark longitudinal stripes. The head is dark rusty red, often with a black stripe on the nose, and the crown may be tawny brown. The chin and chest are pale

Pseudantechinus ningbing, a newly discovered species without tail thickening, photo by P. A. Woolley and D. Walsh.

Three-striped marsupial "mouse" *(Myoictis melas),* photo by P. A. Woolley and D. Walsh.

rufous, and the remainder of the underparts is yellowish gray or whitish. The tail is evenly tapered and covered with long reddish hairs for most of its length; the tip is black above and rufous brown below.

The general body form is similar to that of a small mongoose. The pads are striated. The pouch is only slightly developed, and there are six mammae.

Ziegler (1977) wrote that this marsupial occurred in most rainforests of the lowlands and midmountains of New Guinea but was relatively uncommon. He observed that its external similarities to the largely terrestrial, open forest chipmunks and other sciurids suggested comparable habits and parallel activity patterns. Collins (1973), however, thought that this genus was nocturnal and scansorial in habit. An earlier report stated that one individual, a male, was shot on a recumbent, decayed log in undergrowth, while an accompanying female escaped. A. R. Wallace stated in 1858 that in the Aru Islands these animals "were as destructive as rats to everything eatable in houses." Flannery (1990*b*) questioned that statement but noted that native hunters had indicated that *Myoictis* enters villages at night in pursuit of murid rodents.

DASYUROMORPHIA; DASYURIDAE; **Genus DASYUROIDES**
Spencer, 1896

Kowari

The single species, *D. byrnei,* is known to occur in the south of the Northern Territory of Australia, the southwest of Queensland, and the northeast of South Australia (Aslin 1974). *Dasyuroides* was considered a synonym of *Dasycercus* by Corbet and Hill (1991), Groves (*in* Wilson and Reeder 1993), and Mahoney and Ride (*in* Bannister et al. 1988) but was maintained by Morton, Dickman, and Fletcher (1989).

Head and body length is 135–82 mm, tail length is 110–40 mm, and weight is 70–140 grams, with males being about 30 grams heavier than females (Aslin 1974; Aslin *in* Strahan 1983). The back and sides are grayish with a faint rufous tinge. The underparts are pure creamy white, and the feet are white. Less than the basal half of the tail is rufous; the remainder is densely covered on both the upper and lower surfaces with long black hairs that form distinct dorsal and ventral crests. The soft and dense pelage is composed mainly of underfur with few guard hairs. The body form is strong and stout. The hind feet are very narrow and lack a first toe; the bottoms of the feet are hairy. The tail is not thickened. Pouch development is sufficient to conceal the young completely, and there usually are six (five to seven) mammae (Aslin 1974).

The kowari inhabits desert associations and dry grassland. It may shelter in the hole of another mammal or dig its own burrow, and both sexes construct therein a nest of soft materials (Aslin 1974; Ride 1970). It is primarily terrestrial but climbs well and is capable of vertical leaps of at least 45.7 cm (Collins 1973). In the field, Aslin (1974) observed *Dasyuroides* only after dark; in captivity one individual was largely nocturnal, while another was sporadically active both day and night. Collins (1973) stated that the kowari enjoys sun- and sandbathing. He also reported that the diet consists of insects, arachnids, and probably small vertebrates such as birds, rodents, and lizards. Aslin (1974) reported that one wild individual had fed on *Rattus villosissimus* and some insects. Torpor has been induced experimentally in *Dasyuroides* through a moderate reduction in diet, and it is likely that individuals also enter torpor in the wild in response to declining food supplies (Geiser, Matwiejczyk, and Baudinette 1986).

Aslin (1974) estimated that a population of at least 14 adults occupied an area of less than 750 ha. and suggested that some kind of social aggregation might occur in the wild. Collins (1973) observed that a captive pair or small group could be maintained together but that serious fighting might result if the enclosure were too small. Aslin (1974) noted, however, that aggressive behavior is stylized in such a way as to prevent serious injury in intraspecific conflict. She also reported that each sound made by adults—hissing, chattering, and snorting—is restricted to threat situations and that both sexes use scent produced by sternal and cloacal glands to mark parts of their home range.

The following reproductive information is based on studies of both wild and captive individuals (Aslin 1974, 1980; Collins 1973; Woolley 1971*a*). Each female may have up to four estrous periods a year, which recur at two-month intervals if young are not being suckled. Mating takes place from April to December, mostly from May to July, and females may produce two litters in a season. Gestation is 30–36 days. Litter size is 3–7, having averaged 5.1 in captivity and 5.8 in the wild. The young are 4 mm long at birth, first detach from the nipples at 56 days, are weaned and practically independent at about 100 days, reach sexual maturity at about 235 days, and attain adult weight at 1 year. The young may ride on the mother's back or sides when

Kowari *(Dasyuroides byrnei)*, photo by Constance P. Warner.

they are 2–3 months old, and she remains tolerant of them for a while after weaning. Both sexes are capable of breeding through their fourth year of life. One captive kowari lived for 7 years and 9 months (Marvin L. Jones, Zoological Society of San Diego, pers. comm., 1995).

The IUCN now classifies *D. byrnei* as vulnerable but recognizes it as part of the genus *Dasycercus.* Kennedy (1992) reported that it has declined by 50 to 90 percent, apparently because of fox predation and habitat destruction through grazing of domestic livestock.

DASYUROMORPHIA; DASYURIDAE; **Genus DASYCERCUS**
Peters, 1875

Mulgara

The single species, *D. cristicauda,* inhabits the arid region from the Pilbara in northwestern Western Australia to southwestern Queensland (Ride 1970). Remains found in owl pellets indicate that its range may have extended as far as western New South Wales in the nineteenth or early twentieth century (Ellis 1992). Some authorities include *Dasyuroides* (see account thereof) within *Dasycercus.*

Head and body length is 125–220 mm, tail length is 75–125 mm, and weight is 60–170 grams (Woolley *in* Strahan 1983). Individuals of this genus sometimes exceed *Dasyuroides* in size but usually are smaller. The upper parts vary from buffy to bright red brown, and the underparts are white or creamy. The close and soft pelage consists principally of underfur with few guard hairs. The tail is usually thickened for about two-thirds of its length and is densely covered near the body with coarse, chestnut-colored hairs. Beginning at about the middle of its length the tail is covered with coarse, black hairs that increase in length toward the tip to form a distinct dorsal crest.

This desert dweller is compactly built, with short limbs, ears, and muzzle. The pouch area consists of only slightly developed lateral skin folds. The mammae usually number six or eight, or rarely four.

Dasycercus occupies sandridge or stony desert and spinifex grassland. Woolley (1990) reported two kinds of burrows excavated in the sandhills. One had a grass-lined nest area at a depth of about 0.5 meter that was connected to the surface by one large tunnel and several near vertical popholes. This kind of burrow was found to be occupied by a female and pouch young. The other kind, occupied only by a single male, had up to five entrance holes, a complex system of interconnecting tunnels, and one grass-lined nest at a depth of up to 1 meter. The mulgara is terrestrial but is capable of climbing (Collins 1973). It seems to be both diurnal and nocturnal. Ride (1970) noted that it avoided exposure to heat by remaining in its burrow during the hot part of the day. Collins (1973), however, observed that it was partially diurnal, would bask in the sun whenever the opportunity arose, and would even make use of sun lamps. When sunning itself, it flattens its body against the substrate and the tail twitches sporadically. The mulgara can withstand considerable exposure to both heat and cold. However, captives were found to enter torpor for up to 12 hours, during which metabolic rate and body temperature were greatly reduced (Geiser and Masters 1994). Studies of captives indicate that the physiological adaptations of *Dasycercus* are such that it is able to subsist without drinking water or even eating succulent plants because it is able to extract sufficient water from even a diet of lean meat or mice (Ride 1970).

This genus is relatively uncommon but reportedly increases in numbers when a house mouse plague occurs within its range. *Dasycercus* attacks a mouse with lightning action and then devours it methodically from head to tail, inverting the skin of the mouse in a remarkably neat fashion as it does so. The mulgara also preys on other small vertebrates, arachnids, and insects. It reportedly can skillfully dislodge insects from crevices by means of its tiny forepaws.

Dasycercus can be maintained in captivity in pairs except when the females have young (Collins 1973). They generally do not fight among themselves and appear quite solicitous of each other. According to Woolley (*in* Strahan 1983), however, it is usual to find only a single wild individual in a burrow. Females with up to eight young have been captured between June and December. Mating in captivity has been observed from mid-May to mid-June, and young have been born in late June, July, and August after a gestation period of five to six weeks. The young suckle for 3–4 months and become sexually mature at 10–11 months. Individuals of both sexes have been known to come into

Mulgara *(Dasycercus cristicauda),* photo by Shelley Barker. Insets: undersides of hand and foot, photos from *Report on the Work of the Horn Scientific Expedition to Central Australia, Zoology,* B. Spencer, ed.

breeding condition each year for 6 years, suggesting a fairly long life span.

Archer (1979*a*) considered the mulgara to be a vulnerable or possibly endangered species. It apparently has disappeared or become very rare in Queensland and South Australia. Its decline may be associated with habitat disturbances and predation caused by the European introduction of livestock, cats, foxes, and rabbits (Kennedy 1992). It now is classified as vulnerable by the IUCN. *D. hillieri,* restricted to the vicinity of Lake Eyre and subject to extreme population fluctuations, is recognized as a separate species by the IUCN and is classified as endangered.

DASYUROMORPHIA; DASYURIDAE; **Genus DASYURUS**
E. Geoffroy St.-Hilaire, 1796

Native "Cats," Tiger "Cats," or Quolls

There are six species (Archer 1979*a;* Kirsch and Calaby 1977; Ride 1970; Van Dyck 1987; Ziegler 1977):

D. hallucatus, northern parts of Western Australia and Northern Territory, northern and eastern Queensland;
D. viverrinus, southeastern Australia, Tasmania, Kangaroo Island, King Island;
D. geoffroii, Australia except extreme north;
D. spartacus, southwestern Papua New Guinea;
D. albopunctatus, New Guinea;
D. maculatus, eastern Australia, Tasmania.

These species sometimes were placed in separate genera or subgenera as follows: *D. maculatus* in *Dasyurops* Matschie, 1916; *D. viverrinus* in *Dasyurus* E. Geoffroy St.-Hilaire, 1796; *D. geoffroii* in *Dasyurinus* Matschie, 1916; and *D. hallucatus* and *D. albopunctatus* in *Satanellus* Pocock, 1926. Based on paleontological data, Archer (1982*a*) again recognized *Satanellus* as a full genus. This view was not accepted by Van Dyck (1987), who considered *D. albopunctatus* to be closely related not to *D. hallucatus* but to his newly described *D. spartacus.* The assignment of all of the above species to *Dasyurus* was accepted by Corbet and Hill (1991), Groves (*in* Wilson and Reeder 1993), Mahoney and Ride (*in* Bannister et al. 1988), and Morton, Dickman, and Fletcher (1989).

Measurements are: *D. hallucatus* and *D. albopunctatus,* the two smallest species, head and body length approximately 240–350 mm, tail length 210–310 mm; *D. spartacus,* head and body 305–80 mm, tail 240–85 mm; *D. viverrinus,* head and body 350–450 mm, tail 210–300 mm; *D. geoffroii,* head and body 290–650 mm, tail 270–350 mm; and *D. maculatus,* the largest species, head and body about 400–760 mm, tail 350–560 mm. Weights in a series of specimens of *D. viverrinus* were 850–1,550 grams for males and 600–1,030 grams for females (Green 1967). An adult female *D. geoffroii* weighed 550 grams. Weight in *D. maculatus* usually is about 2–3 kg.

The upper parts are mostly grayish, or olive brown to dark rufous brown. In *D. viverrinus* there is a less common black phase that appears to be unrelated to sex, often in the same litter with individuals of the more common color. Regardless of the basic coloration, all individuals of the genus have prominent white spots or blotches on the back and sides. Only in *D. maculatus* do the spots usually extend well onto the tail. In *D. geoffroii* the face is paler and grayer than the upper parts, the spots extend onto the head, and the terminal half of the tail is black. In *D. hallucatus* and *D. albopunctatus* the tip and entire ventral surface of the tail are dark brown or black. In *D. spartacus* the body color is deep bronze to tan brown with small white spots, and the tail is black with no spots. In all species the underparts are paler than the back, usually yellowish or white. The coat generally is short, being soft and thick in *D. viverrinus* and *D. geoffroii* and coarse with little underfur in *D. hallucatus* and *D. albopunctatus.*

The species *D. viverrinus* differs from all the others in lacking a first toe on the hind foot. The pads of the feet are granulated in *D. viverrinus* and *D. geoffroii* and striated in the other species. In *D. hallucatus* and *D. albopunctatus* pouch development consists only of lateral folds of skin around the 8 mammae. In the other species there is a shallow pouch formed by a flap of skin and usually 6 mammae (sometimes 8 in *D. viverrinus*).

Habitats include dense, moist forest for *D. maculatus,* drier forest and open country for *D. viverrinus,* savannah for *D. geoffroii,* woodland and rocky areas for *D. hallucatus,* low savannah for *D. spartacus,* and a variety of conditions up to an altitude of 3,500 meters for *D. albopunctatus* in New Guinea. All species are primarily terrestrial but able to climb well (Collins 1973). All are nocturnal, but some occasionally are seen by day. *D. maculatus* may follow its prey for over 4 km in one night. By day *D. viverrinus* shelters in rock piles or hollow logs and *D. hallucatus* reportedly nests in hollow trees or even abandoned buildings. Female *D. geoffroii* excavate burrows in which to give birth, the main tunnel descending to a nursery chamber nearly 1 meter below the ground (Serena and Soderquist 1989*a*). All species are predators but also will eat vegetable matter (Ride 1970). The diet of the smaller species contains a high proportion of insects, while *D. maculatus* may take mammals as large as wallabies *(Thylogale).* Native "cats" occasionally raid poultry yards and are therefore disliked by farmers, but they probably also benefit human interests by destroying many mice and insect pests (Ride 1970).

Radio-tracking studies of *D. geoffroii* in Western Australia (Serena and Soderquist 1989*b*) showed females to use home ranges averaging about 337 ha., including a core area of about 90 ha. with numerous dens, where most activity is concentrated. There was usually little or no overlap of these core areas, suggesting that females are intrasexually territorial, but sometimes a daughter shared her mother's core area. Males used much larger ranges, including core areas of at least 400 ha., and there was broad overlap of their core areas with those of both other males and females. Both sexes were found to be essentially solitary and to scent-mark with feces.

In captivity all 6 species can be maintained as pairs except when the female has young (Collins 1973). Most native "cats" appear to be monestrous, winter breeders, producing a single litter between May and August (Ride 1970). There is a lengthy courtship in *D. maculatus,* during which the female may be bitten severely about the head and shoulders. In Victoria the pouch of females of this species enlarges in June and July whether young are born or not. The estrous cycle is 21 days long (Hayssen, Van Tienhoven, and Van Tienhoven 1993), gestation also is 21 days, and litter size is four to six. The young first detach from the nipples at 7 weeks, are one-third grown and independent at 18 weeks, attain sexual maturity at 1 year, and reach full size by 2 years.

Green (1967) collected pouch young of *D. viverrinus* in Tasmania during June and July. A few litters contained 5 young, but most had 6, the same as the number of mammae. Females of this species, however, reportedly some-

Australian native cats: Top, *Dasyurus viverrinus,* photo from U.S. National Zoological Park; Middle, *D. geoffroii;* Bottom, *D. hallucatus,* photos by Shelley Barker.

Australian native cat *(Dasyurus (Satanellus) albopunctatus)*, photo by P. A. Woolley and D. Walsh.

times have 8 mammae and 8 young and occasionally produce up to 24 embryos, far more than can be nurtured. Fletcher (1985) reported that mating activity in captive *D. viverrinus* reaches a peak in late May and early June, that females of this species are polyestrous, and that one female had an estrous cycle of 34 days. There are old and highly doubtful reports that gestation in *D. viverrinus* is as short as 8 days, but Fletcher found the period to vary from 20 to 24 days. The young first detach from the nipples at 8 weeks and become independent by 18 weeks (Collins 1973). There also is evidence of polyestry in *D. geoffroii* (Serena and Soderquist 1990). In that species gestation is 16–23 days, up to 6 young are born, they separate from the mother's nipples by 62 days, and females reach sexual maturity at 1 year (Collins 1973; Ride 1970; Serena and Soderquist 1988). The 6–8 young of *D. hallucatus* measure only 3 mm in length, are weaned after 3 months, and reach full maturity by 10–11 months (Collins 1973).

In a study of *D. hallucatus* in Western Australia, Schmitt et al. (1989) found females to produce a single litter in July or August; lactation continued through November. Most males died during the post-mating period from July to September, evidently because of high testosterone levels and resulting trauma, as has been reported for *Antechinus* (see account thereof). However, males with lower testosterone levels, apparently the less dominant individuals, had increased prospects of survival. Dickman and Braithwaite (1992) determined that a given population of *D. hallucatus* may experience a complete male die-off in some years but not in others. The following longevity data are available for native "cats": individuals of *D. maculatus* have been maintained for 3–4 years in zoos (Collins 1973); maximum known life span of *D. viverrinus* is 6 years and 10 months (Jones 1982); a specimen of *D. geoffroii* was kept in captivity more than 3 years (Archer 1974*b*); and *D. albopunctatus* has been maintained for up to 3 years (Collins 1973).

All species of native "cats" have suffered since the settlement of Australia by Europeans and evidently are continuing to decline. The IUCN classifies *D. geoffroii, D. spartacus, D. albopunctatus,* and *D. maculatus* generally as vulnerable and *D. hallucatus* and *D. viverrinus* as near threatened. The subspecies *D. maculatus gracilis,* of northeastern Queensland, is singled out as endangered. *D. viverrinus* also is listed as endangered by the USDI. *D. geoffroii* survives only in remote parts of southwestern Western Australia, central Australia, and Queensland. During the 1930s *D. viverrinus* was common on the southeastern mainland, being found even in the suburbs of large cities such as Adelaide and Melbourne. It subsequently became very rare in this region (Emison et al. 1975; Grzimek 1975). In Tasmania, however, *D. viverrinus* remains common and even has increased in numbers in some areas (Green 1967). In addition to direct human persecution and clearing of cover, native "cats" have been adversely affected by introduced placental predators and competitors, such as domestic cats and dogs and foxes. During the first decade of the twentieth century a severe epidemic seems to have greatly reduced the populations of native "cats," as well as certain other marsupials, in southeastern Australia and Tasmania, and in some areas there apparently never was a substantial recovery. *D. maculatus* is now uncommon to rare on the mainland, having declined through such factors as destruc-

Large spotted native "cat" *(Dasyurus maculatus)*, photo by C. W. Turner through Basil Marlow.

tion of forests and widespread trapping and poisoning (Mansergh 1984*c*).

DASYUROMORPHIA; DASYURIDAE; **Genus SARCOPHILUS**
E. Geoffroy St.-Hilaire and F. Cuvier, 1837

Tasmanian Devil

The single living species, *S. harrisii,* is now found only in Tasmania. Once it also occupied much of the Australian mainland (Calaby and White 1967), but it probably disappeared before European settlement, through competition with the introduced dingo *(Canis familiaris dingo).* Subfossil remains indicate that it or a closely related species still was present as recently as 3,120 years ago in the Northern Territory (Dawson 1982*a*), 400 years ago in southwestern Australia (Baynes 1982), and 600 years ago in Victoria (Guiler *in* Strahan 1983). Living specimens collected in Victoria in 1912 and 1971 are generally thought to have escaped from captivity (Guiler 1982; Ride 1970).

Head and body length is about 525–800 mm and tail length is about 230–300 mm. A series of specimens of males weighed 5.5–11.8 kg and a series of females weighed 4.1–8.1 kg (Green 1967). The coloration is blackish brown or black except for a white throat patch, usually one or two white patches on the rump and sides, and the pinkish white snout.

The general form, except for the tail, resembles that of a small bear. The head is short, broad, and covered with masses of muscle; the skull and teeth are extremely massive and rugged. The molar teeth are developed into heavy bone crushers in a manner reminiscent of the hyenas. The first toe is not present, and the granular pads of the feet lack striations. The pouch during the breeding season is a completely closed receptacle, unlike that of many dasyurids. There are 4 mammae.

Sarcophilus is found throughout Tasmania, except in areas where cover has been extensively cleared, and is most numerous in coastal heath and sclerophyll forest (Guiler 1970*a*). It is nocturnal, sheltering by day in any available cover, such as caves, hollow logs, wombat holes, or dense bushes. Both sexes make nests of bark, button grass, or leaves. The Tasmanian devil is terrestrial but capable of climbing and occasionally emerges from its lair to bask in the sun (Collins 1973). Its movements are slow and clumsy, and its habit of continually nosing the ground suggests a well-developed sense of smell (Ride 1970). The diet consists of a wide variety of invertebrates and vertebrates, including poisonous snakes, and a small amount of plant material. Guiler (1970*a*) found the main foods to be wallabies (*Macropus* and *Thylogale*), wombats, sheep, and rabbits, most of which were taken as carrion. Both laboratory and field observations indicate that *Sarcophilus* is an inefficient killer that does not usually hunt large, active prey (Buchmann and Guiler 1977). It is, however, an efficient scavenger of carrion, consuming all parts of a carcass, including the fur and bones. It is said formerly to have fed on the remains of animals killed by the larger Tasmanian wolf *(Thylacinus).*

Population densities in two parts of Tasmania were about 3/sq km and 25/sq km, the latter figure being considered abnormally high (Buchmann and Guiler 1977; Guiler 1970*a*). In an area of abundant food home ranges were found to be small, with individuals traveling about 3.2 km during a night. In an area of less food home ranges were larger and nocturnal movement covered about 16 km. There was considerable overlap in home ranges, both between and among sexes, and no evidence of territorial behavior (Guiler 1970*a*).

As in so many animals, disposition of *Sarcophilus* is individually variable; some seem quite vicious, others more tractable. Eric R. Guiler, of the University of Tasmania, reported (pers. comm., 1972) that most of the more than 7,000 Tasmanian devils he handled were docile to the point of being lethargic and could be handled with ease; he said that ferocity has been greatly exaggerated. Captives, however, generally are highly aggressive toward their own kind, and there is severe fighting over food (Buchmann and Guiler 1977). Agonistic encounters are accompanied by much vocalization, starting with low growls and proceeding to a rising and falling vibrato or even loud screeching. Eventually two individuals may form a stable dominant-subordinant relationship with reduced aggression. Observations of wild animals congregated to feed on carcasses indicated considerable agonistic interaction but few serious physical clashes (Pemberton and Renouf 1993). Collins (1973) reported that pairs could be maintained in captivity if the animals were gradually introduced to each other. Turner (1970) observed that adult males would assist females in cleaning the offspring and that after the young were too big to fit in the mother's pouch they might cling to the back of either parent. Guiler (1971*c*) cautioned, however, that at a later stage the father might devour the young.

Sarcophilus is monestrous. Most mating occurs in March, but some may take place a little later. The gestation period is about 31 days and litter size is two to four. The young weigh 0.18–0.29 grams at birth, can first release the

Tasmanian devil *(Sarcophilus harrisii)*, photo from U.S. National Zoological Park.

nipples at 90 days, leave the pouch at about 105 days, and are weaned by 8 months (Collins 1973; Guiler 1970*b*, 1971*c*). Females do not become sexually mature until they are two years old. Maximum known longevity is eight years and two months (Jones 1982).

The Tasmanian devil had become rare by the first decade of the twentieth century, apparently because of persecution by European settlers, destruction of forest habitat, and a severe epidemic in the early twentieth century. There was concern that it, along with the Tasmanian wolf, might be facing extinction. More recently, however, *Sarcophilus* has recovered and even become abundant in many areas. Its increase in numbers seems to have resulted partly from newly available food supplies in the form of carrion from livestock and commercial trapping operations (Green 1967; Guiler 1982; Ride 1970). The Tasmanian devil is sometimes considered a nuisance and has been subject to official control measures. Buchmann and Guiler (1977), however, stated that its reputation as a killer of livestock is unmerited, that sheep probably are eaten only as carrion, and that money spent on control is wasted.

DASYUROMORPHIA; **Family MYRMECOBIIDAE; Genus MYRMECOBIUS**
Waterhouse, 1836

Numbat, or Banded Anteater

The single known genus and species, *Myrmecobius fasciatus*, occurred in suitable areas from southwestern Western Australia, through northern South Australia and extreme southwestern Northern Territory, to southwestern New South Wales (Friend 1989; Ride 1970). This genus sometimes has been placed in the family Dasyuridae, but the sum of available information indicates that it represents a distinct family (Archer and Kirsch 1977). There are Pleistocene records of *M. fasciatus* from Australia (Archer 1984); otherwise the Myrmecobiidae are known only from the Recent.

Head and body length is 175–275 mm, tail length is 130–200 mm, and usual weight is 300–600 grams. In *M. fasciatus fasciatus*, the subspecies that inhabits southwestern Western Australia, the foreparts of the body are grayish brown with some white hairs. In *M. fasciatus rufus*, which occurred farther to the east, the main body color is rich brick red. Both subspecies have six or seven white bars between the midback and the base of the tail, producing an effect of transverse light and dark stripes. There is a dark cheek stripe running through the eye, with a white line above and below. The body pelage is short and coarse, and both the head and hindquarters are remarkably flattened above. The body is larger posteriorly than anteriorly. The tail is long, somewhat bushy, and nonprehensile. It is covered with long, stiff hairs that are sometimes bristly and form a brush. The snout is long and tapering, the mouth small; the slender tongue can be extended at least 100 mm. The tip of the nose is naked. The front legs are comparatively thick and widely spaced; the forefoot has five toes and the hind foot has four, all bearing strong claws. A complex gland on the chest opens onto the skin through a number of conspicuous pores. The female has four mammae but lacks any suggestion of a pouch. When the young are attached to her nipples they are protected only by the long hair on her underparts. She supplies them with milk and warmth, but they are dragged beneath her as she travels.

There may be 4 or 5 upper molar teeth and 5 or 6 lower molars (Archer and Kirsch 1977). Therefore, the total number of teeth may be as high as 52, more than that of any other land mammal. The overall dental formula is: (i 4/3, c 1/1, pm 3/3, m 4–5/5–6) × 2 = 48–52. Most of the teeth are small, delicate, and separated from one another. The size is not constant; the molars on the right and left sides of the skull often vary in length and width. The numbat's bony palate extends farther back than in many mammals and may be associated with the long, extensible tongue. This type of palate is also present in the armadillos (Xenarthra,

Numbat *(Myrmecobius fasciatus)*, photo by L. F. Schick.

Dasypodidae) and the pangolins, or scaly anteaters (Pholidota, Manidae). According to Friend (1989), the palate has no vacuities, the alisphenoid tympanic wing forms virtually the entire floor of the middle ear, processes from the frontal and jugal bones form an incomplete bony bar behind the orbit of the skull, and the lachrymal bone is very large and extends a long way out onto the face.

The numbat inhabits open scrub woodland, generally where eucalyptus trees predominate, and desert associations. There had been some question whether this animal constructs its own burrow. Available evidence now suggests that the female may dig a short tunnel with a chamber at the end where she builds a nest and gives birth to her young (Christensen 1975; Ride 1970). The numbat also is known to shelter by night in a bed of leaves or grass placed in a hollow log or the burrow of some other animal.

Radio-tracking studies by Christensen, Maisey, and Perry (1984) show that the numbat uses logs for overnight shelter during warm weather and changes to burrows when the weather becomes cold and that an individual may utilize many such shelters. One burrow consisted of a tunnel about 1 meter long that opened to a nest chamber about 200 mm across and 100 mm below ground level. These studies indicate that home ranges are large, with some individuals using over 100 ha. during a two-month period.

Unlike most marsupials, *Myrmecobius* is active during the day. It is nimble, climbs readily, and spends much of its time in search of food. It trots and leaps about with jerky movements and carries its tail in line with its body, but with a slight upward curve. When the animal is startled or frightened it may sit bolt upright, flatten its entire body and fluff its tail, or run into a hollow log. When caught, it reportedly emits snuffling and hissing sounds but does not attempt to bite. During cooler months it often basks in the sun.

Termites and ants form the regular diet, and other invertebrates are eaten only occasionally. Termites actually are the preferred food; most of the ants consumed are of small predatory species that rush in when the numbat uncovers a termite nest and are lapped up along with the termites. A captive numbat consumed from 10,000 to 20,000 termites of the smaller species daily. These are swallowed whole, whereas those of the larger species are masticated. Termites are obtained from rotten logs, dead trees, and subsurface soil. The strong foreclaws are used in scratching into soil and decayed wood, and the long snout sometimes functions as a lever. The extensible, cylindrical tongue extracts the termites from crevices in the wood and is manipulated with great speed and dexterity. In southwestern Australia, wandoos *(Eucalyptus redunca elata)* are often attacked by a species of termite, so that the woodland floor is littered with hollow branches from the infested trees. These logs provide both food and shelter for the numbat.

Except during the breeding season the numbat is usually solitary, though Collins (1973) observed that captives could be maintained in pairs. The gestation period is 14 days (Friend 1989). In southwestern Australia the young are born from January to April or May. Litters of two to four have been recorded, but four seems to be the usual number. The young are carried by the female, attached to her nipples, for about four months and then are suckled in a nest for about two more months. Juveniles have been seen foraging for themselves by October. A male in the Taronga Zoo in Australia survived for five years, and its mate was still living after six years in captivity (Strahan 1975).

The numbat is classified as vulnerable by the IUCN and as endangered by the USDI. According to J. A. Friend (1990*a*), the decline of the species began in the east shortly after European settlement around 1800 and moved progressively westward. It now has disappeared entirely from New South Wales, South Australia, and the Northern Territory and survives only in a few suitable areas in extreme southwestern Western Australia. The eastern subspecies, *M. f. rufus*, is apparently extinct. There had been speculation that the main cause of the decline was clearing of land for agriculture, which eliminates the dead and fallen trees from which termites can be obtained; however, the continued abundance of the echidna *(Tachyglossus)* in the same region suggests no shortage of food for a mammal that depends on termites and ants. Probably an overriding factor in the decline of the numbat has been predation by introduced carnivores, especially the red fox *(Vulpes vulpes)*, exacerbated by the loss of vegetative cover to fire and clearing. Unlike the echidna, which can dig into the ground and is protected by its spiny pelage, the numbat depends on cover to escape predators. The numbat actually remained common in many of the arid parts of South Australia and Western Australia until about 1930 but then underwent a drastic population crash in the 1940s and 1950s. This decline coincided both with the establishment of the red fox in the region and with the departure of the aboriginal people from the deserts. These people traditionally had burned off small patches of vegetation throughout the year, but the cessation of this practice led to a buildup of scrub and eventually to huge summer wildfires that destroyed all cover. In addition, introduced rabbits *(Oryctolagus)* increased in the same region and further encouraged numbers of fox and domestic cat *(Felis catus)*. Even in the 1970s the numbat was locally common on reserves in southwestern Western Australia, but populations there underwent another crash at that time and disappeared entirely from some reserves, again apparently because of fox predation. Subsequent control of the fox has allowed a modest increase and successful reintroduction of the numbat in some areas.

DASYUROMORPHIA; **Family THYLACINIDAE; Genus THYLACINUS**
Temminck, 1824

Thylacine, or Tasmanian Wolf or Tiger

The single Recent genus and species, *Thylacinus cynocephalus,* apparently was found only in Tasmania within historical time. There were many reports of animals bearing some resemblance to the thylacine from the Australian mainland during the eighteenth, nineteenth, and twentieth centuries (Heuvelmans 1958; Ride 1970); none have been confirmed, though Archer (1984) noted that there are some hints that a population may have survived in the north until the early twentieth century. During the late Pleistocene and early Recent periods the genus was widespread in Australia and New Guinea, with the latest definite subfossil mainland records dating from just over 3,000 years B.P. The most likely cause for the disappearance of the thylacine from the mainland and New Guinea was competition with domestic dogs introduced by aboriginal human populations starting several thousand years ago (Archer 1974*a;* Dawson 1982*b;* Partridge 1967). These dogs, which in Australia have the name "dingo" *(Canis familiaris dingo)*, became feral and spread over a large region but did not enter Tasmania.

The thylacine often was placed in the family Dasyuridae but subsequently came to be considered distinct from that group and to have closer affinity with three extinct South American families of large predatory marsupials (Kirsch 1977*b*). Those three families, plus the Thylacinidae, were then grouped in the superfamily Borhyaenoidea, which was thought to have originated before the process of continental drift had separated South America, Antarctica, and Australia. More recent studies of morphology, serology, and mitochondrial DNA (Archer 1982*b*, 1984; Krajewski et al. 1992; Sarich, Lowenstein, and Richardson 1982; Szalay 1982) have indicated once again that the Thylacinidae are most closely related to the Dasyuridae, though the two groups are considered very distinct families within the order Dasyuromorphia.

Head and body length of *Thylacinus* is 850–1,300 mm, tail length is 380–650 mm, shoulder height is 350–600 mm, and weight is 15–30 kg (Moeller *in* Grzimek 1990). The up-

Tasmanian wolf *(Thylacinus cynocephalus)*, photo from U.S. National Zoological Park.

per parts are tawny gray or yellowish brown with 13–19 blackish brown transverse bands across the back, rump, and base of the tail; the underparts are paler. The face is gray with some indistinct white markings around the eyes and ears. The fur is short, dense, and coarse. The dental formula is: (i 4/3, c 1/1, pm 3/3, m 4/4) × 2 = 46. The unusual dental features, as noted below, distinguish *Thylacinus* from the Dasyuridae. In addition, the epipubic bones of *Thylacinus* are vestigial and cartilaginous, the clavicles are reduced, and the foramen pseudovale is lacking. Unlike that of the Myrmecobiidae, the palate of *Thylacinus* has large vacuities (Marshall 1984).

With its general external resemblance to a canid, the thylacine represents one of the most remarkable examples of convergent evolution found in mammals. Its scientific name means "pouched dog with a wolf head." The jaws have a remarkably wide gape. Long canine teeth, shearing premolars, and grinding molars show further similarity with the dog family. The overall build is doglike, but the hind end tapers gradually, rather than abruptly, to the rigid tail. The feet also resemble those of dogs, and like the canids, the thylacine is digitigrade, walking on its toes. The pads of the feet are granulated, not striated, and the forefoot leaves a five-toed print. In females there is a rear-opening pouch consisting of a crescent-shaped flap of skin enclosing the four mammae. Despite the superficial appearance of the thylacine, an analysis of its skeletal proportions by Keast (1982) indicated that it is essentially like a large dasyurid and does not have the specialized pursuit adaptations of the placental wolves *(Canis)*.

The preferred habitat of the thylacine probably was open forest or grassland, but its last populations may have occupied dense rainforest in southwestern Tasmania. Its lair reportedly was in a rocky outcrop or a hollow log. Although nocturnal, it was sometimes seen basking in the sun. The thylacine seems to have been mostly solitary but sometimes reportedly hunted in pairs or small family groups. It was said to trot relentlessly after its prey until the victim was exhausted and then to close in with a rush. It fed mainly on kangaroos, wallabies, smaller mammals, and birds. Various calls were reported, including a whine (which may have been for communication), a low growl expressing irritation, and a coughing bark when hunting.

Based on a study of bounty records, Guiler (1961) found that births occurred throughout the year, with a pronounced peak in the summer (December–March). The two to four young were thought to leave the pouch after three months but to remain with the mother until they were about nine months old. Although many individuals were successfully maintained in zoos, the thylacine was never bred in captivity. Record longevity was nearly 13 years (Moeller *in* Grzimek 1990).

As in the case of most large predatory mammals, there is controversy regarding the extent of damage the thylacine caused to domestic livestock. Nonetheless, it was generally considered to be a sheep killer in Tasmania, and intensive private bounty hunting began about 1840. By 1863 the thylacine already had been restricted to the mountains and more inaccessible parts of the island. From 1888 to 1909, 2,184 thylacines were destroyed for payment of a government bounty then in effect, and many more were killed for the private bounties. A rapid decline in the population was noticed about 1905, and by 1914 the thylacine was a rare species. In addition to human hunting, trapping, and poisoning, the demise of the thylacine has been blamed on disease, habitat modification, and increased competition from the settlers' domestic dogs.

The last definite record of a wild individual being killed was in 1930 (Ride 1970). The last captive animal died in 1936 (Paddle 1993). Several organized searches made over the next 40 years failed to find conclusive evidence that the species still existed. A number of recent reports, including the supposed killing of a thylacine in western Tasmania in 1961, were doubted by Brown (1973). Many alleged sightings both in Tasmania and on the mainland of Australia were discussed in Rounsevell and Smith (1982) and Smith (1982). Some persons still have hope that the Tasmanian wolf survives, and a 647,000-ha. game reserve was established in the southwestern part of the island in 1966, partly to protect the species. The thylacine has received complete legal protection since 1938, is designated as endangered by the USDI, and is on appendix 1 of the CITES. It now is classified as extinct by the IUCN.

The geological range of the family Thylacinidae is middle Miocene to Recent in Australasia (Archer 1984).

Order Peramelemorphia

Bandicoots

This order of 2 families with 8 Recent genera and 22 species occurs in Australia, Tasmania, and New Guinea and on certain nearby islands. Until recently the group generally was considered a suborder or superfamily within a larger assemblage of syndactylous marsupials that also included the Diprotodontia. However, Aplin and Archer (1987) explained that there evidently is closer affinity to the Dasyuromorphia and that in any case full ordinal status is warranted. The latter procedure was followed by Groves (*in* Wilson and Reeder 1993) though not by Szalay (1994). Marshall, Case, and Woodburne (1990) accepted this order but used the name Peramelina for it. There also have been suggestions of a direct relationship between the Peramelemorphia and the New World Didelphimorphia (Gordon and Hulbert 1989). A history of the varying systematic outlook on the group was provided by Strahan (*in* Seebeck et al. 1990).

The term "bandicoot" is a corruption of a word in the Telugu language of the people of the eastern Deccan Plateau of India meaning "pig rat," which originally was applied to a large species of rodent *(Bandicota indica)* from India and Sri Lanka. It seems likely that the name was first applied to the Australian marsupials by the explorer Bass in 1799, and the use of the name is of questionable desirability because of confusion with *Bandicota.* It has been suggested that the semantic association of marsupial bandicoots with placental rats has reduced the popularity of the former as subjects of natural history investigation. This situation now seems to have changed; detailed summaries of recent studies of the ecology, behavior, physiology, morphology, and systematics of the Peramelemorphia have been provided by Gordon and Hulbert (1989) and Lyne (1990).

The genera expressing both extremes of size in this order are endemic to New Guinea. The smallest is *Microperoryctes,* with a head and body length of 150–75 mm; the largest is *Peroryctes,* with a head and body length of up to 558 mm and a weight sometimes over 4.7 kg. The muzzle is elongate and pointed. The hindquarters are not as elongated as in the kangaroos (Macropodidae), but the hind limbs are larger than the forelimbs. The tail is hairy, of variable length, and nonprehensile.

The second and third digits of the hind foot are syndactylous, that is, bound together by integument; only the tops of the joints and the nails are separate. The combined toes function somewhat like a single digit and are useful in grooming. The main digit of the hind foot is the fourth; the fifth is well developed in most species but is shorter than the fourth; the first toe of the hind foot lacks a nail and is generally poorly developed, and in the genera *Macrotis* and *Chaeropus* it is absent. Except in *Chaeropus* the forefoot has five digits: the third is the longest, the second is slightly shorter, the fourth is somewhat shorter than the second, and the first and fifth are vestigial. In *Chaeropus* the second and third digits of the forefoot are about the same size, the fourth is vestigial, and the first and fifth are absent. Those digits of the forefoot that are well developed have sharp nails for digging.

Although the Peramelemorphia are syndactylous, like the Diprotodontia, their dentition is polyprotodont, as in the Dasyuromorphia. The incisors are numerous and small; the uppers are flattened and unequal, and the lowers slope forward; the crown of the last lower incisor has two lobes. The canines are slender and pointed. The premolars are narrow and pointed, and the molars are four-sided or triangular in outline with four external cusps. The dental formula is: (i 4–5/3, c 1/1, pm 3/3, m 4/4) × 2 = 46 or 48.

The members of this order are alert and sprightly, generally traveling on all four feet with a galloping action. They are terrestrial and mainly nocturnal, sheltering by day in grassy nests on the surface of the ground. Most species have mixed feeding habits. They drink water in captivity, but dew seems sufficient for their needs in the wild when water is not present.

The Peramelemorphia are the only marsupials in which a chorioallantoic placenta develops after the transient yolk-sac placenta. This chorioallantoic placenta, however, unlike that of placental mammals, lacks villi. It evolved independently from the placenta of the latter group and is probably functionally correlated with a high rate of reproduction; the Peramelemorphia are characterized by the shortest gestation period known in mammals, rapid litter succession in polyestrous females, rapid development of pouch young, minimal parental care, and early sexual maturity (Gordon and Hulbert 1989). Females have 6, 8, or 10 mammae (usually 8) and generally have two to five young per litter. The pouch opens downward and backward. Newly born young of some bandicoots have deciduous claws that are shed soon after the young reach the pouch.

Bandicoots have suffered to varying degrees since the settlement of Australia by Europeans. Several kinds have been adversely affected by such factors as habitat modification and introduced predators. In general, those that live in the coastal forests and woodlands of Australia and in Tasmania have survived well. In contrast, those that live on the inland plains and deserts have declined severely (Aitken 1979).

The geological range of the Peramelemorphia is middle Miocene to Recent in Australasia (Archer 1984).

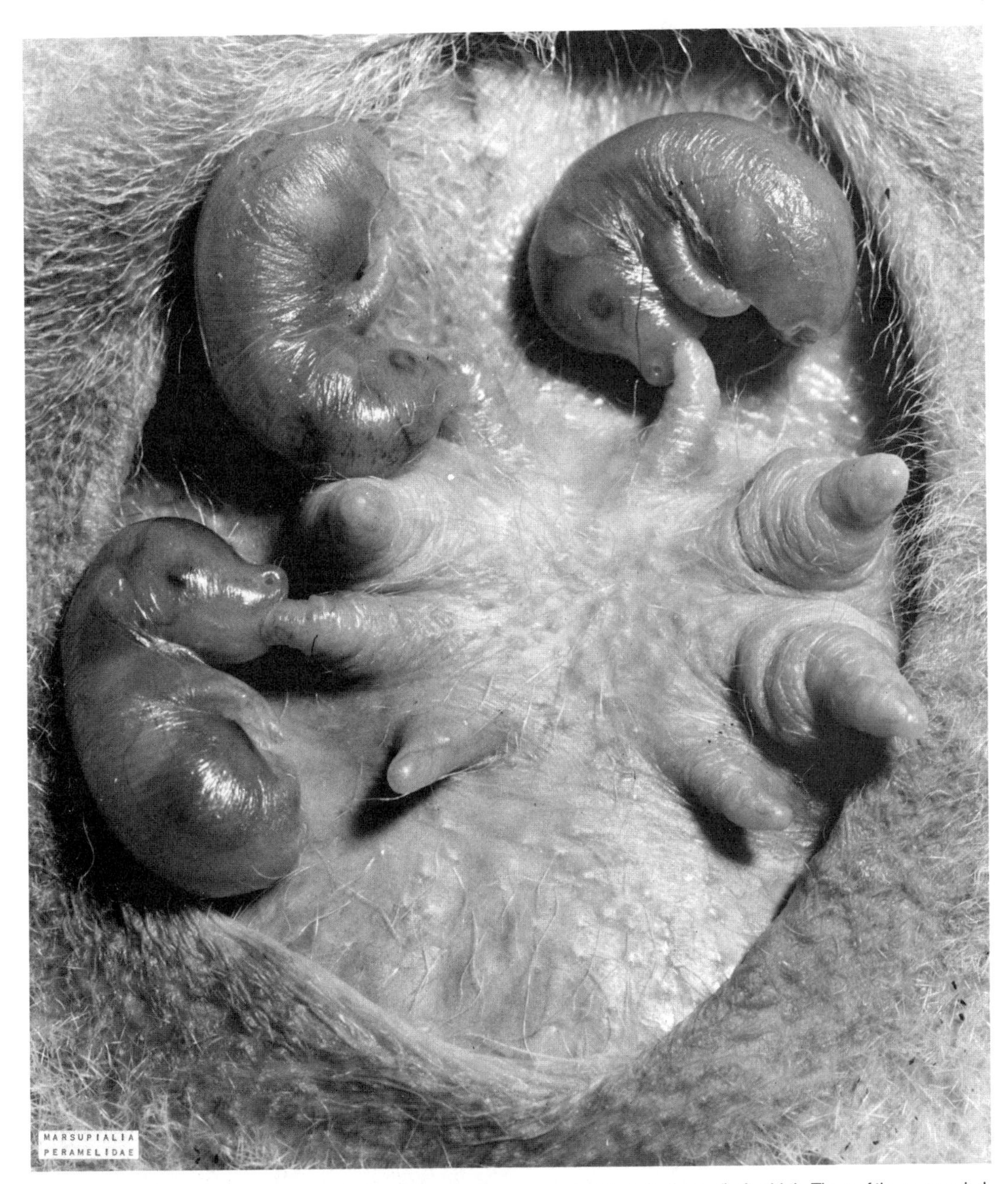

Long-nosed bandicoot *(Perameles nasuta).* Litter of three pouch young at four days (estimated) after birth. Three of the unoccupied teats are still enlarged and were suckled by a previous litter. Photo by A. Gordon Lyne through Australian Commonwealth Scientific and Industrial Research Organization.

PERAMELEMORPHIA; **Family PERAMELIDAE**

Dry-Country Bandicoots

This family of 4 Recent genera and 11 species is found in Australia and southern New Guinea. Until recently this family generally was considered to include the Peroryctidae, but Groves and Flannery (1990) regarded the latter group as a separate family. In addition, Archer and Kirsch (1977) had placed the genus *Macrotis* in still another family, the Thylacomyidae, but such arrangement was not considered appropriate by Groves and Flannery (1990) or Baverstock et al. (1990). Szalay (1994) did not accept the Peroryctidae as a full family and treated the Thylacomyidae as a subfamily of the Peramelidae. Kirsch et al. (1990) supported recognition of the members of Peroryctidae as a monophyletic group but also indicated that the Thylacomyidae are even more distinct from the Peramelidae.

Head and body length is 200–550 mm, tail length is 75–275 mm, and weight is around 200–1,600 grams. Coloration varies considerably. Depending on the genus, the pelage may be soft or coarse, the skull may be narrow or relatively broad, the tail may be relatively long or short,

A young long-nosed bandicoot *(Perameles nasuta)* about 54 days old entering the rear-opening pouch of its mother. The tail of a young already in the pouch shows beneath the young that is entering. Photo by A. Gordon Lyne.

and the ears may be long or short. The dental formula is consistently: (i 5/3, c 1/1, pm 3/3, m 4/4) × 2 = 48.

According to Groves and Flannery (1990), the members of this family occur mainly in dry country, while those of the Peroryctidae are found in rainforest. The morphological distinctions between the two families, however, are not thought to express this difference in habitat. The Peramelidae are characterized as being flat-skulled, while the Peroryctidae are cylindrical-skulled. In the Peramelidae the anteromedial orbital margin slopes diagonally outward, forming a remarkable ridge, or crest, along its edge; the facial extension of the lacrimal bone is very narrow, its suture with the maxilla parallel to the antorbital ridge; the lower margin of the mandible is convex; the auditory bullae (bones housing the inner ear) are complete; and the molar teeth are squared.

The geological range of the Peramelidae is Miocene to Recent in Australasia (Gordon and Hulbert 1989).

PERAMELEMORPHIA; PERAMELIDAE; **Genus MACROTIS**
Reid, 1837

Bilbies, or Rabbit-eared Bandicoots

There are two species (Johnson 1989):

M. lagotis (greater bilby), formerly from Western Australia to southwestern Queensland and New South Wales;
M. leucura (lesser bilby), formerly east-central Western Australia, southern Northern Territory, and northern South Australia.

The generic name *Thylacomys* Owen, 1840, sometimes has been applied to these species, but for technical reasons it is no longer considered valid. Based on cranial and dental morphology, serology, and karyology, Archer and Kirsch (1977) had considered *Macrotis* to represent a distinct family. In accordance with the rules of zoological nomenclature, that family is properly called Thylacomyidae even though the name is based on an incorrect generic name. In any event, Aplin and Archer (1987) suggested that the assignment of *Macrotis* to a separate family may have been premature, and Groves and Flannery (1990) restored the genus to the Peramelidae, noting a particularly close relationship to *Chaeropus*. However, new analyses of DNA by Kirsch et al. (1990) support maintenance of the Thylacomyidae.

According to Johnson (1989), head and body length of *M. lagotis* is 290–550 mm, tail length is 200–290, and weight is 600–2,500 grams; respective figures for *M. leucura* are 200–270 mm, 120–70 mm, and 311–435 grams.

Rabbit-eared bandicoot *(Macrotis lagotis)*, photo by Basil Marlow.

Males are much heavier than females (Johnson and Johnson 1983). The pelage is long, silky, and soft. The upper parts are fawn gray, blue gray, or ash gray, often pale vinaceous along the sides, and the underparts are white. The color of the body fur extends onto the base of the tail. In *M. lagotis* the remainder of the proximal half of the tail is black, and the terminal half is white, with a conspicuous dorsal crest of long hairs. The tail of *M. leucura* is entirely white except for a gray line extending from the body onto the upper surface of the base. The end of the tail bears a prominent dorsal crest of hairs, as in *M. lagotis*, and in both species the extreme tip of the tail is naked.

The general form is light and delicate with a long, tapered muzzle. The ears are very long, pointed, and finely furred. As in all the Peramelemorphia except *Chaeropus*, the forefeet have three functional toes that bear stout, curved claws; the remaining toes are small. On the hind foot the first toe is not present, and in this characteristic both *Macrotis* and *Chaeropus* differ from the other genera of the Peramelemorphia. In *Macrotis*, as in the rest of the order, the second and third hind toes are partially united, the fourth toe is the largest, and the fifth is of moderate size. The pouch opens downward and slightly backward, enclosing eight mammae.

Groves and Flannery (1990) pointed out that *Macrotis* shares many characters with *Chaeropus*. However, the skull of *Macrotis* has an extremely broadened braincase and narrowed snout, the most excessively flattened cranium in the Peramelemorphia, and enormous, pear-shaped auditory bullae. The dental formula is identical to that of the other peramelids. Archer and Kirsch (1977) thought that the teeth of *Macrotis* differed from those of the other peramelids in several ways, particularly with regard to the structures involved in the squaring of the molars. Groves and Flannery (1990), however, noted that while the molars of *Macrotis* have undergone an enormous expansion of the metacone, those of *Chaeropus* show a less modified version of the same condition.

Bilbies live in a variety of mainly dry habitats, including woodland, savannah, shrub grassland, and sparsely vegetated desert. Their distribution seems to have been associated with the presence of suitable soils for burrowing. Unlike most peramelids, bilbies are powerful diggers and live mainly in burrows of their own making. These burrows are characteristic in that they usually descend from a single opening as a fairly steep, ever-widening spiral to a depth of 1–2 meters. According to Ride (1970), *M. leucura* burrows only in sandhills, never on flats, and the burrows of *M. lagotis* in Western Australia were found to be most common in shrub grassland. An area occupied by the latter species had 58 burrows, none more than 168 meters from any other. It seems likely that each animal has a number of burrows within its range.

Bilbies are nocturnal and terrestrial. They progress on all four feet by means of a shuffling gait; hopping is facilitated by the relatively long hind feet. These animals usually do not lie down to sleep; instead they squat on their hind legs and tuck their muzzle between their forelegs; the long ears are laid back and then folded forward over the eyes and along the side of the face. They are largely carnivorous, taking mainly insects but also small vertebrates and some vegetable matter. Conical scratching in the earth around trees and bushes gives evidence of their foraging for subterranean insect larvae. Raw or cooked meat, insects, snails, birds, mice, bread, and cake are readily accepted by captives, but roots and fruit usually are rejected.

These bandicoots are found alone or in small colonies usually consisting of a single adult male, a female, and an independent young (Johnson 1989). Burrows generally seem to have only one occupant (Ride 1970) but reportedly sometimes contain a pair or a female with young. Groups with more than one male or with several females can be maintained in captivity if the enclosure is sufficiently large (Collins 1973). In this situation the males form a rigid dominance hierarchy without severe fighting; the dominant male chases its subordinates away from the burrows but freely shares these places with females (Johnson and Johnson 1983).

According to McCracken (1990), *M. lagotis* is physiologically capable of breeding throughout the year and captives have been observed to do so. The former wild populations in the temperate zone of South Australia had a distinct breeding season from about March to May, but the existing populations in arid regions may breed at any time of year, perhaps depending on rainfall and food availability. Ovulation occurs toward the end of lactation; thus births occur as soon as the young of the previous litter are weaned. The average estrous cycle is 20.6 days and the average gestation is 14 days. Litter size is usually one or two young; triplets are rare. Pouch young develop rapidly and have a pouch life averaging about 80 days. The young are then left in a burrow for about 2 more weeks and suckled at regular intervals. Johnson (1989) added that sexual maturity occurs at about 175–220 days in females and 270–420 days in males. The longevity record for a captive *M. lagotis* is 7 years and 2 months (Collins 1973).

Both *M. leucura* and *M. lagotis* are listed as endangered by the USDI and are on appendix 1 of the CITES. The IUCN now designates *M. leucura* as extinct and *M. lagotis* as vulnerable. *M. leucura* has not definitely been collected since 1931. This species formerly was common but was reduced drastically in the early twentieth century through such factors as trapping for its pelt, predation by introduced foxes, and competition with introduced rabbits for burrows. The same problems beset the once even more abundant *M. lagotis*. This species may even have occupied Victoria (Scarlett 1969) as well as suitable habitat all across the southern half of Australia. However, Southgate (1990) reported that it had disappeared from New South Wales by 1912 and from South Australia by about 1970 and is now restricted to a few isolated colonies in Western Australia, the Northern Territory, and southwestern Queensland; the progression of the decline seems to have coincided with the spread of foxes and rabbits. Johnson (1989) suggested that its decline in arid areas is associated with the disappearance of the traditional aboriginal practice of burning off vegetation gradually, thereby producing a patchwork of habitat types in different stages of regeneration. The current regime of large wildfires results in a homogeneity of habitats that is less favorable to the species. Thornback and Jenkins (1982) noted that its numbers may be kept at very low densities by the depletion of food supplies by grazing cattle.

PERAMELEMORPHIA; PERAMELIDAE; **Genus CHAEROPUS**
Ogilby, 1838

Pig-footed Bandicoot

The single species, *C. ecaudatus*, formerly occurred in suitable localities from southwestern and west-central Western Australia, through South Australia and the southern part of the Northern Territory, to southwestern New South Wales and western Victoria (Baynes 1984; Ride 1970). The name *ecaudatus* is an unfortunate misnomer as the original description was based on a mutilated specimen from which the tail was missing.

Pig-footed bandicoot *(Chaeropus ecaudatus)*, photo of mounted specimen by E. F. Aitken.

The head and body length is 230–60 mm and the tail length is 100–140 mm. The hair, though coarse, is not spiny. The upper parts are grizzled gray tinged with fawn, or almost orange brown, and the underparts are white or pale fawn. The tail is gray or fawn below and on the sides and black above. There is an inconspicuous crest above, with a few white hairs toward the tip.

The body form is light and slender. The head is broad with a long, sharply pointed snout. The ears are long and narrow. The limbs are long, slender, and peculiarly developed. The first and fifth digits of the forefeet are absent, the fourth is minute, and the second and third are well developed with long, sharp nails and conspicuous pads under their tips. On the hind feet the first digit is absent, the fifth is barely evident, the fourth is large and long and bears a short, stout nail and a well-developed pad, and the second and third are small and united. The forefeet thus resemble those of Artiodactyla, though in pigs the functional digits are the third and fourth rather than the second and third. The hind feet show some resemblance to those of Perissodactyla, except that in the horses the functional digit is the third, not the fourth.

Ride (1970) described the habitat as sclerophyll woodland, mallee, heath, and grassland. A nest of grass, sticks, and leaves is constructed in a hollow in the ground and lined with soft grasses. *Chaeropus* is cursorial; the gait was described by G. Krefft as resembling that of a "broken-down hack in a canter, apparently dragging the hind quarters after it." It is said to squat in the open with its ears laid back and to run for shelter in hollow logs and trees when chased. *Chaeropus* may not be strictly nocturnal, as on several occasions individuals were seen out in daylight. The genus is omnivorous. Breeding occurs in May or June. The maximum recorded litter size is only two, but females possess eight mammae (Collins 1973).

This genus may well be extinct. The last specimen collected was taken at Lake Eyre in South Australia in 1907. There were unconfirmed reports from central Australia in the 1920s. The decline is thought to have been caused by introduced predators, such as foxes and house cats, destruction of habitat by domestic livestock, and competition with introduced rabbits (Thornback and Jenkins 1982). According to Friend (1990*b*), there is extensive evidence that *Chaeropus* actually persisted in the interior of Western Australia well into the twentieth century. Its final disappearance may not have been until the 1930s or 1940s and may have been associated with the same factors that affected *Isoodon* (see account thereof). The pig-footed bandicoot is designated as extinct by the IUCN and endangered by the USDI and is on appendix 1 of the CITES.

PERAMELEMORPHIA; PERAMELIDAE; **Genus ISOODON**
Desmarest, 1817

Short-nosed Bandicoots

There are four species (Close, Murray, and Briscoe 1990; Ellis, Wilson, and Hamilton 1991; Groves *in* Wilson and Reeder 1993; Lyne and Mort 1981; Ride 1970; Seebeck et al. 1990):

I. macrourus, southern New Guinea, northeastern Western Australia, northern Northern Territory, eastern Queensland, eastern New South Wales;
I. obesulus, southeastern New South Wales, southern Victoria, southeastern South Australia and nearby Kangaroo Island and Nuyts Archipelago, southwestern Western Australia and nearby Recherche Archipelago, Tasmania, West Sister Island in Bass Strait;
I. auratus, northern and eastern Western Australia and nearby Barrow Island, Northern Territory, inland South Australia, western New South Wales;
I. peninsulae, Cape York Peninsula of northern Queensland.

The name *Thylacis* Illiger, 1811, often used instead of *Isoodon* for these species, is considered here to be a synonym of *Perameles. I. peninsulae* was treated as an extremely disjunct population of *I. obesulus* by Groves (*in* Wilson and Reeder 1993) and Seebeck et al. (1990), but Close, Murray, and Briscoe (1990), using chromosome and electrophoretic analysis, found it to be the most distinctive member of *Isoodon.*

In *I. macrourus* head and body length is 300–470 mm, tail length is 80–215 mm, and weight is 500–3,100 grams.

Short-nosed bandicoot *(Isoodon macrourus)*, photo by Stanley Breeden.

In *I. obesulus* head and body length is 280–360 mm, tail length is 90–140 mm, and weight is 400–1,600 grams. In *I. auratus* head and body length is 210–95 mm, tail length is 88–121 mm, and weight is 267–670 grams. The pelage in this genus is coarse and glossy with distinct hairs. The upper parts of *Isoodon* generally are a fine mixture of blackish brown and orange or yellow, and the underparts are yellowish gray, yellowish brown, or white.

An elongated snout is characteristic of all bandicoots, but in this genus the head is relatively broader than in *Perameles* and the jaws are slightly shorter and stouter. The ears are short and rounded. The well-developed pouch opens backward, as in other bandicoots, and the pouch contains eight mammae.

Short-nosed bandicoots inhabit open woodland, thick grass along the edges of swamps and rivers, and thick scrub on dry ridges. They are active and inoffensive animals and are often mistaken for large rats or rabbits from a distance. When walking they move the forelegs and hind legs separately, and they do not bound when running. They often make long tunnels through the grass. Their nests of sticks, leaves, and grass, sometimes mixed with earth, are located on the ground or in a hollow log. These bandicoots level off the surrounding vegetation in making their nests. They are terrestrial and generally nocturnal, but captive *I. macrourus* have been found to be active during part of the day (Collins 1973). *I. obesulus* seems to detect the approach of bad weather, as it will enlarge its nest before heavy rains. After entering or leaving its nest, it closes the opening behind it. This species also has been known to construct a short burrow (Kirsch 1968).

Isoodon feeds in definite areas, the location changing from time to time, as evidenced by scratch marks in the ground. *I. obesulus* uses a rapid movement of its forefeet to crush its living prey. This species feeds on a variety of items, at least in captivity, but prefers insects and worms. Individuals that were not fed for several days refused potatoes, carrots, and turnips despite their hunger.

Population densities of *I. obesulus* have been reported as 0.75–2.0 individuals per 10 ha. in Tasmania (Ride 1970) and about 1.5 per ha. in the Franklin Islands (Copley et al. 1990). Home ranges in this species were found to be 4.05–6.48 ha. for four males and 2.31 ha. for one female. Home ranges of *I. macrourus* in Queensland were 1.7–5.2 ha. for males and 0.9–2.1 ha. for females (Stodart 1977). The species *I. obesulus* has been reported to be highly aggressive and territorial, with probably very little overlap in home ranges of individuals of the same sex. Territories of males have been reported to be larger than those of females, with each male's territory probably overlapping those of several females (Ride 1970). However, recent field studies suggest that the degree of range exclusion in this species may be a function of food availability and population density (Broughton and Dickman 1991). *Isoodon* is solitary in the wild, apparently coming together only to mate (Stodart 1977). Two individuals can rarely be kept together in captivity as they are intolerant of each other and pugnacious. Two males should never be placed in the same enclosure (Collins 1973). They fight with open mouths, but their long claws usually inflict greater injury than their teeth.

Short-nosed bandicoots are polyestrous, with the estrous cycle averaging 22 days in *I. macrourus* (Gemmell 1988; Lyne 1976). Breeding reportedly can take place year-round for *I. macrourus* in Queensland and New South Wales and in May–February for *I. obesulus* in Tasmania (Ride 1970). Several litters can be produced by each female during one season, there usually being two or three in *I. obesulus* in Tasmania (Ride 1970) and four in *I. macrourus* (Friend 1990). In a population of *I. obesulus* in the Franklin Islands, off South Australia, breeding occurred throughout the year and females produced up to five litters during the period (Copley et al. 1990). A captive *I. macrourus* bore eight litters totaling at least 32 young in 17 months (Collins 1973). The gestation period in *I. macrourus* has been calculated as from 12 days and 8 hours to 12 days and 11 hours, the shortest recorded for any mammal (Lyne 1974). In *I. macrourus* there are up to 7 young per litter, with reported averages being 2.7 (Friend 1990) and 3.4 (Stodart 1977). In *I. obesulus* there are up to 5 young (Ride 1970), though Copley et al. (1990) reported an average ranging from 1.6 in the summer to 2.4 in the spring. According to Collins (1973), probably no more than 4 young of any one litter survive, so that four small, unused mammae are immediately available for the next litter and reproduction can proceed at a rapid rate, just as in *Perameles*. The young of *I. macrourus* leave the pouch after 7–8 weeks, are weaned at 8–10 weeks, become sexually mature at only about 4 months, and reach adult size at about 1 year (Collins 1973; Friend 1990). Gemmell (1990) reported longevity of *I. macrourus* to be up to 28 months in captivity, but Lobert and Lee (1990) found wild *I. obesulus* to live up to 4 years.

I. auratus now is classified as vulnerable by the IUCN. It formerly occupied much of northern and interior Australia but now is known to survive only in remote parts of the Kimberley area of northern Western Australia, on nearby Augustus Island, and on Barrow Island farther to the west. Its decline may be associated with habitat destruction by introduced grazing animals and predation by introduced cats, foxes, and pigs. However, the main period of decline took place in the arid interior from about the 1930s to the 1950s, when there was a large-scale movement of aboriginal people from their native lands to permanent settlements and cattle stations. In their former nomadic movements these people had burned off vegetation gradually, thereby producing a mosaic of habitats that provided abundant food and cover to small mammals. Disappearance of these conditions led to massive destruction of habitat by wildfires started by lightning and the subsequent loss of habitat diversity. In contrast, *I. macrourus*, which occurs predominantly in zones of higher rainfall and greater vegetative cover, has not suffered such a substantial decline. *I. obesulus* has lost 50–90 percent of its original habitat; the subspecies *I. o. nauticus*, of South Australia, is classified as vulnerable by the IUCN, and the subspecies *I. o. obesulus*, of eastern mainland Australia, and *I. o. fusciventer*, of Western Australia, are designated as near threatened. *I. peninsulae*, also designated as near threatened, appears to be uncommon, but its status is not well understood (Ashley et al. 1990; Friend 1990*b*; Gordon, Hall, and Atherton 1990; Johnson and Southgate 1990; Kemper 1990; Kennedy 1992).

PERAMELEMORPHIA; PERAMELIDAE; **Genus PERAMELES**
E. Geoffroy St.-Hilaire, 1803

Long-nosed Bandicoots

There are four species (Kirsch and Calaby 1977; Ride 1970; Seebeck et al. 1990):

P. nasuta, eastern Queensland, New South Wales, and Victoria;
P. gunnii, southern Victoria, Tasmania;
P. bougainville, from islands in Shark Bay at west edge of Western Australia across arid parts of southern Australia to western New South Wales and Victoria;
P. eremiana, southern Northern Territory, northern South Australia, east-central Western Australia.

Head and body length is 200–425 mm and tail length is 75 to about 170 mm. Weights are 500–1,900 grams in *P. nasuta*, 450–900 grams in *P. gunnii*, and 190–250 grams in *P. bougainville* (Burbidge *in* Strahan 1983; Seebeck *in* Strahan 1983; Seebeck et al. 1990; Stodart *in* Strahan 1983). The sleek-looking pelage is composed chiefly of coarse, distinct hairs. The upper parts may be drab, light brown with a slight pinkish tinge, dull orange, yellowish brown, grayish brown, or gray. Black or black-tipped hairs are often interspersed with lighter ones. In all species except *P. nasuta* there are transverse or diagonal dark and light bars on the back and rump, forming in some cases an elaborate pattern. The underparts are white or whitish. These bandicoots have long, tapered snouts and conspicuous pointed ears.

Habitats include rainforest and sclerophyll forest for *P. nasuta*, woodland and open country with good ground cover for *P. gunnii*, heaths and dune vegetation for *P. bougainville*, and spinifex grassland for *P. eremiana* (Ride 1970). Collins (1973) wrote that in heavy vegetation long-nosed bandicoots construct a nest consisting of an oval mound of twigs, leaves, and humus on the surface of the ground. In more open areas they excavate a nest chamber, line it with plant fibers, and then cover it with a mound of twigs and leaves. Abandoned rabbit burrows, rock piles, and hollow logs are also used as nest sites. Long-nosed bandicoots are nocturnal, terrestrial, and highly active. Their rapid running has been described as a kind of gallop, and they have been seen to jump straight up into the air and then to take off immediately in another direction.

The members of this genus are largely insectivorous, though they also feed on worms, snails, lizards, mice, and plant material. *P. nasuta* scratches and digs in the ground for insects and grubs, including weevils found on fruit tree roots and the larvae of the scarabaeid beetle, which feeds on the grass roots of lawns and pastures. Stodart (1977) observed that the claws are used to make a hole big enough for the snout to enter.

An investigated population of *P. gunnii* seemed variable in density, declining in 13 months from about 15/ha. to 4.5/ha. (Ride 1970). Home range in this species has been reported as about 1–4 ha. for females and 18–40 ha. for males. There may be considerable overlap in female ranges, and presumably several are included within the home range of each male. The species *P. nasuta* has been reported to be extremely solitary, with males acting aggressively when coming into contact with one another (Ride 1970). Stodart (1977) found that in enclosures this species was not gregarious, individuals tended to ignore each other, and there was no territorial defense. Collins (1973) wrote that pairs or family groups could be kept together if the enclosure were sufficiently large but that two or more males should not be placed in the same enclosure.

Long-nosed bandicoots are polyestrous, with the average estrous cycle in *P. nasuta* lasting 21 days (Lyne 1976). In some areas reproduction is said to occur year-round. Stodart (1977), however, reported that while males can breed throughout the year, females are inactive during the autumn. She also observed that there was a 62- to 63-day interval from one period of female receptivity through pregnancy to the next period of receptivity. In Tasmania *P. gunnii* breeds from May to February, each female producing three to four litters per season at intervals of 60 days (Ride 1970; Stodart 1977). Robinson et al. (1978) caught a female *P. bougainville* with two pouch young on Bernier Island, Western Australia, in April. Gestation in *P. nasuta* has been timed at about 12.5 days. Litter size in both *P. nasuta* and *P. gunnii* is usually two or three but can be as small as one or as large as five. Since females have eight mammae, the number of young seems rather low, but actually there is an advantage, considering the short interval between births. A new litter is born at the same time that the previous one is weaned. The mammae used by the previous litter have greatly expanded and are too large for the newborn to grasp. Because usually at least four other mammae have not been used, however, the newborn are assured nourishment (Collins 1973). Young of *P. gunnii* leave the pouch at 48–53 days and are weaned at 59–61 days. Females are sexually mature by 3 months and can bear one or two litters in the same breeding season in which they were born. Adult size and male sexual maturity are attained at 4–6 months (Collins 1973; Ride 1970). In *P. nasuta* the young are carried in the pouch 50–54 days, then remain briefly in the nest, and begin to forage with the female at 62–63 days (Stodart 1977). In this species sexual maturity comes at 4 months for females and 5 months for males

Long-nosed bandicoot *(Perameles nasuta)*, photo by Stanley Breeden.

(Ride 1970). Captive specimens of *P. gunnii* have lived for at least 3 years (Collins 1973).

The species *P. eremiana* and *P. bougainville* are listed as endangered by the USDI, and *P. bougainville* also is on appendix 1 of the CITES. The IUCN designates *P. bougainville* generally as endangered, the subspecies *P. b. bougainville*, of eastern Australia, as extinct, *P. gunnii* generally as vulnerable, the mainland population of *P. gunnii* as critically endangered, and *P. eremiana* as extinct. These species declined severely following European settlement, apparently because of clearing of natural vegetation and other habitat modification, inadvertent destruction in poisoning and trapping efforts to control introduced rabbits, and the spread of introduced predators, such as the fox, domestic dog, and domestic cat (Aitken 1979; Menkhorst and Seebeck 1990). Apparently the most drastic losses in the inland arid country were coincident with the large-scale departure of aboriginal people from the 1930s to the 1950s. The gradual, patchwork burning practiced by these people has been replaced by intensive, lightning-caused wildfires that destroy habitat diversity (Johnson and Southgate 1990).

P. eremiana has not been collected since a specimen was taken in Western Australia in 1943. *P. bougainville*, formerly occurring all across southern Australia, had disappeared from New South Wales by the 1860s and from South Australia and the mainland of Western Australia by about 1930. At present it is known to survive only on Bernier and Dorre islands in Shark Bay, Western Australia, where it is common but highly vulnerable to the potential introduction of predators (Friend 1990*b*; Kemper 1990). *P. nasuta* may have fared better, but *P. gunnii* apparently is declining in Tasmania (Robinson, Sherwin, and Brown 1991) and is nearly extinct on the mainland. It formerly occupied a large part of Victoria, but since the 1960s it has been essentially restricted to suburban gardens and lightly farmed areas around the city of Hamilton in the southwestern part of the state (Emison et al. 1978; Minta, Clark, and Goldstraw 1989). Survival in this area may depend on the maintenance of gardens and the willingness of the gardeners to accept the digging activity of the bandicoot. However, the population seems to be in an uncontrollable decline, perhaps associated with loss of genetic viability. Numbers were calculated at 633 individuals in 1985 (Sherwin and Brown 1990), 246 in 1988 (Dufty 1991), and fewer than 100 in 1992 (Robinson, Murray, and Sherwin 1993). Recent reintroduction projects have established small populations at two other sites (Reading et al. 1992).

PERAMELEMORPHIA; **Family PERORYCTIDAE**

Rainforest Bandicoots

This family of 4 Recent genera and 11 species is found in New Guinea, on a number of nearby islands, and in extreme northeastern Australia. Until recently this group generally was considered a part of the Peramelidae (see account thereof), but Groves and Flannery (1990) designated it a separate family.

Head and body length is 150–558 mm, tail length is 50–335 mm, and weight is around 200–4,700 grams. The pelage may be soft or spiny depending on the genus, and coloration varies considerably. The tail may be relatively long or short, but the ears are short in all genera. The dental formula is mostly the same as in the Peramelidae, but in the genera *Echymipera* and *Rhynchomeles* there are only four upper incisors on each side.

Groves and Flannery (1990) distinguished this family on the basis of characters of the skull and teeth. Whereas the Peramelidae are flat-skulled, the Peroryctidae are "cylindrical skulled," the crania being much higher and subcylindrical in cross section in both the rostral and the neurocranial portion. In the Peroryctidae the foramen rotundum is prolonged into a tube; the alveolar plate in the molar region is reduced, so that the zygomatic arch swings forward low over the posterior molars; the infraorbital fossa is low and compressed dorsoventrally; the main palatal vacuities are long and narrow and do not tend to fuse across the midline; the mesopterygoid fossa is broad and parallel-sided; the antorbital surface is flattened; the gonial hooks are directed posteriad; the superior temporal lines are nearly parallel throughout, not converging further back; and the auditory bullae are incomplete.

Gordon and Hulbert (1989) noted that fossils of *Echymipera* and *Peroryctes* have been recovered in Australia from sites as old as middle and late Miocene. Flannery (1990*b*) indicated that there are no paleontological

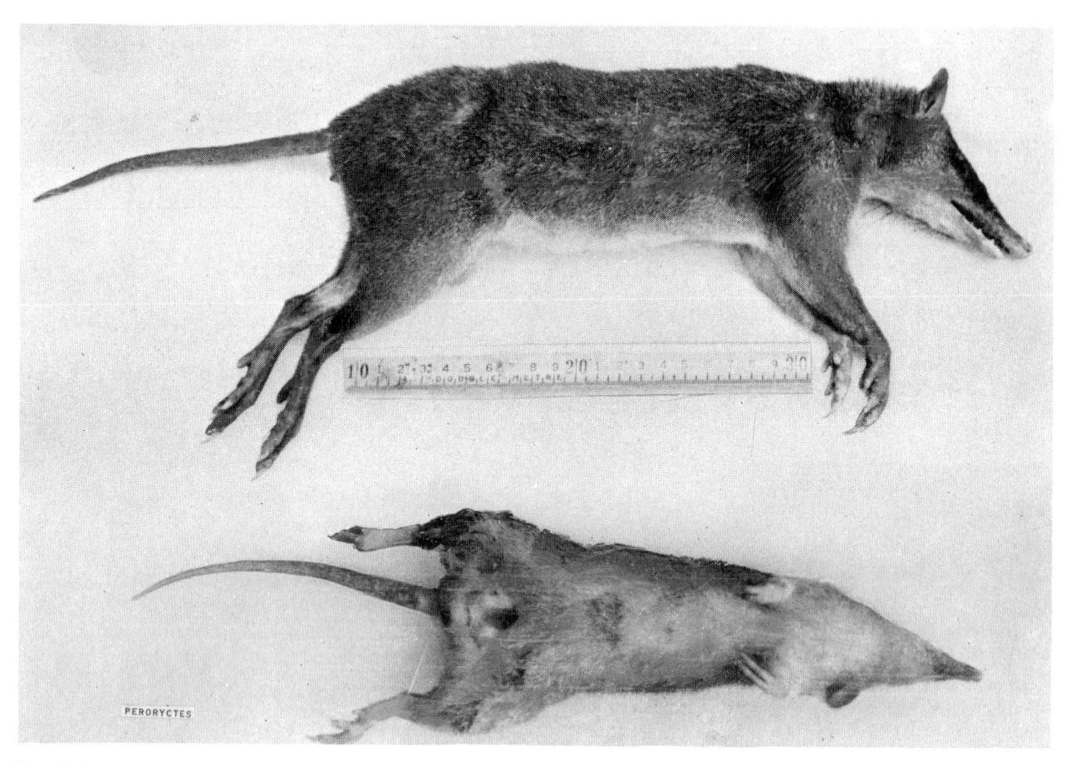

New Guinean bandicoot (*Peroryctes* sp.), ventral view of pouch of the female and the long claws, photo from American Museum of Natural History, Archbold Expedition Collection.

data for New Guinea but that the Peroryctidae appear to be a derived group that originated and radiated almost exclusively there. All species can be found in rainforest and seem to be insectivorous and/or omnivorous.

PERAMELEMORPHIA; PERORYCTIDAE; **Genus ECHYMIPERA**
Lesson, 1842

New Guinean Spiny Bandicoots

There are five species (Collins 1973; Flannery 1990*a*, 1995; George and Maynes 1990; Menzies 1990; Ziegler 1977):

E. kalubu, New Guinea and many nearby islands, Bismarck Archipelago;
E. rufescens, New Guinea and Misool Island (Mysol) to west, Kei and Aru islands, D'Entrecasteaux Archipelago, Cape York Peninsula of Queensland;
E. clara, northern New Guinea, Japen Island;
E. echinista, known only from Western Province of Papua New Guinea;
E. davidi, known only from Kiriwina Island in the Trobriand Islands off southeastern Papua New Guinea.

The species *E. rufescens* had long been known in Australia only by a single specimen collected in 1932, but in 1970 five more individuals were captured (Hulbert, Gordon, and Dawson 1971), and subsequently the species was found over a large part of the Cape York Peninsula (Gordon and Lawrie 1977).

Head and body length is about 200–500 mm and tail length is 50–125 mm. Often the tail is missing, perhaps bitten off, or it may detach easily. Weight is 500–2,000 grams in *E. rufescens* (Gordon *in* Strahan 1983), 450–1,500 grams in *E. kalubu,* and 825–1,700 grams in *E. clara* (Flannery 1990*b*). The upper parts are bright reddish brown, dark coppery brown, black mixed with yellow, or black interspersed with tawny; the underparts usually are buffy or brownish. The entire pelage is stiff and spiny. The snout is comparatively long and sharp. There are only four upper incisor teeth on each side of the jaw. There are three pairs of mammae in *E. kalubu* and four pairs in *E. rufescens* (Lidicker and Ziegler 1968).

These bandicoots generally inhabit rainforest. Flannery (1990*b*) listed the following altitudinal ranges: *E. kalubu,* 0–2,000 meters; *E. rufescens,* 0–1,200 meters; and *E. clara,* 300–1,200 meters. They are apparently terrestrial, nocturnal, and omnivorous (Collins 1973). Van Deusen and Keith (1966) reported that *E. clara* feeds in part on the fruit of *Ficus* and *Pandanus* and in turn forms an important food staple for the native people within its range.

In a spool-and-line tracking study Anderson et al. (1988) found *E. kalubu* to forage for fallen fruit and to root in the forest floor for earthworms and grubs. Average nightly movement was 344 meters. Individuals maintained regular home ranges; two males utilized 1.0 ha. and 2.1 ha. over a period of three nights. There were several nests within each range, located in hollow logs, leaf piles, and shallow burrows. Ranges overlapped, though there were probably exclusive core areas.

Available evidence indicates that these bandicoots are solitary and highly intolerant of their own kind. According to Flannery (1990*b*), breeding of *E. kalubu* occurs throughout the year, average birth interval is probably about 120 days, one to three pouch young are usually present, and fe-

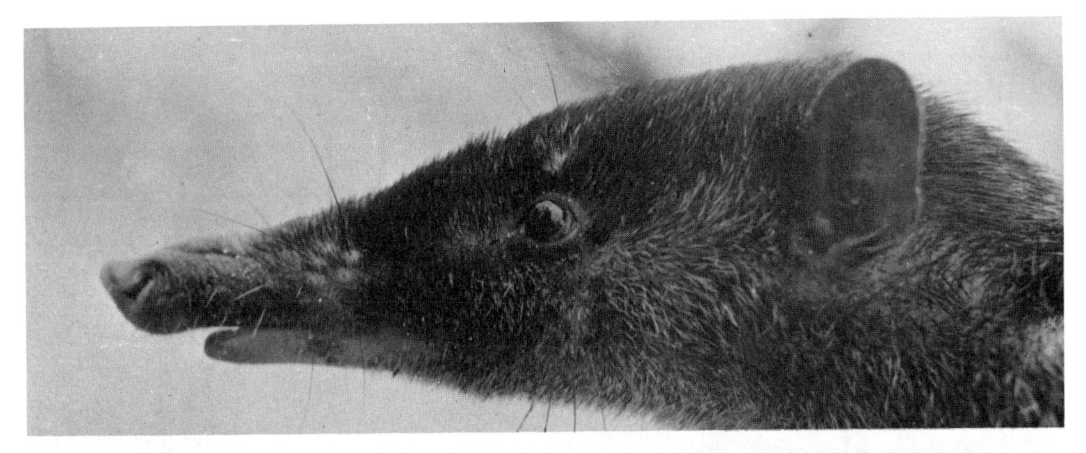

New Guinean spiny bandicoot *(Echymipera rufescens):* Top, photo by G. Gordon; Bottom, photo by Sten Bergman.

males begin breeding at a very early age; pouch young of *E. rufescens* were recorded in New Guinea in May, October, March, and August. On the basis of negative evidence, Gordon and Lawrie (1977) suggested that *E. rufescens* in Australia might be a seasonal breeder, with anestrus during the dry season. A specimen of *E. rufescens* lived in captivity for two years and nine months (Jones 1982).

G. G. George (1979) considered *E. clara* to be threatened in Papua New Guinea because of constant hunting and expanded agricultural activity. Kennedy (1992) designated *E. clara, E. rufescens,* and *E. echinista* as potentially vulnerable because of their restricted ranges.

PERAMELEMORPHIA; PERORYCTIDAE; **Genus RHYNCHOMELES**
Thomas, 1920

Seram Island Long-nosed Bandicoot

The single species, *R. prattorum,* is known only by seven specimens from Seram Island, located in the South Moluccas between New Guinea and Sulawesi. These specimens are in the British Museum of Natural History. The species is named after the collectors, Felix Pratt and his sons Charles and Joseph. Holocene fossils recently discovered on Halmahera Island in the North Moluccas share some similarities with *Rhynchomeles* and suggest that the genus may once have been more widespread (Flannery 1995).

Head and body length is 245–330 mm and tail length is 105–30 mm (Flannery 1995). The fur is crisp, glossy, and completely nonspinous. There is little underfur. This bandicoot is dark chocolate brown above and below with a patch of white on the chest. The head is somewhat lighter than the back, and there may be a whitish area on the forelimbs. The tail is blackish brown and almost naked.

As the generic name suggests, the muzzle is long because of the elongation of the nasal bones of the skull. The ears are small and oval in shape. As in *Echymipera,* there is no fifth upper incisor. Groves and Flannery (1990) noted that *Rhynchomeles* differs from *Echymipera* in only a few features but is characterized by the extreme length and slenderness of its muzzle, its very small last molar, and its generally well-spaced cheek teeth.

All known specimens were obtained during February 1920 in heavy jungle in very precipitous limestone country on Mount Manusela, at an elevation of 1,800 meters

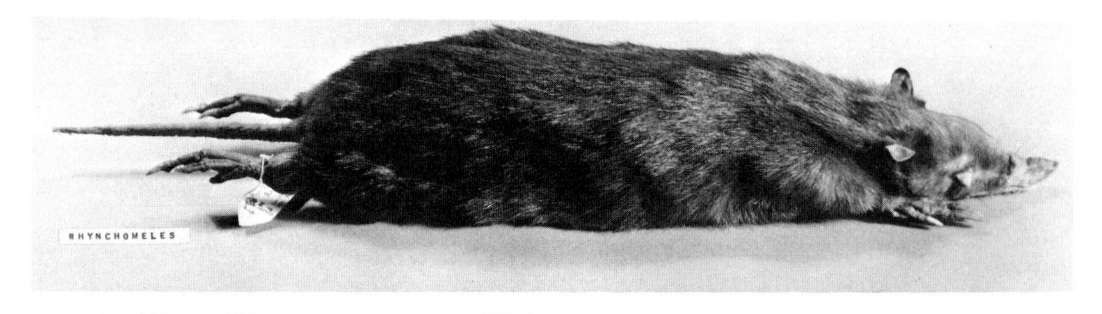

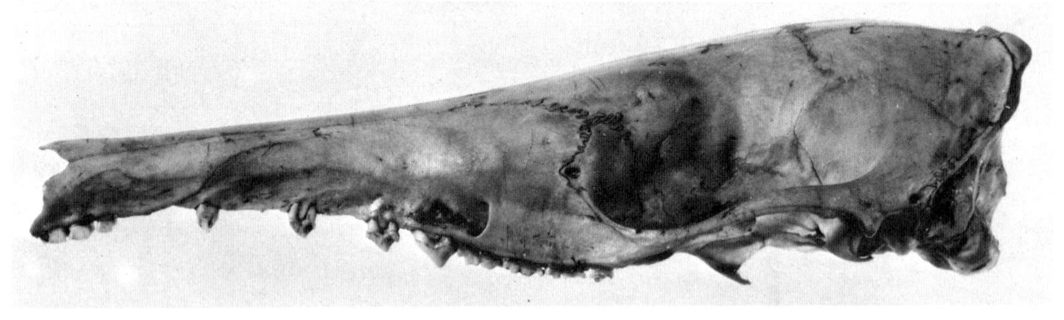

Seram Island long-nosed bandicoot *(Rhynchomeles prattorum)*, photos from British Museum (Natural History).

(Flannery 1995). Kennedy (1992) referred to *Rhynchomeles* as very rare and recommended an IUCN classification of vulnerable. However, the current IUCN designation is "data deficient." Archer (1984) stated that the genus is probably now extinct, but Flannery (1995) doubted that and even suggested that it may be locally common. Kitchener et al. (1993) were unable to locate it during an expedition in 1991, but conversations with local villagers indicated that it might survive in certain areas of undisturbed montane forest.

PERAMELEMORPHIA; PERORYCTIDAE; **Genus MICROPERORYCTES**
Stein, 1932

New Guinean Mouse Bandicoots

There are three species (Groves and Flannery 1990; Ziegler 1977):

M. murina, known only by three specimens taken in the Weyland Mountains of western New Guinea and one collected on the Vogelkop Peninsula at the extreme western tip of the island;
M. papuensis, southeastern Papua New Guinea;
M. longicauda, Central Range of New Guinea.

Microperoryctes long was considered to comprise only the species *M. murina,* but recent investigations prompted the transfer of *M. papuensis* and *M. longicauda* to this genus from *Peroryctes* (Flannery 1990*b;* Groves *in* Wilson and Reeder 1993; Groves and Flannery 1990).

Head and body length of *M. murina* is 150–75 mm and tail length is 105–10 mm. The three specimens from the Weyland Mountains have a dark gray, unpatterned coloration above and below and a grayish white scrotum. The feet have scattered white hairs. The short-haired tail is dark fuscous above and below. The specimen from the Vogelkop differs in having light grayish buff fur, a dark gray middorsal stripe on the back, and a white tip on the tail. In *M. longicauda* head and body length is 239–303 mm, tail length is 141–258 mm, and weight is 350–670 grams; in *M. papuensis* head and body length is 175–200 mm, tail length is 135–55 mm, and weight is 137–84 grams (Aplin and Woolley 1993; Flannery 1990*b;* Lidicker and Ziegler 1968). *M. longicauda* is reddish brown or pale brown speckled with black above, and some subspecies have a dark middorsal line, paired lateral rump stripes, and/or dark eye stripes; the underparts are rufous or buff. In *M. papuensis* the upper parts are dark with a prominent black middorsal line; there also are dark lateral rump stripes and eye stripes as in some subspecies of *M. longicauda,* and the underparts are rich orange buff.

Microperoryctes is characterized by a broadened braincase but a very narrow snout, a highly fenestrated palate with two pairs of vacuities that are shifted forward, straight posterior palatal margins, a deeply wavy coronal suture, small and incomplete auditory bullae lacking anterior spurs, trituberculate molars equal in size to or slightly larger than premolars, granular soles of feet, small ears, a soft and dense pelage, and only three pairs of mammae (Gordon and Hulbert 1989; Groves and Flannery 1990).

Ziegler (1977) stated that *M. murina* lives in moss forest at an altitude of 1,900–2,500 meters. He observed that its short, soft fur suggests a semifossorial existence. He also listed an altitudinal range of 1,000–4,500 meters for *M. longicauda.* The hind foot seems to be adapted for running in *M. longicauda.* Dwyer (1983) collected a 170-gram immature male *M. longicauda* in June. According to Flannery (1990*b*), *M. longicauda* inhabits primary forests and has been taken in nests among roots and at the base of a tree; a female taken in March had four furred young and a female taken in October had three unfurred young. Aplin and Woolley (1993) reported that most specimens of *M. papuensis* have been taken in lower montane forest or associated secondary growth at 1,200–1,450 meters but that

New Guinean mouse bandicoot *(Microperoryctes longicauda)*, photo by J. I. Menzies.

New Guinean bandicoot *(Peroryctes raffrayana)*, photo by Pavel German.

the species also has been found at elevations of up to 2,650 meters. Limited data suggest that *M. papuensis* breeds year-round and litter size is one. Kennedy (1992) designated *M. murina* as potentially vulnerable because of its apparent rarity and restricted range.

PERAMELEMORPHIA; PERORYCTIDAE; **Genus PERORYCTES**
Thomas, 1906

New Guinean Bandicoots

There are two species (Flannery 1995; Ziegler 1977):

P. raffrayana, New Guinea and Japen Island to northwest;
P. broadbenti (giant bandicoot), southeastern New Guinea.

Kirsch and Calaby (1977) did not separate *P. broadbenti* from *P. raffrayana*, but Ziegler (1977) argued that the two are distinct and seem to occur together in southeastern Papua New Guinea. The latter arrangement was accepted by Groves and Flannery (1990), who also transferred the species *P. longicauda* and *P. papuensis* to the genus *Microperoryctes*. Kirsch et al. (1990) agreed that *P. longicauda* is not immediately related to the other species of *Peroryctes* but suggested that it has closer affinity to *Echymipera*.

Measurements (based in part on Lidicker and Ziegler 1968) are: *P. raffrayana*, head and body length 175–384 mm, tail length 110–230 mm; and *P. broadbenti*, head and body 394–558 mm, tail 118–335 mm. Flannery (1990*b*) listed weights of 650–1,000 grams for *P. raffrayana*, and George and Maynes (1990) referred to a specimen of *P. broadbenti* that weighed 4.9 kg. In *P. raffrayana* the upper parts are dark brown with a slight mixture of black, and the underparts are white, brownish yellow, or buff. *P. broadbenti* also is dark brown dorsally but has reddish buff on the flanks and is near white ventrally. The fur of *Peroryctes* is comparatively soft and long, not spinous as in *Echymipera*. Flannery (1990*b*) stated that *Peroryctes* has much longer and narrower hind feet than do other peroryctids and can be further differentiated from *Microperoryctes* by having a single pair of posterior palatal vacuities rather than two.

New Guinean bandicoots dwell mainly in upland forests. Ziegler (1977) gave altitudinal ranges of 60–3,900 meters for *P. raffrayana* and 0–2,700 meters for *P. broadbenti*. He also noted that *P. raffrayana* is rare below 500 meters, occurs primarily in dense forest, and avoids grassland. The hind foot seems to be adapted for leaping in *P. raffrayana*. Collins (1973) wrote that New Guinean bandicoots are terrestrial, nocturnal, and apparently solitary. Flannery (1990*b*) reported that *P. raffrayana* apparently is a rare inhabitant of undisturbed forest, that it may feed on fruit, and that some specimens have been taken from nests; a female with a pouch young was found in August or September. Dwyer (1983) collected a 600-gram female *P. raffrayana* with a 13-gram pouch young in December. A specimen of *P. raffrayana* lived in captivity for three years and three months (Jones 1982).

According to George and Maynes (1990), the great size of *P. broadbenti* renders it vulnerable to hunting pressure. It occupies a very restricted range, is presumably rare, and may already be extinct, though there remain extensive tracts of suitable habitat with a sparse human population. *P. raffrayana* is more widespread and common but also is subject to hunting.

Order Notoryctemorphia

NOTORYCTEMORPHIA; **Family NOTORYCTIDAE; Genus NOTORYCTES**
Stirling, 1891

Marsupial "Mole"

This order contains one family, the Notoryctidae, with the single known genus and species *Notoryctes typhlops*, which occupies parts of northern and east-central Western Australia, southern Northern Territory, and western South Australia (Corbett 1975). A second species, *N. caurinus*, was described from the coast of northern Western Australia in 1920. Since then, most authorities have considered it part of *N. typhlops*, but Johnson and Walton (1989) stated that there is no substantive basis for synonymy. *N. caurinus* was accepted by Walton (*in* Bannister et al. 1988) and Groves (*in* Wilson and Reeder 1993) but not by Corbet and Hill (1991). In any case, over the years *Notoryctes* has been placed in various orders of other Australian mammals either at the familial or the subordinal level. However, studies of morphology, serology, and karyology (Archer 1984; Calaby et al. 1974; Kirsch 1977*b*) did not provide a clear idea of the affinities of this unusual mammal with the other marsupial groups. Finally, Aplin and Archer (1987) proposed that it be placed in a full order. The same order was accepted by Marshall, Case, and Woodburne (1990). Based on an analysis of DNA, Westerman (1991) concluded that *Notoryctes* is not closely related to other marsupial groups and belongs in its own order.

Notoryctes has about the size and general proportions of placental moles (Talpidae), with a thick, powerful, and somewhat elongate body. Head and body length is 90–180 mm and tail length is 12–26 mm. A female (head and body 114 mm) weighed approximately 66 grams. Another specimen weighed 40.5 grams (Howe 1975). The coloration has been reported to vary from almost white through pinkish cinnamon to rich golden red. However, Howe (1975) observed that the reddish color probably results from iron staining. A live specimen lost this coloration after a week in captivity, then again became red after its terrarium was filled with red sand. In life the pelage, which consists almost entirely of underfur, is remarkably iridescent, fine, and silky. The short, cylindrical, stumpy tail is hard and leathery, marked by a series of distinct rings, and terminates in a horny knob. There is a horny shield on the nose. The vertebrae in the neck are fused, enhancing the rigidity of the body as the animal digs. The third and fourth digits of the

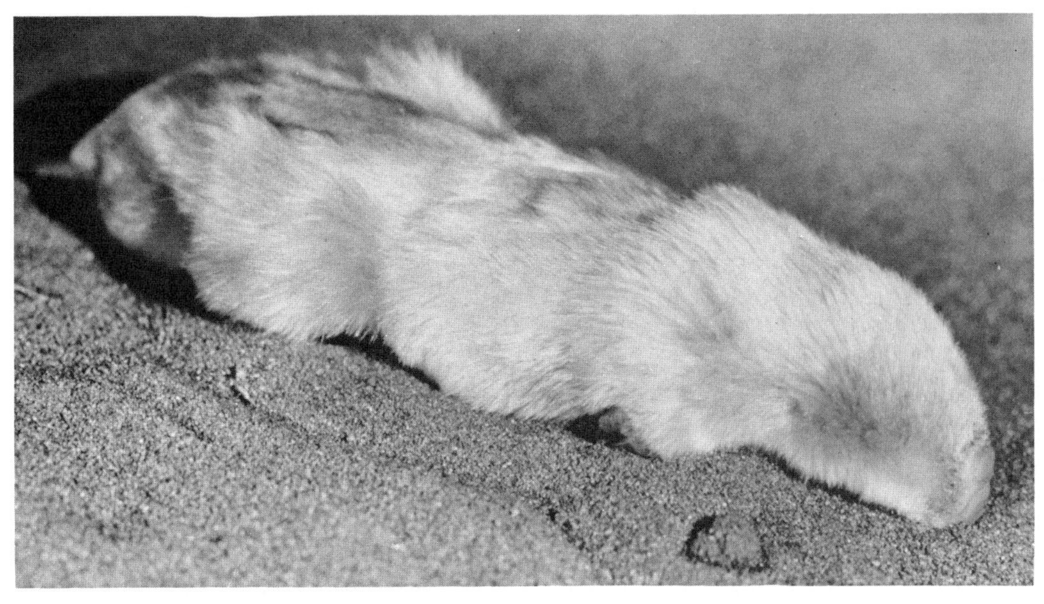

Marsupial "mole" *(Notoryctes typhlops)*, photo by D. Roff.

Marsupial "mole" *(Notoryctes typhlops)*. Inset shows broad flat surface of claws. Photos of mounted specimen from Royal Scottish Museum.

forefoot are greatly enlarged and bear large, triangular claws that form a cleft spade or scoop. The remaining three digits of the forefoot are small, but the first and second bear claws and are opposed to the third and fourth. The middle three digits of the hind foot also possess enlarged claws. The only external indication of ears is a small opening beneath the fur on either side of the head. The eyes of *Notoryctes* are vestigial, measure 1 mm in diameter, and are hidden under the skin. There is no lens or pupil, and the optic nerve to the brain is reduced. The female has a distinct pouch that opens posteriorly and contains two mammae. The testes of the male are situated between the skin and the abdominal wall (K. A. Johnson *in* Strahan 1983), and there is no scrotum.

The skull is conical in shape with an expanded occipital region and overhanging nasal bones. The sutures between the bones of the posterior half of the braincase are obliterated. The variable dental formula is: (i 3–4/3, c 1/1, pm 2/2–3, m 4/4) × 2 = 40–44. The incisors, canines, and premolars are simple and often blunt, though the last upper premolar is bicuspidate; the upper molars are tritubercular. The teeth are well separated from each other. The epipubic (marsupium) bones are reduced.

The discovery of this molelike marsupial in 1888 created a stir among mammalogists comparable to the sensation that accompanied the discovery of the duck-billed platypus. The marsupial "mole" affords an interesting example of evolutionary parallel between marsupial and placental mammals. *Notoryctes* resembles the golden moles (order Insectivora, family Chrysochloridae) in general body form, burrowing habits, texture of fur, and even external features of the brain (Kirsch 1977*b*).

The habitat of *Notoryctes* is sandridge desert with acacias and shrubs, especially along river flats (Corbett 1975; Ride 1970). The marsupial "mole" apparently is not as subterranean as the true moles (Talpidae). It may burrow a relatively short distance and then emerge and shuffle along the surface, propelling itself mainly by the hind feet (Ride 1970). It leaves a peculiar triple track, produced by the two hind feet and the tail. When burrowing, it often moves along at about 8 cm below the surface but then may dip vertically to depths exceeding 2.5 meters (K. A. Johnson *in* Strahan 1983). The shielded snout acts as a bore, the foreclaws function as scoops, and the hind feet throw up the sand, which then falls in behind. Thus, *Notoryctes* practically "swims" through the ground and generally does not leave a long-lasting tunnel as the true moles do. Females, however, probably construct a deep, permanent burrow in which to bear young.

Notoryctes is reported to be active, timorous, apparently solitary, and both diurnal and nocturnal (Collins 1973). One captive proved to be extremely active and moved continuously about its enclosure in search of food. Its nose was always held downward. It fell asleep suddenly on several occasions and awoke just as suddenly to resume its feverish activity. Despite the appearance of being highly nervous, it did not seem to resent handling; it would even consume milk rapidly while being held and then would suddenly fall asleep again. This individual fed ravenously on earthworms, but some other captives have refused them (Collins 1973). Actually, it is unlikely that they are part of the natural diet since only one rare species of earthworm occurs in the range of *Notoryctes*. Available information indicates that the preferred diet consists of the larvae of certain beetles and moths (Corbett 1975; Howe 1975). Examination of the digestive tracts of 10 specimens revealed mostly ants, some other insect remains, and some seeds (Winkel and Humphrey-Smith 1987). One captive made low-intensity, sharp squeaking noises when restrained (Howe 1975). The only information on reproduction is that in the Northern Territory females give birth about November, that the pouch of one contained a single young, and that there is a museum specimen with two sucklings (Collins 1973; Johnson and Walton 1989; Ride 1970).

The IUCN recognizes *N. typhlops* and *N. caurinus* as separate species and now classifies both as endangered. *Notoryctes* is thought to have declined by at least 50 percent in the last decade because of loss of habitat, and a similar rate of decline is projected over the next 10 years. Surprisingly, there was once an active trade in the skin of this small and poorly known creature. Several thousand pelts are said to have been brought by Aborigines to trading posts in the Northern Territory from about 1900 to 1920 and later sold on the market for about two British pounds each (then about U.S. $9). Subsequently, the collection of specimens has remained relatively steady at about 5–15 every 10 years (Johnson and Walton 1989).

Geologically this order is known only from the Recent in Australia, and no fossils can be assigned directly to it (Kirsch 1977*a*). However, Aplin and Archer (1987) noted the recent discovery of fossils from the middle Miocene of Queensland, which suggest a very distant relationship between *Notoryctes* and the Peramelemorphia.

Order Diprotodontia

Koala, Wombats, Possums, Wallabies, and Kangaroos

This order of 10 Recent families, 40 genera, and 131 species is found throughout Australia, Tasmania, and New Guinea and on many islands of the East Indies from Sulawesi to the Solomons. Although this group once was generally considered to form only a marsupial suborder or superfamily, there long has been recognition of its unity and there has been increasing acceptance of its ordinal status. Aplin and Archer (1987), whose sequence is followed here, designated two suborders: Vombatiformes, with the families Phascolarctidae and Vombatidae; and Phalangerida, with the superfamily Phalangeroidea for the family Phalangeridae, the superfamily Macropodoidea for the Potoroidae and Macropodidae, the superfamily Burramyoidea for the Burramyidae, the superfamily Petauroidea for the Pseudocheiridae and Petauridae, and the superfamily Tarsipedoidea for the Tarsipedidae and Acrobatidae. The position of the Macropodoidea in this sequence is questionable; although there is evidence of common ancestry of that group and the Phalangeroidea (Hume et al. 1989), Flannery (1989*b*) suggested that the Macropodoidea may represent an evolutionary branch separate from all other components of the Phalangerida. Marshall, Case, and Woodburne (1990) accepted the same order and subordinal division but indicated certain differences with respect to superfamilies and families. Szalay (1994) considered the Diprotodontia a semiorder of the Syndactyla, which he regarded as also including what are here designated the Peramelemorphia and Notoryctemorphia.

The Diprotodontia are the largest and most diverse marsupial order, with extensive variation in size and shape, but there are several critical characters common to the group. As the name of the order implies, the dentition is diprotodont; the two middle incisor teeth of the lower jaw are greatly enlarged and project forward. Usually there are no other remaining lower incisors or canine, but if any are present they are very small, and there is a gap between the incisors and the lower cheek teeth. A second major unifying character is syndactyly; the second and third digits of the hind foot are joined together by integument, are relatively small, and retain claws (see Hall 1987 for a detailed discussion of the condition and its phylogenetic significance). No other marsupial order is both diprotodont and syndactylous. Other diagnostic characters of the Diprotodontia include reduction of the upper incisors to three or fewer, selenodont (having crescent-shaped cusps) upper molar teeth, a fasciculus aberrans (connection between the cerebral hemispheres) and large neocortex in the brain, a superficial thymus gland, an expanded squamosal epitympanic wing in the roof of the tympanic bullae, and an unusually complex morphology of the glenoid fossa in the basicranial region of the skull (Aplin 1987; Aplin and Archer 1987; Archer 1984). Serological investigations also support the content of the Diprotodontia as set forth herein (Baverstock 1984; Baverstock, Birrell, and Krieg 1987).

The order is predominantly herbivorous, though some members consume invertebrates and even small vertebrates. The smallest species, *Acrobates pygmaeus,* may weigh less than 15 grams, is adapted for gliding like a flying squirrel, and eats mainly insects. The largest living species, *Macropus rufus,* weighs up to 100 kg, is structured for a leaping mode of progression, and like many placental bovids is a grazer. *Diprotodon optatum,* which became extinct about 10,000 years ago, had the size and general appearance of a modern hippopotamus. Other fossils extend the geological range of the Diprotodontia back to the late Oligocene and show that the order was highly diverse even then.

DIPROTODONTIA; **Family PHASCOLARCTIDAE; Genus PHASCOLARCTOS**
De Blainville, 1816

Koala

The single living genus and species, *Phascolarctos cinereus,* has a modern range extending from southeastern Queensland through eastern New South Wales and Victoria to southeastern South Australia (Ride 1970). During the late Pleistocene the koala also occurred in southwestern Western Australia, where suitable habitat still seems to exist. *Phascolarctos* often has been placed in the family Phalangeridae, but it is now considered to represent a distinct and very primitive family with some affinity to the Vombatidae (Archer 1984; Harding, Carrick, and Shorey 1987; Kirsch 1977*b*).

Head and body length is 600–850 mm, the tail is vestigial, and weight is 4–15 kg. Ride (1970) stated that average weight in Victoria was about 10.4 kg for males and 8.2 kg for females. The dense, woolly fur is grayish above and whitish below. The rump often is dappled, and the ears are fringed with white.

The compact body, large head and nose, and big, hairy ears produce a comical, appealing appearance. Cheek pouches and a 1.8- to 2.5-meter caecum aid in digesting the bulky, fibrous eucalyptus leaves that form the main diet.

Koalas *(Phascolarctos cinereus)*, photo by Garth Grant-Thomson through Frank W. Lane. Insets: undersurfaces—A. Left foot; B. Left hand, photos from *The Mammals of South Australia*, F. Wood Jones.

Both the forefoot and the hind foot have five digits, all strongly clawed except the first digit of the hind foot, which is short and greatly broadened. The second and third digits of the hind foot are relatively small and partly syndactylous but have separate claws. The first and second digits of the forefoot are opposable to the other three. The palms and soles are granular. Females have two mammae; their marsupium opens in the rear and extends upward and forward.

The skull is massive and flattened on the sides. The tympanic bullae are elongated and flattened from side to side. The dental formula is: (i 3/1, c 1/0, pm 1/1, m 4/4) $\times$ 2 = 30. The lower incisor and first upper incisor are large, the canine is small, and the molars are blunt and tubercular.

The koala is confined to eucalyptus forests. It is largely nocturnal and arboreal, coming to the ground only to move between food trees or to lick up soil or gravel, which serves as a digestive aid. However, although it is often viewed as a sluggish animal and usually does move at a sedate pace, its relatively long legs can propel it rapidly over the ground or up the trunk of a tree (Lee and Carrick 1989). In addition to the leaves and young bark of about 12 species of eucalyptus, or "gum," tree, the koala may also eat mistletoe and box *(Tristania)* leaves. The animal has a characteristic eucalyptuslike odor.

In a study in Victoria, Mitchell (1990*a*) found koalas to be inactive for more than 20 hours a day, to be usually solitary, and to occupy restricted home ranges centered on a few large food trees for periods of years. Home range size averaged 1.70 ha. for adult males and 1.18 ha. for females. There was extensive overlap between male and female ranges, with pairs tending to use the same trees in their common areas. In a study of the introduced koala population on Kangaroo Island, South Australia, Eberhard (1978) found each individual to remain in an area of about 1.0–2.5 ha. containing a few favorite food trees. Home ranges of adults were largely separate, but there was some overlap between sexes. Of 943 sightings only 11 percent were of pairs and only 1 percent of three animals. Trespass into an occupied tree led to a savage attack by the resident. Males uttered a loud call during the breeding season and apparently demarcated a territory through this vocalization and the production of a powerful scent from a sternal gland. Mitchell (1990*b*) distinguished a series of sounds, including bellows, most frequently emitted at night and probably helping to space the dominant males, and various wails and screams associated with both sexual and antagonistic encounters; he also observed urine marking as well as rubbing of trees with the sternal glands. Gordon, McGreevy, and Lawrie (1990) suggested that the males in any given established population tend to be mostly more than five years or less than two years of age, with the older group comprising dominant residents and the younger group made up of immature animals that will soon disperse to seek openings in other populations. In contrast, Mitchell and Martin (1990) indicated that young females tend to remain in the vicinity of their mothers.

During the breeding season males may attempt to defend a territory containing several females. The season reportedly extends from September to January in New South Wales and from November to February in Victoria. Eberhard (1978) reported that on Kangaroo Island births occurred from late December to early April, with a peak in February. The following natural history data were summarized by Eberhard (1978), Handasyde et al. (1990), Martin and Handasyde (1990), and Martin and Lee (1984). Females are seasonally polyestrous, with an estrous cycle of about 35 days, and usually breed once every year. Loss of a pouch young during the breeding season is followed by further ovulations and sometimes another birth. The gestation period is variable but averages about 35 days, and the usual litter size is 1, though twins have been recorded regularly. The young weigh as little as 0.36 gram at birth, have a pouch life of 5–7 months, and are weaned at 6–12 months. Toward the end of their pouch life for a period of up to six weeks the young feed regularly on material passed through the mother's digestive tract. Both sexes may attain sexual maturity at 2 years but may not begin to breed that soon. Full physical maturity is reached during or after the fourth year in females and the fifth year in males, and the latter probably do little mating before that time. Reproductively active females as old as 12 years have been observed in the wild. One female aged 17–18 years was captured, and there have been reports that captive koalas can live 20 years.

Until the early twentieth century the koala seems to have numbered in the millions in southeastern Australia despite grievous losses from forest fires and epidemics. In the early twentieth century, clearing of woodland habitat and increased access to its range, combined with a demand for its beautiful, warm, durable fur, signaled the end of the vast koala populations (Eberhard 1978; Grzimek 1975; Ride 1970). The commercial kill increased until in 1924 more than 2 million koala skins were exported. By the end of that year the species appeared to have been exterminated in South Australia and nearly wiped out in Victoria and New South Wales. There still was a huge population in Queensland, but in 1927 the state government, yielding to pressure from economic interests, allowed an open season, and nearly 600,000 more pelts were exported. Subsequent public outcry both in Australia and abroad resulted in legal protection, but only a few thousand scattered animals were left.

Since the 1920s, intensive conservation efforts, including the breeding and transplantation of thousands of individuals by the Victoria Fisheries and Wildlife Division, have allowed partial recovery of the species in some parts of southeastern Australia. However, even though the koala is now legally protected from direct killing by people, its habitat is being reduced and fragmented by settlement, road construction, fires, logging, and massive cutting of forests for production of woodchips (mostly for export to Japanese paper mills). More than half of the tall to medium-sized trees, on which the koala depends for survival, have already been destroyed (Phillips 1990). The IUCN classifies the koala as near threatened. The Australasian Marsupial and Monotreme Specialist Group of the IUCN Species Survival Commission has designated the koala as "potentially vulnerable" and indicated that the species has lost 50–90 percent of its habitat (Kennedy 1992). Another problem is the widespread prevalence of the bacterial pathogen *Chlamydia*, which infects the reproductive tract of females, thereby reducing fertility and thus restricting the size of some populations but is not considered an overriding threat to the survival of the species as a whole (Martin and Handasyde 1990). There is much controversy regarding the precise status of this highly popular mammal and what conservation measures should be undertaken. Overall estimates of numbers remaining in the wild range from about 40,000 to more than 400,000 (Payne 1995; Thompson 1995).

The geological range of the family Phascolarctidae is middle Miocene to Recent in Australia (Archer 1984; Kirsch 1977*a*).

DIPROTODONTIA; **Family VOMBATIDAE**

Wombats

This family of two Recent genera and three species is known from eastern and southern Australia, Tasmania, and islands of Bass Strait. The sequence of genera presented here follows that of Kirsch and Calaby (1977). The name Phascolomyidae has sometimes been used for this family.

Wombats resemble small bears in general appearance. The body is thick and heavy; the head and body length of adults is about 700–1,200 mm, and the weight is 15–35 kg. The muzzle is naked and the pelage is coarse in *Vombatus*, but in *Lasiorhinus* the muzzle is haired and the pelage is soft. Underfur is almost lacking. The eyes are small. The limbs are short, equal in length or nearly so, and extremely strong. There are five digits on all the limbs, and all have claws except the first toe, which is vestigial. The second and third digits of the hind foot are partly united by skin. The marsupium opens posteriorly and encloses a single pair of mammae. Wombats have traces of cheek pouches. They also have a group of glands of unusual structure inside the stomach that may be associated with digestion of a special type of plant food. The skull is massive and flattened.

The dentition is remarkably similar to that of rodents. Wombats also resemble rodents in their manner of feeding and in the rapid side-to-side movements of their jaws. All the teeth of wombats are rootless and ever-growing, which compensates for wear. The two rootless incisors in each jaw are large and strong, have enamel on their front and lateral surfaces only, and are separated by a wide space from the cheek teeth. The premolars are small, single-lobed, and close to the molars; the molar teeth have two lobes and are rather high-crowned. The dental formula is: (i 1/1, c 0/0, pm 1/1, m 4/4) $\times$ 2 = 24.

Wombats are shy, timid, and difficult to observe in the wild; they are sometimes active during the day but are considered nocturnal. They live in burrows and are rapid, powerful diggers. Under certain conditions they may construct burrow systems more than 30 meters in length. Wombats dig with their front feet, thrusting the soil out with the hind feet, and use their strong incisors to cut such obstructions as roots. The burrow entrance normally is a low arch that fits the body. A grass or bark nest is located near the end of the burrow. A shallow resting place is usually excavated against a tree or log near the mouth of the burrow as a site for sunbathing. Paths often lead from burrows to feeding areas. Wombats eat mainly grasses, roots, bark, and fungi.

Wombats do well in captivity and often become interesting and affectionate pets. They have suffered serious reduction in numbers and range. People are their chief enemy. Colonies have been exterminated near settled areas because of damage to crops and because domestic livestock may injure a leg by breaking through into the burrows. Wombats also have been destroyed in the campaign against rabbits, as these introduced pests often shelter in wombat burrows. There are many regions in Australia where these unusual mammals could reside without disturbing human developments. Adequate protection would probably enable wombats to maintain a stable population. It is surprising how little detailed information is available regarding them.

The geological range of this family is Miocene to Recent in Australia (Kirsch 1977*a*).

DIPROTODONTIA; VOMBATIDAE; **Genus VOMBATUS**
E. Geoffroy St.-Hilaire, 1803

Common, or Coarse-haired, Wombat

The single species, *V. ursinus,* formerly occupied the coastal region from southeastern Queensland to southeastern South Australia, as well as Tasmania and the islands of Bass Strait (Ride 1970). The mainland populations sometimes have been designated as a separate species with the name *V. hirsutus.*

The head and body length is 700–1,200 mm and the tail is a mere stub. Adults generally weigh from 15 kg to about 35 kg. The general coloration is yellowish buff, silver gray, light gray, dark brown, or black. The fur in this genus is coarse and harsh, not soft and silky as in *Lasiorhinus.*

Vombatus also differs from *Lasiorhinus* in skull and dental features and in its hairless nose and somewhat rounded ears. In the fine-haired wombats the nose is haired and the ears are relatively pointed. The teeth are rootless, that is, they grow continuously from pulpy bases. Both genera of wombats are the only living marsupials that possess two rootless incisors in each jaw, an arrangement also seen in rodents. In the lesser curvature of the stomach is a peculiar gland patch that is also found in the koala *(Phascolarctos)* and the beaver *(Castor)*. The pouch of the female opens posteriorly and contains two mammae.

The common wombat occurs in upland forests, especially in rocky areas. It constructs holes that have a single entrance but may branch into a complex network of tunnels under the surface (Ride 1970). The nesting chamber, lined with vegetation, usually is placed about 2–4 meters from the entrance. This wombat is nocturnal except for occasional bouts of basking in the sun (Collins 1973). It feeds mainly on grass, roots, and fungi, apparently preferring fresh seed stems; it uses its forefeet to tear and grasp pieces of vegetation. It also has been seen foraging among sea refuse along the shore.

This wombat is quick in its movements and can run rapidly, at least for short distances. When touched, especially near the hindquarters, it kicks backward with both hind feet, and when annoyed it may emit a hissing growl. Often, however, it becomes a playful and affectionate pet. Individuals formerly occurring on King Island in Bass Strait were reportedly domesticated by fishermen. These animals, in a reversal of normal routine, would feed in forests during the day and return in the evening to the cabins that served as their retreat.

According to McIlroy (*in* Strahan 1983), population densities of 0.3/ha. occur in areas of favorable habitat. Home ranges measure 5–23 ha., contain a number of burrow systems, and overlap extensively. Collins (1973) wrote that the common wombat is solitary except during the breeding season and that keeping more than one individual in an enclosure often results in fighting and injury. He added, however, that pairs and groups of compatible individuals have been maintained together successfully. Ride (1970) cited a study of a wild population in Victoria in

Coarse-haired wombat *(Vombatus ursinus)*, photo from New York Zoological Society.

which usually only a single wombat occupied each burrow, but individuals were sociable and would visit one another's burrows. Wells (1989) indicated that territories are maintained by rubbing scent on logs and branches and by depositing feces along trails. There are a number of vocalizations, the most common being a harsh cough.

Ride (1970) also stated that births probably occur in late autumn (about April–June in Australia) and that the young are independent by the following summer (December–March). According to a report from the Perth Zoo, however, an adult pair were together in late November 1968, the female was found to have a pouch young on 30 July 1969, the young left the pouch first on 12 October 1969 and permanently by 29 October 1969, and the mother and offspring had to be separated in May 1970 (Collins 1973). The estrous cycle is 32–34 days (Hayssen, Van Tienhoven, and Van Tienhoven 1993). Wells (1989) reported that gestation lasts approximately 21 days and that the young are fully weaned at 12 months. In Tasmania, births were found to occur throughout the year, though about 48 percent took place from October to January (Green and Rainbird 1987). A single young is the usual case, but twins are known to occur. Sexual maturity is attained after 2 years (McIlroy *in* Strahan 1983). This genus appears to be capable of long life in captivity; the record longevity is 26 years and 1 month (Collins 1973).

Although the common wombat has declined through human persecution, some of its mainland and Tasmanian populations probably are secure in their mountain forest habitat (Ride 1970). By the late nineteenth century, however, the species had been exterminated from all islands of Bass Strait except Flinders Island, where it still occurs. The subspecies there, *V. ursinus ursinus*, is classified as vulnerable by the IUCN.

DIPROTODONTIA; VOMBATIDAE; **Genus LASIORHINUS**
Gray, 1863

Hairy-nosed, or Soft-furred, Wombats

There are two species (Kirsch and Calaby 1977; Ride 1970):

L. latifrons, southeastern Western Australia to southeastern South Australia;
L. krefftii, formerly east-central and southeastern Queensland, south-central New South Wales.

The population of *L. krefftii* in east-central Queensland sometimes is considered a distinct species with the name *L. barnardi*. The population of *L. krefftii* in southeastern Queensland, along the Moonie River, has been called *L. gillespiei*, but that name is a synonym of *L. krefftii*.

Head and body length is 772 mm to about 1,000 mm, tail length is 25–60 mm, and weight is 19–32 kg (Gordon *in* Strahan 1983; Wells *in* Strahan 1983). The upper parts are usually dappled with gray and black or brown; the cheeks, neck, and chest are often white, and the remainder of the underparts is gray.

This genus is distinguished from *Vombatus* by skull, dental, and external characters. The specific name *latifrons* refers to the great width of the anterior part of the skull. Externally the fine-haired wombat can be differentiated from *Vombatus* by its haired nose and relatively longer, more pointed ears. The fur in *Lasiorhinus* is soft and silky, not coarse and harsh as in *Vombatus*. The marsupial pouch is well developed and opens to the rear; there are two mammae.

These wombats occupy relatively dry country—savannah woodland, grassland, and steppe with low shrubs. They construct complex tunnel systems comprising a large number of separate burrows that join together to form a warren; the entrances are connected above ground by a network of trails, and additional trails radiate out to feeding areas and other warrens (Ride 1970). The burrows of one colony in the early twentieth century covered an area 800 meters long and about 80 meters wide. Wells (*in* Strahan 1983) pointed out that while these wombats appear to be slow and bumbling, they are exceedingly alert and can run as fast as 40 km/hr for short distances. They are said to be more docile in captivity than *Vombatus*. They are nocturnal and feed mainly on grass.

Wells (1978) found the density of a population of *L. latifrons* in South Australia to be 1/4.8 ha. Home ranges of individuals varied from 2.51 ha. to 4.18 ha. and seemed centered on a warren. There was some evidence of territorial marking and defense. Ride (1970) observed that despite the gregarious appearance suggested by the warrens, hairy-nosed wombats seem basically solitary, with each individual possessing its own burrow and feeding area. Wells (*in* Strahan 1983), however, stated that each warren system is inhabited by 5–10 wombats, with about equal numbers of each sex. Johnson and Crossman (1991) confirmed the latter position but noted that individuals of the same sex kept to separate ranges when they left the burrow system to feed. They also found that, in contrast to the situation in *Phascolarctos* (see account thereof) and most other mammals that have been studied, dispersal from the natal colony was common for young females but rare for young males.

The young of *L. latifrons* are born in the spring (starting in October) in South Australia (Wells 1978). Crowcroft (1977) found gestation for two births of this species to be 20–21 and 21–22 days. There usually is one young, but twins have been reported. The young first leave the pouch at approximately 8 months, vacate it permanently at 9 months, are weaned at 12 months, may reach full size by 2 years, and are usually sexually mature at 3 years (Wells 1989). A captive specimen of *L. latifrons* lived for 24 years and 6 months (Jones 1982).

Hairy-nosed wombats declined in numbers and distribution following persecution by European settlers. Ride (1970) expressed serious concern about the steadily contracting range of *L. latifrons* in South Australia. Aitken (1971*a*), however, thought that while the distribution of this species had become somewhat fragmented, it still was abundant over large areas. He stated that its density had been reduced in the late nineteenth century, probably because of competition from introduced rabbits. He added that the recent human take has been negligible but there was some killing based on the unfounded belief that the wombat damaged fences, competed with livestock for grazing, and dug burrows that allowed rabbits (considered pests) to find shelter.

The status of *L. krefftii* is worse. The once plentiful population in the Riverina area of south-central New South Wales disappeared long ago (Ride 1970). The population in southeastern Queensland, formerly called *L. gillespiei*, is thought to have been extinct since about 1900 (Goodwin and Goodwin 1973). The only remaining population of *L. krefftii*, in east-central Queensland, seems to have been long restricted to a small area of about 15.5 sq km. In 1971 most of this area was protected by being declared a nation-

Soft-furred wombat, or hairy-nosed wombat *(Lasiorhinus latifrons)*, photo by Ernest P. Walker.

al park, but the habitat may be jeopardized by cattle grazing in the park (Thornback and Jenkins 1982). The population fell to only about 25 individuals in 1981 but since has increase to about 70 and has been found to remain genetically viable (Taylor, Sherwin, and Wayne 1994). *L. krefftii* is classified as critically endangered by the IUCN and as endangered by the USDI and is on appendix 1 of the CITES.

DIPROTODONTIA; **Family PHALANGERIDAE**

Possums and Cuscuses

This family of 6 living genera and 22 species occurs in Australia, Tasmania, and New Guinea and on islands from Sulawesi to the Solomons. The family sometimes has been considered to be much larger and to include all the genera that herein, in accordance with Aplin and Archer (1987) and Kirsch and Calaby (1977), are allocated to the families Phascolarctidae, Burramyidae, Pseudocheiridae, Petauridae, Tarsipedidae, and Acrobatidae. The sequence of genera presented here basically follows that of Flannery, Archer, and Maynes (1987), who recognized two subfamilies: Ailuropinae, with the single genus *Ailurops;* and Phalangerinae, with *Strigocuscus, Trichosurus, Spilocuscus,* and *Phalanger. Wyulda,* considered only a subgenus of *Trichosurus* by those authorities, is here given generic rank. Kirsch (1977*b*) used an arrangement by which the species here assigned to *Trichosurus* and *Wyulda* were placed in the subfamily Trichosurinae and those assigned to all other genera were placed in the subfamily Phalangerinae.

Head and body length is 320–650 mm and tail length is 240–610 mm. The soft, dense pelage is woolly in all genera except *Wyulda.* All four limbs have five digits, and all the digits except the first toe of the hind foot have strong claws; this first toe is clawless but opposable and provides for a firm grip on branches. The second and third digits of the hind foot are partly syndactylous, being united by skin at the top joint, but the nails are divided and serve as hair combs. The largest digits of the hind foot are the fourth and fifth. In the genera other than *Trichosurus* and *Wyulda* the first and second digits of the forefoot are opposable to the other three. The well-developed marsupium opens to the front, and there are two to four mammae.

Flannery, Archer, and Maynes (1987) listed the following diagnostic characters of the Phalangeridae: the mastoid wing on the rear face of the cranium is reduced in size and the rear face of the cranium above the mastoid is composed of an extension of the squamosal; almost the entire plantar surface of the hind foot is covered by a large striated pad; the distal portion of the tail is nearly naked and at least part is covered in scales or tubercles; and the upper second premolar is greatly reduced or lost but the upper first premolar is large. The dental formula of the family is: (i 3/1–2, c 1/0–1, pm 2–3/3, m 5/5) × 2 = 40–46 (Archer 1984). The first incisors are long and stout. The second and third incisors, if present, are vestigial, as are the lower canine and some of the premolars. The molars have sharp cutting edges. The skull is broad and flattened.

The brush-tailed possum *(Trichosurus vulpecula)* has an unusual set of glands: both sexes possess two pairs of anal glands, one pair producing oil and the other pair producing cells. The former is a scent gland, but the function of the latter is not known. The cell-producing glands do not liquefy their secretion, which eventually appears in the urine as cells. This phenomenon has not been reported for placental mammals.

The geological range of this family is late Oligocene to Recent in Australasia (Archer 1984).

Cuscus (*Phalanger* sp.), photo by Bernhard Grzimek.

Ailurops ursinus, photo by Colin P. Groves.

DIPROTODONTIA; PHALANGERIDAE; **Genus AILUROPS**
Wagler, 1830

Large Celebes Cuscus

The single species, *A. ursinus*, occurs on Sulawesi and the nearby Togian, Peleng, and Talaud islands. This species usually had been assigned to the genus *Phalanger*, but Flannery, Archer, and Maynes (1987) and George (1987) considered it to represent its own distinct genus and subfamily.

According to Tate (1945), one specimen had a head and body length of 564 mm and a tail length of 542 mm. Miller and Hollister (1922) indicated that weight is about 7 kg. For another specimen, a male, Flannery and Schouten (1994) listed a head and body length of 610 mm, a tail length of 580 mm, and a weight of 10 kg. The upper parts vary greatly in color, being black, grayish, or brown, and the underparts are usually whitish. The limbs are relatively long, the feet and claws very large, the rostrum unusually short and broad, the rhinarium large and naked, the canine tooth short, and the third upper molar large. George (1987) noted that the pelage consists of fine but wiry underfur and coarse, bristly guard hairs and that the ears are short and well furred internally.

Flannery, Archer, and Maynes (1987) listed 10 cranial characters that distinguish *Ailurops* from all other phalangerids. In the others, the mastoid and ectotympanic are usually broadly continuous, but in *Ailurops* these bones are separated by a deep and continuous groove. In the others the entire basicranial region, particularly the squamosal and mastoid, are much more pneumatized than in *Ailurops.* In the others the mastoid is restricted to a thin ventral band on the posterior face of the cranium, but in *Ailurops* the mastoid has an extensive wing on the rear face of the cranium. The third upper incisor of the others is reduced in size, but in *Ailurops* this tooth is larger than the second upper incisor. The first upper premolar is double-rooted in the others but single-rooted in *Ailurops.*

Although *Ailurops* is a distinctive and relatively large animal, up to 1,200 mm in total length, and has the most northwesterly distribution of any marsupial, almost nothing is known of its status and natural history. Kennedy (1992) stated that it occurs in rainforests and designated it as "potentially vulnerable" because it is presumed to be hunted as game. Flannery and Schouten (1994) cited information that it is most abundant at around 400 meters in elevation, is largely diurnal and folivorous, is generally found in pairs, and occurs at an estimated density of one pair per 400 ha.

DIPROTODONTIA; PHALANGERIDAE; **Genus STRIGOCUSCUS**
Gray, 1861

Ground Cuscuses

There are two species (Corbet and Hill 1992; Flannery and Schouten 1994):

S. celebensis, Sulawesi and the islands of Sangihe and Siau to the north and Muna to the southeast;
S. pelengensis, Peleng and Taliabu islands to the east of Sulawesi.

Strigocuscus usually has been placed in *Phalanger,* but Flannery, Archer, and Maynes (1987) regarded the two as generically distinct. Those authorities also tentatively considered *Strigocuscus* to include the species that here is designated *Phalanger ornatus* in accordance with Corbet and Hill (1992), Flannery and Schouten (1994), George (1987), and Groves (1987*a* and *in* Wilson and Reeder 1993) as well as the species that here is designated *Phalanger gymnotis* in accordance with Flannery and Schouten (1994) and Springer et al. (1990). Flannery, Archer, and Maynes (1987) tentatively included still another species, *S. mimicus* from southern New Guinea and the eastern Cape York Peninsula of Queensland, in *Strigocuscus,* but that species was regarded as part of *Phalanger orientalis* by George (1987), Groves (1987*a* and *in* Wilson and Reeder 1993), and Menzies and Pernetta (1986), and it was included in *Phalanger intercastellanus* by Flannery and Schouten (1994). Finally, Menzies and Pernetta (1986) regarded *S. pelengensis* as a subspecies of *S. celebensis,* and Flannery, Archer, and Maynes (1987) included it in the genus *Phalanger,* but the species was reassigned to *Strigocuscus* by Flannery and Schouten (1994).

In *S. celebensis* head and body length is 294–380 mm and tail length is 270–373 mm (Groves 1987*a*). In *S. pelengensis* head and body length is 350–70 mm, tail length is 245–300 mm, and weight is 1,070–1,150 grams (Flannery and Schouten 1994). *S. celebensis* is pale buff in color, and *S. pelengensis* is clear tawny brown to reddish above and paler brown or yellowish below.

According to Flannery, Archer, and Maynes (1987), characters of the skull indicate that *Strigocuscus* is more closely related to *Trichosurus* and *Wyulda* than to *Phalanger.* In *Strigocuscus, Trichosurus,* and *Wyulda* the rostrum is relatively narrower than in other phalangerids, the lachrymal is retracted from the face, the ectotympanic is almost totally excluded from the anterior face of the postglenoid process, and the third upper premolar tooth is set at a more oblique angle relative to the molar row than it is

Ground cuscus *(Strigocuscus pelengensis),* photo by Tim Flannery.

in other phalangerids. George (1987) indicated that *Strigocuscus* is characterized by the very large size of the third upper premolar, a widening of the zygomatic arches at the orbits, and short paroccipital processes. Flannery (1995) noted that *Strigocuscus* also shows affinity to *Trichosurus* by the lack of a dorsal stripe and the presence of hairs on the portion of the tail that is usually naked in *Phalanger.*

According to Flannery (1995), *Strigocuscus* is arboreal and sometimes is found in secondary forests and gardens. Individuals have been noted sleeping in the crowns of coconut palms. *S. celebensis* has been reported to be nocturnal and frugivorous and to usually occur in pairs. *S. pelengensis* evidently breeds throughout the year and normally rears a single young.

DIPROTODONTIA; PHALANGERIDAE; **Genus TRICHOSURUS**
Lesson, 1828

Brush-tailed Possums

There are three species (Flannery and Schouten 1994; Kerle, McKay, and Sharman 1991; McKay *in* Bannister et al. 1988):

T. vulpecula, throughout suitable areas of Australia, including Tasmania and Kangaroo and Barrow islands;
T. johnstonii, between Koomboolomba and Kuranda in northeastern Queensland;
T. caninus, mountainous districts of southeastern Queensland, eastern New South Wales, and eastern Victoria.

Flannery, Archer, and Maynes (1987) considered *Wyulda* (see account thereof) a subgenus of *Trichosurus. T. arnhemensis,* found in extreme northern parts of Western Australia and Northern Territory and on Barrow Island, was long considered a separate species and was maintained as such by Flannery and Schouten (1994) and Groves (*in* Wilson and Reeder 1993). However, morphological, karyological, electrophoretic, and ecological studies by Kerle, McKay, and Sharman (1991) indicate that it is best treated as a subspecies of *T. vulpecula.*

Head and body length is 320–580 mm, tail length is 240–350 mm, and weight is 1.3–5.0 kg. Crawley (1973) found the average adult weight of an introduced population of *T. vulpecula* in New Zealand to be 2.46 kg for males and 2.33 kg for females. The coloration of *T. vulpecula* is extremely variable, with gradations within the main color phases—gray, brown, black and white, and cream. Ride (1970) noted that a black phase was particularly common in Tasmania. There also is sexual dimorphism in color; adult males usually blend to reddish across the shoulders. In *T. caninus* the general color is dark gray or black with a glossy sheen (Ride 1970). The coat is thick, woolly, and soft. The tail is well furred and prehensile, and at least the terminal portion is naked on the lower side. In the subspecies *arnhemensis* the tail is less well haired than in other populations of *T. vulpecula* and appears thinner (Ride 1970). Females of the genus have a well-developed pouch that opens forward and encloses two mammae.

Brush-tailed possums have a well-developed glandular area on the chest. This brown-pigmented, sternal patch is more prominent in the male than in the female. Observations of animals in the field and in captivity have revealed that these areas are rubbed against stumps or fallen logs to form "scent posts." Copious amounts of an anal gland secretion with a strong musky odor are also produced and may have a function in demarcating territory, as well as being a defense mechanism.

All species are nocturnal and arboreal and usually nest in tree hollows. They generally occupy forest habitat, but

Brush-tailed possum *(Trichosurus vulpecula)* male about 159 days old, photo by A. Gordon Lyne.

T. vulpecula also may be found in areas devoid of trees, where it shelters by day in caves or burrows of other animals. It is even found in semideserts of central Australia, where it takes shelter in eucalyptus trees along watercourses. Resident populations occur in most cities in parks and suburban gardens, where the animals shelter on the roofs of houses. The diet consists mostly of young shoots, leaves, flowers, fruits, and seeds. Insects also are eaten, and there are well-documented records of *T. vulpecula* killing young birds. Harvie (1973) found that pasture species of grass and clover formed about 30 percent of the diet of an introduced population of *T. vulpecula* in New Zealand.

Because *T. vulpecula* has long been established in New Zealand and is considered to be of much economic importance, it has been studied there at least as much as in Australia. Some reported population densities for both countries are: vicinity of Canberra, 1/0.67 ha.; dry woodland and sclerophyll forest of southeastern Australia, 1/2.2–2.7 ha.; gardens in Australia, as high as 1/0.72 ha.; North Island of New Zealand, 1/0.16 ha. in August and 1/0.09 ha. in November; and certain parts of New Zealand, 1/0.124 ha. and 1/0.206 ha. (Crawley 1973; How 1978; Ward 1978). Home range in this species was found to average 3 ha. for males and 1 ha. for females in the Canberra vicinity and 0.81 ha. for males and 0.46 ha. for females on the North Island of New Zealand (Crawley 1973). Ward (1978) found nightly ranges of four individuals in New Zealand usually to be under 1,000 sq meters and annual home ranges to be 0.28–3.21 ha. For the species *T. caninus* in southeastern Australia, density was 1/3.3 ha. and home range was 7.67 ha. for males and 4.85 ha. for females (How 1978).

Information on social structure in *Trichosurus* is still limited. According to Collins (1973), two captive males cannot be kept in the same enclosure, but one male and several females can coexist. There is strong evidence that wild *T. caninus* are paired, with an adult male and female sharing a home range (How 1978). The species *T. vulpecula* seems basically solitary. In its low-density populations a territorial system may function, with individuals of the same sex maintaining discrete ranges. In its high-density populations, however, as in parts of New Zealand, there is considerable overlap of home range both among the same sex and between sexes, and territorial behavior is not evident. Spacing between individuals is maintained through agonistic encounters and olfactory and vocal communication (Crawley 1973; How 1978).

The mating call of *T. vulpecula* is a loud, rolling, guttural sound terminating in a series of staccato "ka-ka-ka" syllables. When mildly disturbed, all species make a metallic clicking sound and then, if further aroused, a series of harsh exhalations with the mouth open. Further disturbance causes the animal to rise on its hind limbs with the forelimbs raised and outstretched. The possum then extends itself to full height and screams. All species are powerful fighters.

Females are polyestrous; the estrous cycle averages about 25 days in *T. vulpecula* (Van Deusen and Jones 1967) and about 26.4 days in *T. caninus* (Smith and How 1973). Breeding may extend throughout the year, but generally for *T. vulpecula* in Australia there are two distinct seasons or peaks, one in spring and one in autumn. In some areas a female may produce a litter in both seasons, but usually most females give birth during the autumn (How 1978). The subspecies *T. vulpecula arnhemensis* in the tropics of extreme northern Australia has been found to breed continuously throughout the year, with females evidently conceiving before weaning the pouch young or returning to estrus in about 10 days if the pouch young is lost (Kerle and Howe 1992). In New Zealand it is extremely rare for a female to have two litters in one year (Crawley 1973). For *T. vulpecula* in New South Wales, Smith, Brown, and Frith (1969) found 89.6 percent of births to occur in autumn (late March–June) and 9.1 percent to occur in spring (September–November). For *T. caninus* in New South Wales, How (1976) found 87.5 percent of the births to take place from February to May, mostly in March and April, with a few scattered over the rest of the year. For an introduced population of *T. vulpecula* in New Zealand, Crawley (1973) reported a single well-defined breeding season from April to July, with a few births also occurring later.

Gestation averages about 17.5 days in *T. vulpecula* (Van Deusen and Jones 1967) and 16.2 days in *T. caninus* (Smith and How 1973). Usually a single young is produced; of 60 births of *T. vulpecula* recorded at the London Zoo, 6 resulted in twins and 1 in triplets (Collins 1973). Development appears to take place more rapidly in *T. vulpecula* (How 1976, 1978). The young leave the pouch at about 4–5 months, are weaned by 6–7 months, and separate from the mother at 8–18 months. Females generally attain sexual maturity and begin to breed at 9–12 months, though in some populations this phase is delayed 1 or even 2 years. In *T. caninus* the young emerge from the pouch at 6–7 months, weaning occurs at 8–11 months, dispersion occurs at 18–36 months, and females do not reach sexual maturity until 24–36 months. Males of *T. vulpecula* are sexually mature by the end of their second year of life (Crawley 1973; Smith, Brown, and Frith 1969). In *T. vulpecula* mortality is high among dispersing young; only 25 percent reach the age of 1 year (How 1978). Adult mortality has been found to be relatively low—20 percent in one study—and life expectancy is 6–7 years. In wild populations individuals of *T. vulpecula* over 13 years old and of *T. caninus* up to 17 years have been found (How *in* Strahan 1983). *Trichosurus* also does well in captivity; one specimen of *T. vulpecula* survived for 14 years and 8 months (Jones 1982).

Because of its fecundity and adaptability to a variety of conditions, *T. vulpecula* has been compared to the Virginia opossum *(Didelphis virginiana)* in North America. Both species seem able to live near people and even to expand their range in the face of human development. However, the subspecies *T. vulpecula hypoleucus,* restricted to extreme southwestern Western Australia, is designated as near threatened by the IUCN. Ride (1970) wrote that Australians had more contact with *T. vulpecula* than with any other native mammal. This species can find shelter in artificial structures and can subsist on garden vegetation, but unlike the Virginia opossum, it now is considered by some to be a major pest. In Australia it is said not only to damage flowers, fruit trees, and buildings but also to adversely affect regenerating eucalyptus forests and introduced pine plantations and to carry diseases that are potentially harmful to humans and livestock (How 1976; McKay and Winter 1989; Ride 1970). However, this possum has been taken extensively in Australia for the fur trade, with more than a million pelts going on the market in some years. At present, commercial hunting is restricted to Tasmania and serves to suppress populations that are alleged to be damaging forestry and agriculture. As recently as 1980 more than 250,000 skins were exported with a value of about U.S. $6 each, but subsequently prices fell and the market declined. In 1990 and 1991 exports totaled less than 15,000 skins at a value of about $2 each (Callister 1991).

A desire to share in the fur market led to the introduction of *T. vulpecula* in New Zealand, first in 1840 and more extensively in the 1890s. It adapted well, greatly increased in range and numbers, and now often is treated more as a pest than as a fur bearer (Crawley 1973; Fitzgerald 1976, 1978; Gilmore 1977; Harvie 1973). It reportedly damages gardens, orchards, crops, pasture, exotic tree plantations, and remaining native forests. All legal restrictions on killing it in New Zealand were removed in 1947, a bounty was in effect from 1951 to 1960, and the species has been subject to intensive programs of poisoning, trapping, and shooting. Its overall numbers, however, do not seem to have been greatly affected. From 1962 to 1974 the annual number of skins exported from New Zealand ranged from 346,000 to 1,605,000.

DIPROTODONTIA; PHALANGERIDAE; **Genus WYULDA** *Alexander, 1919*

Scaly-tailed Possum

The single species, *W. squamicaudata,* is restricted to the Kimberley Division in northern Western Australia (Ride 1970). It was known only from four specimens until 1965, when eight more were collected (Fry 1971). It subsequently has been observed with more regularity. Flannery, Archer, and Maynes (1987) considered *Wyulda* a subgenus of *Trichosurus.* That position was followed by Corbet and Hill (1991) but not Groves (*in* Wilson and Reeder 1993) or McKay (*in* Bannister et al. 1988), and *Wyulda* was restored to generic rank by Flannery and Schouten (1994).

Head and body length is 290–395 mm, tail length is about 250–325 mm, and adult weight range is 1.4–2.0 kg (Humphreys et al. 1984). The pelage is short, soft, fine, and dense. The general dorsal color is pale or dark ashy gray. A dark stripe, obscure or distinct, runs along the middorsal line from the shoulders to the rump. The nape, shoulders, and rump of one specimen were mottled with buff, and its throat and chest were gray. The sides and back are usually the same color, though the sides are somewhat paler. The underparts are creamy white.

The prehensile tail is densely furred at the base and has nonoverlapping, thick scales for the remainder of its length. Short, bristly hairs are present around the scales. This is the only member of the family Phalangeridae with a tail of this kind. The head is short and wide. The claws are short and not strongly curved.

This possum inhabits areas with trees and rocks in broken sandstone country (Ride 1970). It is nocturnal and scansorial, apparently sheltering by day among the rocks and emerging at night to feed on leaves, blossoms, fruit, insects, and possibly small vertebrates (Collins 1973). After leaving the rocks, it reportedly climbs the nearest tree and then crosses from tree to tree without coming down. It appears to be solitary, and three female specimens each carried a single young (Ride 1970). One population occurred at a density of about 1/ha. The young are born from March to August and are weaned sometime after 8 months; males do not reach sexual maturity until past 18 months, and females not until their third year (Humphreys et al. 1984). A captive individual was said to be gentle and affectionate and to make a chittering noise like a bird (Fry 1971). It was still alive after being maintained in a private home for 4 years and 4 months (Collins 1973).

The genus has been designated as endangered by the USDI since 1970, though Winter (1979) stated that it is probably more plentiful than previously thought, and Thornback and Jenkins (1982) reported no evidence of any major threat. More recently, Flannery and Schouten (1994) reaffirmed that *Wyulda* should be regarded as endangered. While it reportedly is abundant at one site, it has a very

Scaly-tailed possum *(Wyulda squamicaudata)*, photo by A. G. Wells / National Photographic Index of Australian Wildlife.

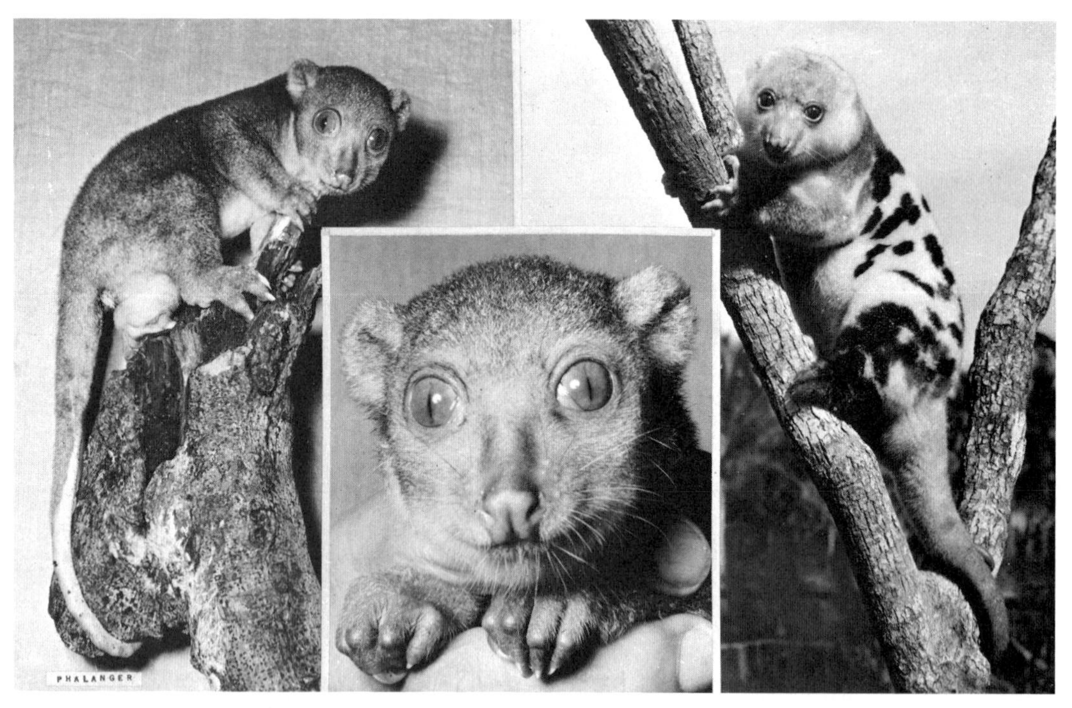

Spotted Cuscus *(Spilocuscus maculatus)*, photos by Sten Bergman.

limited and patchy distribution and may be declining as its habitat is disrupted by the expanding pastoral industry. The IUCN currently designates it as near threatened.

DIPROTODONTIA; PHALANGERIDAE; **Genus SPILOCUSCUS**
Gray, 1861

Spotted Cuscuses

There are four species (Feiler 1978*a*, 1978*b*; Flannery 1990*b*; Flannery, Archer, and Maynes 1987; Flannery and Calaby 1987; Flannery and Schouten 1994; G. G. George 1979, 1987):

S. rufoniger, northern New Guinea;
S. maculatus, New Guinea and nearby islands, Bismarck Archipelago, Seram and nearby islands, Cape York Peninsula of Queensland;
S. kraemeri, Manus and some small nearby islands in the Admiralty Group northeast of New Guinea;
S. papuensis, Waigeo Island off the western tip of New Guinea.

Spilocuscus usually has been placed in *Phalanger*, but Flannery, Archer, and Maynes (1987) and George (1987) regarded the two as generically distinct. Those authorities also suggested that *S. papuensis* is a species distinct from *S. maculatus*; that view was followed by Corbet and Hill (1991) but not Groves (*in* Wilson and Reeder 1993).

Head and body length is 338–640 mm, tail length is 315–590 mm, and weight is 2.0–7.0 kg (Flannery 1995; Flannery and Calaby 1987; Flannery and Schouten 1994; Winter *in* Strahan 1983). On average females are larger than males. Except for *S. papuensis*, the sexes also are colored differently. In *S. maculatus* the adult males, unless completely white, are gray spotted with white above and white below. The females are uniformly gray and usually unspotted. In *S. rufoniger* and the much smaller *S. kraemeri* there is a dark saddle on the back of females but only an area of mottling or spots on the males. In *S. papuensis* both sexes are marked with very small spots. The young go through a sequence of color changes. The fur is dense and woolly, the snout is short, and the ears are almost invisible. Flannery, Archer, and Maynes (1987) listed the following unique characters that distinguish *Spilocuscus* from related genera: there is sexual dichromatism as noted above; in both sexes the frontal bones of the skull are markedly convex and accommodate a large sinus that does not open into the nasal cavity; there is a well-developed protocone on the first upper molar; and the alisphenoid and basoccipital consistently form a more extensive suture that is developed earlier in life than it is in other phalangerids. Females have four mammae.

Winter (*in* Strahan 1983) wrote that *S. maculatus* occurs mainly in rainforest from sea level to an elevation of 820 meters but also has been seen in mangroves and open forests. It does not use a den or nest but is reputed to build a small sleeping platform of leaves. It is mainly nocturnal and arboreal, climbs slowly and deliberately, maintains a strong grip with its feet, and bounds at the speed of a fast human walk when on the ground. Under natural conditions it has been observed to eat fruit, flowers, and leaves, but captives have readily taken dog food and chickens. Males are aggressive and cannot be housed together in captivity. There apparently is an extended breeding season in both New Guinea and Australia. Although three young have been recorded from a pouch, it is likely that only one is reared. A specimen of *S. maculatus* lived for more than 11 years in captivity.

All species of *Spilocuscus* are on appendix 2 of the CITES (though under the single name *S. maculatus*). The IUCN designates *S. rufoniger* as endangered. According to Thornback and Jenkins (1982), that species is known only by 18 specimens, is subject to intensive hunting, and may be losing its rainforest habitat to logging and agricultural expansion. Flannery and Schouten (1994) reported that a few more specimens had been collected recently but considered the species to be seriously endangered by hunting for use as food and disruption of habitat. There is a disjunct population of *S. maculatus* on Salayar Island, south of Sulawesi, where it may have been introduced by human agency (George 1987). *S. kraemeri* apparently was brought to the area from which it is currently known only 1,000–2,000 years ago. Since it probably could not have evolved its distinctive characters over so short a period, there may be an as yet undiscovered parent population (Flannery 1995; Flannery and Schouten 1994).

DIPROTODONTIA; PHALANGERIDAE; **Genus PHALANGER**
Storr, 1780

Cuscuses

There are 11 named species (Colgan et al. 1993; Feiler 1978*a*, 1978*b*; Flannery 1987, 1990*b*; Flannery, Archer, and Maynes 1987; Flannery and Boeadi 1995; Flannery and Schouten 1994; G. G. George 1979, 1987; Groves 1976, 1987*a*, 1987*b*; Kirsch and Calaby 1977; Laurie and Hill 1954; Menzies and Pernetta 1986; Ride 1970; Springer et al. 1990; Ziegler 1977):

P. rothschildi, Obi and Bisa islands in the northern Moluccas;
P. ornatus, Halmahera, Bacan (Batjan), Ternate, Tidore, and possibly Morotai islands in the Northern Moluccas;
P. alexandrae, Gebe Island in the Northern Moluccas;
P. matanim, known only by five specimens from the mountains of western Papua New Guinea;
P. orientalis, Molucca Islands, Buru, Timor, Seram, northern New Guinea and many nearby islands, Bismarck Archipelago, Solomon Islands;
P. intercastellanus, southern New Guinea, Aru Islands, D'Entrecasteaux Islands, Louisiade Archipelago, eastern Cape York Peninsula of Queensland;
P. vestitus, highlands of western, central, and eastern New Guinea;
P. gymnotis, New Guinea and Aru Islands to southwest;
P. carmelitae, highlands of central and eastern New Guinea;
P. sericeus, highlands of central and eastern New Guinea;
P. lullulae, Woodlark Island east of New Guinea and (possibly through introduction) nearby Alcester and Madau islands.

Flannery (1995) and Flannery and Schouten (1994) indicated the existence of two undescribed species, one on Mount Karimui in east-central New Guinea, which may be closely related to *P. intercastellanus,* and one on Gebe and perhaps throughout the northern Molucca Islands, which has been referred mistakenly to *P. ornatus. Phalanger* often has been considered to include the genera *Ailurops, Strigocuscus,* and *Spilocuscus* (see accounts thereof). Flannery, Archer, and Maynes (1987) referred *P. ornatus* and *P. gymnotis* to *Strigocuscus,* but Flannery and Schouten (1994) reassigned both to *Phalanger* based in part on the molecular analyses of Springer et al. (1990). *P. vestitus* sometimes has been referred to as *P. interpositus* and includes *P. permixtio* of east-central New Guinea as a synonym. George (1987) regarded *P. leucippus* of mainland New Guinea as a species distinct from *P. gymnotis.*

Gray cuscus *(Phalanger mimicus)*, photo by Queensland National Parks and Wildlife Service / National Photographic Index of Australian Wildlife.

Head and body length is 325–600 mm, tail length is 240–610 mm, and weight is 1,045–4,850 grams (Flannery 1990*b*; Flannery and Schouten 1994; Lidicker and Ziegler 1968). The fur in most forms is thick and woolly. Coloration in the genus ranges from white, reds, and buffs through various shades of brown to light grays and different intensities of black. Some members of this genus have suffusions of yellow or tawny over the shoulder region, and others have dark dorsal stripes that extend from the head to the rump. In *P. vestitus* some individuals are pale silvery brown and short-haired, others are dark brown and long-haired, and still others represent intermediate forms.

These are heavy and rather powerfully built animals. Their yellow-rimmed protruding eyes, bright yellow noses, inconspicuous ears, and prehensile tails give them a somewhat monkeylike appearance. The terminal portion of the tail is covered with scales and lacks hair. The fingers are not of equal length; the claws are long, stout, and curved; and the soles are naked and striated. Females have four mammae in a well-developed pouch.

Cuscuses inhabit mainly tropical forests and thick scrub. They are arboreal animals with strongly prehensile tails but sometimes descend to the ground. They are nocturnal, resting by day curled up in the thick foliage of a vine tangle, in a tree or tree hollow, under tree roots, or among rocks. Cuscuses are slow-moving and somewhat sluggish, resembling the slow loris *(Nycticebus)* in their movements.

Cuscus *(Phalanger gymnotis)*, photo by Pavel German.

Even when handled gently cuscuses usually emit a penetrating musk odor. Snarls and barks are the sounds they make. Their diet consists mainly of fruits and leaves but also includes insects, small vertebrates, and birds' eggs. Cuscuses are solitary but can be maintained in pairs in roomy enclosures (Collins 1973). The breeding season appears to be extensive and may last throughout the year in some species; females have up to three young, though they may rear only one (Flannery 1995; Winter *in* Strahan 1983).

According to Flannery (1990*b*), *P. gymnotis* is common in many parts of New Guinea from sea level to 2,700 meters but avoids swampy areas and floodplains. It usually rests during the day in burrows under tree roots, in caves, or in human-made tunnels. It is capable of climbing and tends to feed in trees at night. It reportedly eats a wide variety of fruit in the wild, though captives have killed small vertebrates. Native people say that females carry fruit back to the den in their pouch. Severe fighting evidently is frequent, even between opposite sexes when females were in an anestrous condition. Breeding apparently is continuous throughout the year, and there is usually a single pouch young. A captive was known to be at least 11 years old.

Recent archeological research indicates that many of the island populations of *P. orientalis* were introduced through human agency in prehistoric time. Based on electrophoretic and morphological analyses, Colgan et al. (1993) suggested that the species was introduced to New Ireland from New Britain between 10,000 and 19,000 years ago and then spread to the Solomon Islands between 2,000 and 6,000 years ago. Flannery, Archer, and Maynes (1987) noted that *P. orientalis* probably was introduced on Timor about 4,000–5,000 years ago.

The species *P. orientalis* is on appendix 2 of the CITES. The IUCN classifies *P. matanim* as endangered and *P. rothschildi* and *P. vestitus* as vulnerable. All are declining because of loss of forest habitat and hunting by people for use as food. The restricted population of *P. intercastellanus* in Australia is designated as near threatened by the IUCN. Thornback and Jenkins (1982) wrote that *P. lullulae* had not been collected since 1953 and that its limited habitat is unprotected and suitable for logging or agriculture. However, Flannery and Schouten (1994) reported that it recently has been rediscovered in abundance on Woodlark Island.

DIPROTODONTIA; **Family POTOROIDAE**

"Rat"-Kangaroos

This family of 5 Recent genera and 10 species occurs in Australia, especially the southern and eastern parts, and on a few nearby islands and in Tasmania. In the past the Potoroidae often were regarded only as a subfamily of the Macropodidae, but there now is general recognition that the two are distinct families (Calaby and Richardson *in* Bannister et al. 1988; Corbet and Hill 1991; Groves *in* Wilson and Reeder 1993; Strahan 1983). The sequence of genera presented herein follows that of Flannery (1989*b*), who, like most other authorities, recognized two living subfamilies: Hypsiprymnodontinae, with the single genus *Hypsiprymnodon;* and Potoroinae with *Potorous, Bettongia, Caloprymnus,* and *Aepyprymnus.* Szalay (1994) continued to include the Potoroinae as a subfamily of the Macropodidae but considered the Hypsiprymnodontinae a completely separate family.

Seebeck and Rose (1989) provided a detailed description of the Potoroidae. Head and body length is 153–415 mm, tail length is 123–387 mm, and weight is 360–3,500 grams. The body is covered with dense fur. The upper parts range in color from dark chocolate brown, through browns and grays, to rufous and sandy; the underparts are usually pale. Unlike in the Macropodidae, coat color is usually uniform with no stripes or other markings. All living genera are compact animals with variously elongate muzzles, short and rounded ears, and short but muscular forelegs bearing small paws with short, forward-pointing spatulate claws. The hind limbs are well developed and heavily muscled, and the hind feet are elongate. In *Hypsiprymnodon* the proportional difference between the forelimbs and the hind limbs is less than in the other genera. Also, whereas all the other genera have lost the first digit of the hind foot, it is retained by *Hypsiprymnodon.* The second and third digits of the hind foot are syndactylous (united by integument). The fourth toe is the largest. The long tail is furred in most genera but naked and scaly in *Hypsiprymnodon;* it is prehensile in *Bettongia.* The stomach is simple in *Hypsiprymnodon* and complex in all other genera that have been studied. The female reproductive tract of potoroids differs from that of most macropodids in the presence of an anterior vaginal expansion.

The skull is short and broad in most genera but elongate and narrow in some species of *Potorous.* In all genera except *Hypsiprymnodon* the squamosal bone has wide contact with the frontal bone, thereby separating the parietal and alisphenoid bones. In the Macropodidae parietal-alisphenoid contact is usual. In the Potoroidae the masseteric canal of the mandible is confluent with the inferior dental canal and extends farther anteriorly than it does in the Macropodidae. The dental formula in most genera is: (i 3/1, c 1/0, pm 1/1, m 4/4) $\times$ 2 = 30; in *Hypsiprymnodon* the incisors are 3/2. The dentition is typically diprotodont, the pair of lower incisors being enlarged and projecting forward. There are two deciduous premolars on each side of both jaws, but these are shed near maturity and replaced by a single large, permanent tooth. This premolar is bladelike and has vertical corrugations. Unlike in the Macropodidae, there is no forward progression of molar teeth. The molars are mostly quadrituberculate, and they decrease in size from the anterior to the posterior.

Habitats range from hummock grassland to tropical rainforest, but most species are found in forest or woodland in southern Australia. Most are mainly nocturnal. Although omnivorous, many species now are known to feed primarily on fungi. Reproduction generally is continuous, mating is promiscuous, and there is little evidence of territoriality. The reproductive biology of the Potoroidae is much like that of the Macropodidae, but the "rat"-kangaroos generally have more than one young per year and have a relatively shorter pouch life. Despite their fecundity, potoroids are reduced in numbers over most of their range because of the effects of European settlement. Only 3 of the 10 modern species remain relatively common (Rose 1989; Seebeck, Bennett, and Scotts 1989).

The known geological range of the Potoroidae is middle Miocene to Recent in Australia.

DIPROTODONTIA; POTOROIDAE; **Genus HYPSIPRYMNODON**
Ramsay, 1876

Musky "Rat"-kangaroo

The single species, *H. moschatus,* occupies approximately 320 km of the coast of northeastern Queensland (Calaby 1971).

This is the smallest of the "rat"-kangaroos: head and body length is 208–341 mm, tail length is 123–65 mm, and weight is 337–680 grams (Johnson and Strahan 1982). The pelage is close, crisp, and velvety and consists mainly of underfur. The general coloration is rich brown or rusty gray, brightest on the back and palest on the underparts. Some

Musky "rat"-kangaroo *(Hypsiprymnodon moschatus)*, photo by P. M. Johnson.

individuals have a white area on the throat that extends as a narrow line to the chest.

This genus is unique among the Potoroidae in that the hind foot has a well-developed first digit that is movable and clawless but nonopposable to the other digits. The tail, which is used to gather nesting material, also differs from that of all other members of the family in being almost completely naked and scaly; only the extreme base is hairy. The muzzle is naked, and the ears are rounded, thin, and naked except at their posterior base. The limbs are more equally proportioned than those of other "rat"-kangaroos. The claws are quite small, weak, and unequal in length. The females have four mammae and a well-developed pouch. The specific name refers to the musky scent emitted by both sexes.

This "rat"-kangaroo occurs in rainforests, often in dense vegetation bordering rivers and lakes. It seems to be fairly common but is shy, quick in its movements, and difficult to observe in its dense habitat. It differs from other members of the Potoroidae in being completely diurnal (Seebeck and Rose 1989). One individual was seen sunbathing, lying spread-eagle on a fallen tree trunk. Most observers state that *Hypsiprymnodon* runs on all four limbs instead of hopping on its hind feet. According to Johnson and Strahan (1982), adults have been seen to climb on fallen branches and horizontal trees, and juveniles have ascended a thin branch inclined at about 45°. At night and in the middle of the day the musky "rat"-kangaroo sleeps in a nest in a clump of vines or between the plank buttresses of a large tree.

The diet is different from that of other "rat"-kangaroos. It consists to a great extent of insects and worms; the berries of a palm *(Ptychosperma)* and tuberous roots also are eaten. This animal sits on its haunches while eating. Food is obtained by turning over debris and by digging.

Johnson and Strahan (1982) stated that *Hypsiprymnodon* appears to be solitary, but feeding aggregations of up to three individuals have been observed. Breeding occurs from February to July (the rainy season), and usually two young are born. They leave the pouch after about 21 weeks, and then for several more weeks they spend a considerable part of the day in the nest. Females attain sexual maturity at just over a year.

The range of the musky "rat"-kangaroo has been affected by clearing of rainforest for agricultural development. Where it remains, however, it seems relatively common (Seebeck, Bennett, and Scotts 1989). Kennedy (1992) considered it "potentially vulnerable" because of its restricted rainforest habitat.

DIPROTODONTIA; POTOROIDAE; **Genus POTOROUS**
Desmarest, 1804

Potoroos

Three species now are recognized (Johnston and Sharman 1976, 1977; Kirsch and Calaby 1977; Seebeck and Johnston 1980):

P. tridactylus (long-nosed potoroo), southeastern Queensland, coastal New South Wales, Victoria, southeastern South Australia, southwestern Western Australia, islands of Bass Strait, Tasmania;
P. longipes, eastern Victoria;
P. platyops (broad-faced potoroo), southwestern Western Australia.

The subspecies *P. tridactylus apicalis* of Tasmania and the islands in Bass Strait was formerly considered a distinct species. Subfossil specimens indicate that the range of *P. platyops* formerly extended to southeastern South Australia and that *P. longipes* also occurred in southeastern New South Wales, but it is not clear whether these records date from historical time (Seebeck 1992*b;* Seebeck and Rose 1989).

Head and body length is 243–415 mm, tail length is 198–325 mm, and weight is 660–2,200 grams (Johnston *in* Strahan 1983; Kitchener *in* Strahan 1983; Seebeck *in* Strahan 1983). The pelage, at least in *P. tridactylus,* is straight, soft, and loose. The upper parts are grayish or brownish, and the underparts are grayish or whitish. The tail of *P. tridactylus* is often tipped with white. The muzzle is elongated in *P. tridactylus* and shortened in *P. platyops.* Females have four mammae located in a well-developed pouch that opens forward. The hind feet of *Potorous* are shorter than the head.

Following three years of field and laboratory observations, Buchmann and Guiler (1974) were able to clear up the uncertainty regarding potoroos' means of locomotion. There are three principal methods: (1) the "quadrupedal crawl," a slow (20–23 meters per minute) plantigrade movement in which the hind feet are employed for the main forward propulsion and then the weight is shifted to the forefeet, which is used in leisurely feeding and foraging; (2) the "bipedal hop," a series of synchronous digitigrade thrusts by the hind limbs, moving the animal 30–56 meters per minute, used for escape and chasing in aggression; and (3) "jumping," a single bound 2.5 meters long and 1.5 meters high resulting from a powerful thrust of the hind legs, used in initial escape or aggression.

Potoroos inhabit dense grassland or low, thick scrub especially in damp places (Ride 1970). They are nocturnal but occasionally may engage in early morning basking in the sun (Collins 1973). During the day potoroos shelter in shallow "squats," usually excavated at the base of a tussock or under dense shrubs, and do not construct complex nests (Seebeck, Bennett, and Scotts 1989). *P. tridactylus* digs small holes in the ground when feeding. These holes are not round like those made by bandicoots. Well-defined trails lead from one feeding site to another. The overall home range of *P. tridactylus* is relatively large, perhaps because fungi form a major food source. One study in Tasmania found that fungi were of substantial importance from May to December and constituted more than 70 percent of the food in May and June (Guiler 1971*a*). *P. tridactylus* also depends on insects more than do most potoroids, especially in summer (Guiler 1971*b*). In addition, the diet includes grass, roots, and other forms of vegetation.

P. tridactylus has been found at population densities of 0.2–2.55/ha. (Seebeck, Bennett, and Scotts 1989). Its home range in Victoria averages only about 2.0 ha. for males and 1.5 ha. for females, but those of *P. longipes* may exceed 10.0 ha. Male ranges may overlap those of several females, but female ranges are often exclusive (Seebeck and Rose 1989). Observations of *P. tridactylus* indicate that males are territorial and defend a small part of their home range but tend to avoid conflict under natural conditions. They may be kept in captivity together with other males and several females, but when a female is in estrus the males will fight fiercely, with one eventually establishing dominance (Collins 1973; Ride 1970; Russell 1974*b*).

Potoroos (*Potorous* sp.), photo by Ernest P. Walker.

Potoroos are polyestrous, with an estrous cycle of about 42 days, and apparently lack a well-defined breeding season. In Tasmania *P. tridactylus* bears young throughout most of the year, with peaks from July (winter) to January (summer). Each female is able to breed twice a year. Nondelayed gestation is 38 days and normal litter size is 1. Four days after the season's first young is born the female mates again, but because of embryonic diapause development is arrested and parturition does not occur for about 4.5 months unless the first young is lost. The newborn measures 14.7–16.1 mm in length, is able to detach from the nipple at 55 days, and leaves the pouch at about 130 days. Females are able to breed at 1 year (Bryant 1989; Collins 1973; Ride 1970). Reproduction in *P. longipes* is similar, but pouch life is 140–50 days and sexual maturity may not be attained until 2 years (Seebeck 1992*a*). A wild specimen of *P. tridactylus* is known to have lived at least 7 years and 4 months (Guiler and Kitchener 1967), and captives have reached an age of at least 12 years (Johnston *in* Strahan 1983).

The species *P. platyops* of southwestern Australia apparently has been extinct since about 1875 and may have declined considerably prior to the coming of European settlers; it is designated extinct by the IUCN. The subspecies of *P. tridactylus* in southwestern Australia, *P. t. gilbertii*, also may be extinct, none having been collected in more than 80 years (Calaby 1971). Recently, however, there have been unconfirmed sightings in southwestern Australia of a mammal that might represent *Potorous* (Thornback and Jenkins 1982). The IUCN now treats *gilbertii* as a separate and living species and classifies it as critically endangered; it is thought to number fewer than 50 individuals and to be declining. On the mainland of southeastern Australia *P. tridactylus* apparently has declined because of its dependence on dense vegetation and its resultant vulnerability to clearing operations and brush fires, but it is not as rare as once thought. It no longer occurs in South Australia but is widespread from Victoria to southern Queensland and remains common in much of Tasmania. In Bass Strait it still is present, though rare, on Flinders and King islands, but it seems to have disappeared from Clarke Island, which is heavily grazed by livestock and introduced rabbits (Calaby 1971; Poole 1979; Seebeck 1981). It was considered "potentially vulnerable" by Kennedy (1992). The IUCN designates the eastern mainland subspecies, *P. tridactylus tridactylus*, as vulnerable, noting that its total number is less than 10,000. *P. longipes*, with fewer than 2,500, is classified as endangered by the IUCN. It apparently is very rare, and its restricted habitat is threatened by logging (Thornback and Jenkins 1982).

DIPROTODONTIA; POTOROIDAE; **Genus BETTONGIA**
Gray, 1837

Short-nosed "Rat"-kangaroos

There are four species (Kirsch and Calaby 1977; Ride 1970; Sharman et al. 1980):

B. penicillata (woylie), formerly from southwestern Western Australia to central New South Wales;
B. tropica, eastern Queensland;
B. gaimardi, southeastern Queensland, coastal New South Wales, southern Victoria, Tasmania;
B. lesueur (boodie), Western Australia, including certain coastal islands, to southwestern New South Wales.

Another species, *B. cuniculus*, formerly was recognized from Tasmania but is now included in *B. gaimardi*. Sharman et al. (1980) concluded that there is no chromosomal basis for the specific distinction of *B. tropica* from *B. penicillata*. This position was accepted by Groves (*in* Wilson and Reeder 1993) but not by Calaby and Richardson (*in* Bannister et al. 1988), Corbet and Hill (1991), or Flannery (1989*a*).

Head and body length is 280–450 mm and tail length is 250–330 mm. Weight is 1.1–1.6 kg in *B. penicillata* and 1.2–2.25 kg in *B. gaimardi* (Christensen *in* Strahan 1983; K. A. Johnson *in* Strahan 1983; Rose *in* Strahan 1983). The upper parts are buffy gray to grayish brown; the underparts are paler. The tail is crested in all species except *B. lesueur* and usually is white-tipped in *B. gaimardi* and *B. lesueur*. The unworn adult pelage in *B. penicillata* is often crisp or even harsh; in *B. lesueur* it is soft and dense. The tip of the muzzle is naked and flesh-colored and the ears are short and rounded. The hind feet are longer than the head. Females have four mammae and a well-developed pouch.

Habitats include grassland, heath, and sclerophyll woodland. These "rat"-kangaroos are nocturnal. They construct nests, usually of grass but sometimes using sticks or bark, and they carry nesting material in the curled-up tip of their tail. The nests generally are located at the base of a grass tussock or overhanging bush. There sometimes is an earth excavation at the base, but *B. lesueur* constructs a large burrow for community dwelling and may modify a rabbit warren for its own use. This species is entirely bipedal in locomotion, never using its forefeet, even when moving slowly (Ride 1970). The claws of all species are used for digging. When fleeing, *B. penicillata* travels with head held low, back arched, and the tail brush displayed conspicuously. Most species seem primarily herbivorous, feeding on roots, tubers, seeds, and legume pods, but there are some reports of feeding on marine refuse, carrion, and meat (Collins 1973; Ride 1970). There is increasing evidence that *Bettongia* and some other potoroids depend to a large extent on fungi (Seebeck, Bennett, and Scotts 1989). In Tasmania, sporocarps of mycorrhizal fungi make up the bulk of the diet of *B. gaimardi* year-round, though other foods also are thought to be necessary for nutritional balance (R. J. Taylor 1992, 1993*a*).

B. penicillata has been found at population densities of 0.07–0.45/ha. and in home ranges averaging 35 ha. for males and 20 ha. for females (Seebeck, Bennett, and Scotts 1989), an unusually large area for an animal of its size. Taylor (1993*b*) found an average of 61 ha. for individuals of *B. gaimardi* tracked longer than two months, with the ranges of males being generally larger than those of females.

In *B. lesueur* a male and several females form a social group and occupy a burrow system. Large warrens may have 40–50 occupants (Seebeck, Bennett, and Scott 1989). Males are aggressive toward one another and seem to defend groups of females but not a particular territory. Females generally are amicable but sometimes will establish a territory and exclude other females. They are polyestrous, with the modal length of the estrous cycle being 23 days, and can produce up to three litters per year. Breeding may go on throughout the year in some areas, but in the Bernier Island population of *B. lesueur* most births occur between February and September. The modal length of undelayed gestation is 21 days and the usual litter size is one, though twins occasionally occur. Just after one young is born the female mates again, but because of embryonic diapause development is delayed, and parturition of the second young does not take place for about four months un-

Short-nosed "rat"-kangaroo *(Bettongia lesueur)*, photo from New York Zoological Society. Inset: *B. penicillata*, photo from *A Natural History of the Mammalia*, G. R. Waterhouse.

less the first young is lost. The young weigh about 0.317 grams at birth, leave the pouch permanently at about 115 days, and attain adult size at 280 days. Females apparently are capable of giving birth at about 200 days (Ride 1970; Tyndale-Biscoe 1968). Reproductive information on other species is more limited, but in *B. gaimardi* and *B. penicillata* the estrous cycle, gestation period, embryonic diapause, and pouch life are nearly the same as in *B. lesueur* (Rose 1978, 1987). Record longevity for the genus is held by a specimen of *B. gaimardi* that was still living after 11 years and 10 months in captivity (Jones 1982).

Modern human agency seems to have harmed *Bettongia* more than any other polytypic genus of marsupials. All four species are listed as endangered by the USDI and are on appendix 1 of the CITES. The IUCN designates *B. tropica* as endangered, *B. lesueur* generally as vulnerable, the subspecies *B. l. graii* of mainland Australia as extinct, *B. penicillata* generally as conservation dependent, the subspecies *B. p. penicillata* of eastern Australia as extinct, the mainland subspecies *B. gaimardi gaimardi* as extinct, and the Tasmanian subspecies *B. g. cuniculus* as near threatened. *B. tropica* had been known from only six specimens, the latest collected in 1932, and was thought to be possibly extinct, but a population recently was discovered in the Davies Creek National Park in northeastern Queensland (Poole 1979). The species *B. gaimardi* has not been recorded from the mainland since 1910; it still is found in reasonable numbers in many localities of Tasmania but may be jeopardized there by logging and poisoning (Calaby 1971; Ride 1970; Rose 1986).

B. lesueur once was among the most widely distributed of native Australian mammals, occurring in all mainland states except perhaps Queensland. Although the last specimen in New South Wales was collected in 1892, *B. lesueur* remained common in parts of central and southwestern Australia until the 1930s. By the early 1960s it had disappeared completely from the mainland and become restricted to Barrow, Boodie, Bernier, and Dorre islands off the west coast. Its drastic decline apparently resulted from competition with introduced rabbits for burrows and food, habitat disruption and disturbance by domestic livestock, predation by introduced foxes, and direct killing by people. Until recently the remnant island populations were thought to be viable and well protected (Australian National Parks and Wildlife Service 1978; Calaby 1971; Poole 1979). However, surveys by Short and Turner (1993) revealed that the species had been wiped out on Boodie Island, apparently as an inadvertent consequence of a recent poisoning campaign against *Rattus rattus*. There were about 5,000 individuals left in the other three island populations, but they were considered seriously endangered, especially because of the potential introduction of cats or other predators.

Another severe decline has been suffered by *B. penicillata*, which formerly occurred all across the southern part of the continent, including certain islands and the northwestern corner of Victoria. Its range may even have extended well into the Northern Territory and Queensland (Seebeck, Bennett, and Scotts 1989). It was described as "very abundant" in New South Wales in 1839–40 but disappeared from that state shortly thereafter. It was common in South Australia at the end of the nineteenth century but was gone from there by 1923. Today it is restricted to a few tracts of woodland at the extreme southwestern tip of the continent. The primary cause of its decline seems to have been clearing of brush for agricultural development (Australian National Parks and Wildlife Service 1978; Calaby 1971; Poole 1979; Ride 1970). Recently, *B. penicillata* was reintroduced on several islands off the coast of South Australia (Delroy et al. 1986).

DIPROTODONTIA; POTOROIDAE; **Genus CALOPRYMNUS**
Thomas, 1888

Desert "Rat"-kangaroo

The single species, *C. campestris,* was described in 1843 on the basis of three specimens from an unknown locality in South Australia. There were no further records until 1931, when a specimen was collected in the northeastern corner of South Australia. A subsequent survey indicated that in the early 1930s the species occurred in an area about 650 km north to south and 250 km wide in the Lake Eyre Basin in northeastern South Australia and southwestern Queensland. In addition, apparently Recent cave remains were found in southeastern Western Australia, and late Pleistocene fossils were located in New South Wales (Australian National Parks and Wildlife Service 1978; Calaby 1971; Ride 1970).

Head and body length is 254–82 mm, tail length is 297–377 mm, and weight is 637–1,060 grams (Smith *in* Strahan 1983). The pelage is soft and dense. The coloration of the upper parts is clear pale yellowish ochre, which matches that of clay pans and plains; the underparts are whitish. The ears are longer and narrower than those of any other "rat"-kangaroo, and the muzzle is naked. The long, cylindrical tail is evenly short-haired without a trace of a crest. There is a well-defined neck gland, at least in the skin of the type specimen. The most conspicuous feature, however, is the relatively enormous hind foot. The forelimb is small and delicate; the bones of its three segments weigh only 1 gram, whereas the bones of the hind limb weigh 12 grams.

A peculiar feature of the hopping gait of *Caloprymnus* is that the feet are not brought down in line with one another; rather, the right toe mark registers well in front of the left toe mark. This "rat"-kangaroo seldom dodges or doubles back when moving rapidly and is noted for endurance rather than speed. A young adult male exhausted two galloping horses in a 20-km run. The gait, when the animal moves on all four limbs and uses the tail as a support, is normal for the Potoroidae.

The area known to have been inhabited consists mainly of gibber plains, clay pans, and sandridges. There is a sparse cover of saltbush and other shrubs. *Caloprymnus* is a nest builder, not a burrower, and shelters in simple leaf and grass nests in scratched-out excavations despite the glaring heat of its habitat. It has the unusual habit of protruding its head through a gap in the roof of the nest to observe its surroundings. It apparently is nocturnal (Collins 1973). It has been reported to feed mainly on the foliage and stems of plants and to feed less on roots than do the other "rat"-kangaroos. However, recent examination of the stomach and colon contents of a preserved specimen revealed extensive remains of beetles (Dixon 1988).

Females with a single pouch young have been taken in June, August, and December. The breeding season is apparently irregular: one female had a small, naked young in her pouch at the same time that two other females were carrying well-furred and almost independent "joeys." According to Jones (1982), long ago a specimen lived in captivity for 13 years.

The desert "rat"-kangaroo apparently was rare for many years but then became fairly common in its restricted range when severely dry conditions abated in the early 1930s. By 1935, however, it again was rare, and there have been no reliable records since that year. Possibly it is now extinct, or perhaps a small population still survives, awaiting the time when it again may increase in response to proper conditions. The species is designated as extinct by the IUCN and as endangered by the USDI and is on appendix 1 of the CITES.

DIPROTODONTIA; POTOROIDAE; **Genus AEPYPRYMNUS**
Garrod, 1875

Rufous "Rat"-kangaroo

The single species, *A. rufescens,* formerly occurred from northeastern Queensland to northeastern Victoria (Ride 1970). Remains have been found in cave deposits in southwestern Victoria and on Flinders Island, near Tasmania (Calaby 1971).

Desert "rat"-kangaroo *(Caloprymnus campestris)*, from *The Red Centre*, Hedley H. Finlayson (Angus & Robertson, publishers).

Rufous "rat"-kangaroo *(Aepyprymnus rufescens)*, photo by John Warham.

This is the largest of the "rat"-kangaroos. Head and body length is 380–520 mm and tail length is 350–400 mm. According to Johnson (1978), adults stand 350 mm tall, and weight in a series of specimens ranged from 2.27 to 2.72 kg for males and 1.36 to 3.60 kg for females. The pelage is crisp and often harsh. The upper parts, grizzled in appearance, are rufescent gray, and the underparts are whitish. The tail is thick, evenly haired, and not crested. This genus can be distinguished by its ruddy color, black-backed ears, whitish hip stripe (usually not distinct), and hairy muzzle. Females have four mammae in a well-developed pouch.

At present this "rat"-kangaroo is found from sea level to the tops of plateaus, in open forest and woodland with a dense grass floor. It is nocturnal and constructs nests in which it shelters during the day. These nests consist of a shallow excavation lined and covered with grass or bark (Seebeck, Bennett, and Scotts 1989). An individual constructs one or two clusters of nests that provide protection from the sun and concealment from predators (Wallis et al. 1989). Although not particularly fast, *Aepyprymnus* is wonderfully agile, adept at dodging, and difficult to approach on foot. On horseback, however, a person may get quite close. When startled it usually seeks shelter in a hollow log, if available. It has little fear of people at night, is often attracted by a bush camp, and, if not molested by dogs, can be enticed up to a tent door to receive scraps of food. When taken young and treated kindly, it becomes tame and responsive. In the wild its diet consists mainly of grasses (Johnson 1978), but in captivity it accepts a variety of foods and apparently thrives on them. Like most of the coastal species, it has little resistance to drought; during dry periods it excavates holes in creek beds to reach the water level.

The rufous "rat"-kangaroo is generally solitary, though a female often is accompanied by a nearly full-grown offspring and loose feeding aggregations sometimes form. Captive males are extremely aggressive toward one another. The species is polyestrous, with an estrous cycle of about 21–25 days. Breeding apparently occurs over most of the year. In the Dawson Valley of central coastal Queensland females with pouch young were taken from January to March. Near Killarney in southeastern Queensland three females with pouch young were caught in June, and local residents reported finding pouch young all year. Females in a captive colony at Canberra had large pouch young in February, July, and August and smaller ones in July, August, and December. The gestation period is 22–24 days. Litters usually consist of a single offspring, but twins occasionally occur. The pouch young releases the nipple at 7–8 weeks, vacates the pouch permanently at about 16 weeks, and then remains with the mother for another 7 weeks. Sexual maturity is attained at about 11 months in females and 12–13 months in males (P. M. Johnson 1980; P. M. Johnson *in* Strahan 1983; Moors 1975; Rose 1989). One specimen reportedly lived more than eight years in captivity (Collins 1973).

The rufous "rat"-kangaroo still is widely distributed along the coast from northeastern Queensland to central New South Wales. It is fairly common in much of this region and seems able to coexist with grazing beef cattle (Calaby 1971). Farther south the species has disappeared, the last record in Victoria dating 1905. Its decline may have resulted from predation by introduced foxes. Because of the continuing pressures on its habitat it was designated as "potentially vulnerable" by Kennedy (1992).

DIPROTODONTIA; **Family MACROPODIDAE**

Wallabies and Kangaroos

This family of 12 Recent genera and 61 species is native to Australia, Tasmania, New Guinea, some nearby islands,

Western gray kangaroo *(Macropus fuliginosus)* and her young, about 12 months old, photo by Mary Eleanor Browning.

and the Bismarck Archipelago. The Potoroidae, or "rat"-kangaroos (see account thereof), sometimes have been considered a subfamily of the Macropodidae but now are generally considered a separate family. Flannery (1983, 1989*b*), whose phylogenetic sequence is followed here, recognized two living macropodid subfamilies: Sthenurinae, with the single living genus *Lagostrophus,* and Macropodinae for the other 11 genera. Analysis of albumin immunologic relationships (Baverstock et al. 1989) supports the same basic division but suggests certain sequential alternatives, especially an association of *Petrogale* with *Dorcopsulus.* A somewhat different arrangement, suggested by the work of Kirsch (1977*b*), would put the genus *Onychogalea* in a position intermediate to the Potoroidae and Macropodidae, while the genera *Macropus* and *Wallabia* would form one main line at the opposite end of the family from the Potoroidae; the genus *Lagorchestes* would form another main line, and all the other genera would compose a fourth branch of the Macropodidae.

The adults in this family vary in head and body length from less than 300 mm to as much as 1,600 mm in *Macropus.* Full-grown individuals of the latter genus weigh as much as 100 kg (Hume et al. 1989). The head is rather small in relation to the body, and the ears are relatively large. The tail is usually long, thick at the base, hairy, and nonprehensile. It is used as a prop or additional leg, as a balancing organ when leaping, and sometimes for thrust. In all Recent members of this family except *Dendrolagus* the hind limbs are markedly larger and stronger than the forelimbs. The forelimbs are small with five unequal digits. The hind foot is lengthened and narrowed in all genera, hence the family name Macropodidae, meaning "large foot." Digit 1 of the hind foot is lacking in all genera, the small digits 2 and 3 are united by skin (the syndactylous condition), digit 4 is long and strong, and digit 5 is moderately long. The well-developed marsupial pouch opens forward and encloses four mammae.

In kangaroos and wallabies the dentition is suitable for a grazing or browsing diet. The first upper incisors are prominent and with the other incisors form a U- or V-shaped arcade; the two remaining lower incisors are very large and forward-projecting (the diprotodont condition), and except in the Sthenurinae their tips fit within the upper incisor arcade and press on a pad on the front of the palate; there are no lower canines, and the upper canines are small or absent, thus leaving a space between the incisors and cheek teeth; the premolars are narrow and bladelike; and the molars are broad, generally high-crowned, rectangular teeth with two transverse ridges separated by a deep trough crossed by a longitudinal ridge that is weakly formed in the browsing species but strong in grazers (Hume et al. 1989). In many macropodid genera, as in elephants, manatees, and certain pigs, an anterior migration of the molariform teeth occurs throughout life, making room for the late-erupting rear molars. The fourth molar may not erupt until well after adulthood is attained. The dental formula in this family is: (i 3/1, c 0–1/0, pm 2/2, m 4/4) × 2 = 32 or 34.

Most species are nocturnal, but they may sunbathe on warm afternoons and some are active at intervals during

Tree kangaroo (*Dendrolagus goodfellowi*), photo from Zoological Society of London.

the day. All the members of this family except *Dendrolagus* progress rapidly by leaps and bounds, using only the hind limbs. Representatives of the genus *Macropus* illustrate the peak of development of the jumping mode of progression. Macropodids commonly move on all four limbs and often use the tail as a support when feeding, progressing slowly, or standing erect.

The members of this family are mainly grazers or browsers, feeding on many kinds of plant material. The occurrence of ruminantlike bacterial digestion in the kangaroos and wallabies enables them to colonize areas that would be nutritionally unfavorable to most other large mammals. In this kind of digestion the food is fermented by a dense bacterial population in the esophagus, stomach, and upper portion of the small intestine, thus providing the available energy for chemical breakdown of foods over a longer period of time and enhancing the uptake of nitrogen and other nutrients.

Females of most macropodid genera usually give birth to one young at a time. A phenomenon known as embryonic diapause, or delayed birth, approximately equivalent to delayed implantation in placental mammals, occurs in some, perhaps most, members of this family. Following the birth of one young the female often mates again. Usually this mating occurs only a day or two after the birth, but in some species it is late in the pouch life of the young. In one species, *Wallabia bicolor*, the mating takes place just before the young is born (Kaufmann 1974). The development of the embryo resulting from this mating is arrested at about the 100-cell stage. The embryo remains in this state of diapause until the first young nears the end of its pouch life, dies, or is abandoned. At such a time the embryo resumes development, and a second young is soon produced. This reproductive approach seems well adapted to the variable, often severe climate of inland Australia. Should adverse conditions lead to the death of a pouch young or force the female to discard it, a successor is in reserve for another try at rearing during the season. If conditions are consistently favorable, both offspring can be raised, the second being born shortly after the first permanently vacates the pouch. Other views regarding embryonic diapause are that it functions simply to prevent two young from crowding the pouch at the same time or that it helps to ensure synchrony between the embryo and the uterus (Russell 1974*a*). A feature unique to the macropodids and potoroids is the capability of females to produce milk in one mammary gland that has a very different nutritional composition from that of the other gland when young at different stages of development are being reared (Merchant 1989).

Modern humans have greatly affected many species of this family. Perhaps the single most important instance was the introduction of vast herds of domestic livestock, which cropped grasslands, thereby eliminating the cover needed by smaller macropodids. Some of the larger species of *Macropus*, however, appear to have at least temporarily benefited from the elimination of tall, dry grass by livestock and the resultant production of more favorable vegetation (Newsome 1975).

The geological range of the family Macropodidae is Miocene to Recent in Australasia (Kirsch 1977*a*).

DIPROTODONTIA; MACROPODIDAE; **Genus LAGOSTROPHUS**
Thomas, 1887

Banded Hare Wallaby, or Munning

The single species, *L. fasciatus*, occupied Western Australia, including islands in Shark Bay. It may also have been present in historical time in South Australia, where its remains have been found at an archeological site on the lower Murray River (Calaby 1971).

Head and body length is 400–460 mm and tail length is 320–400 mm. Prince (*in* Strahan 1983) listed weight as 1.3–2.1 kg, occasionally up to 3.0 kg. The fur is thick, soft, and long. *Lagostrophus* can be readily distinguished by its banded color pattern. The dark transverse bands on the posterior half of the body contrast sharply with the general grayish coloration. The underparts are buffy white with gray hair bases, and the hands, feet, and tail are gray. This species, like those in the genus *Lagorchestes*, is referred to as a hare wallaby because of its harelike speed, its jumping ability, and its habit of crouching in a "form." The muzzle in *Lagostrophus* is rather long and pointed, and the nose is naked rather than hairy. The tail is evenly haired throughout, except for an inconspicuous pencil of longer hairs at the tip. The claws of the hind feet are hidden by the fur. Unlike the condition in all other genera of the Macropodidae, the lower incisors of *Lagostrophus* occlude with the upper incisors, not with a pad between the upper incisors (Hume et al. 1989). Additional technical characters of the dentition that distinguish *Lagostrophus* were listed by Flannery (1983, 1989*b*).

On the mainland this marsupial inhabits prickly thickets on the flats and the edges of swamps. On the islands it lives in thickets of a thorny species of *Acacia*. Runs and "forms" are made in these tangled masses. The species is nocturnal, emerging at night from retreats in the scrub to feed on various plants and fruits.

Banded hare wallaby *(Lagostrophus fasciatus)*, photo by A. G. Wells / National Photographic Index of Australian Wildlife.

The banded hare wallaby is gregarious, congregating in spaces under the low-hanging limbs of bushes in dense thickets (Ride 1970). According to Collins (1973), available data indicate that *Lagostrophus* is a seasonal breeder, with a reproductive peak in the first half of the year. A postpartum estrus may occur, followed by embryonic diapause while the first young still is suckling. Normal litter size is one, but indications of twin offspring were present in one specimen. The Australian National Parks and Wildlife Service (1978) stated that in the island populations young animals of all ages can be observed at any one time, thus suggesting that the breeding season is an extended one, occurring at least from February to August. According to Prince (*in* Strahan 1983), young spend about six months in the pouch and are weaned after another three months; both sexes are capable of breeding in the first year but usually do not do so until the second year.

The banded hare wallaby has not been recorded on the mainland since 1906. Its disappearance may have been associated with clearing of vegetation for agriculture, competition for food with rabbits and livestock, and predation by introduced foxes. It still is found on Bernier and Dorre islands, where it is now well protected, and its populations fluctuate between high numbers and relative scarcity (Ride 1970; Thornback and Jenkins 1982). Surveys by Short and Turner (1992) indicated minimum numbers of about 3,900 on Bernier and 3,800 on Dorre Island. Earlier, when sheep were brought temporarily to Bernier Island, the number of banded hare wallabies was reduced severely. A population on nearby Dirk Hartog Island disappeared entirely in the 1920s following the establishment of sheep. In the 1970s an unsuccessful attempt was made to reintroduce *L. fasciatus* on that island (Poole 1979); further efforts are being planned (Kennedy 1992; Short et al. 1992). The species is designated as endangered by the USDI and is on appendix 1 of the CITES. The IUCN classifies the mainland subspecies *L. f. albipilis* as extinct and the island subspecies *L. f. fasciatus* as vulnerable.

DIPROTODONTIA; MACROPODIDAE; **Genus DORCOPSIS**
Schlegel and Müller, 1842

New Guinean Forest Wallabies

Four species now are recognized (Groves and Flannery 1989):

D. luctuosa, coastal lowlands of southeastern New Guinea;
D. muelleri, lowlands of western New Guinea, including Misool, Salawatti, and Japen islands;
D. hageni, lowlands of northern New Guinea;
D. atrata, Goodenough Island off southeastern New Guinea.

The genus *Dorcopsulus* (see account thereof) sometimes has been considered a subgenus or synonym of *Dorcopsis.* The name *D. veterum* sometimes has been used in place of *D. muelleri,* but the latter designation was recommended by George and Schürer (1978) and accepted by Groves and

New Guinean forest wallaby (*Dorcopsis* sp.), photo from Zoological Society of London.

Flannery (1989). Groves (*in* Wilson and Reeder 1993) reported the range of *D. muelleri* to extend to the Aru Islands. Skeletal remains indicate that *D. muelleri* or a closely related species was present on Halmahera Island in the North Moluccas until at least 1,870 years B.P. (Flannery, Bellwood, et al. 1995).

Head and body length is 340–970 mm and tail length is 270–550 mm. Flannery (1990*b*) listed weights of 3.6–11.6 kg for *D. luctuosa,* 5.0 kg for *D. muelleri,* and 5.0–6.0 kg for *D. hageni.* Adults of *D. atrata,* a medium-sized species, weigh about 3.9–7.5 kg. The pelage is short and sparse in *D. hageni* but long and thick in the other species. In *D. muelleri* the coloration of the upper parts is dull brown or blackish brown tipped with light buff and the underparts are gray or white. *D. hageni* is light brown or fuscous above with a narrow white dorsal stripe and grayish white below. *D. luctuosa* is dark gray above and drab gray to creamy orange below. In *D. atrata* the upper parts are black or blackish brown and the underparts are also blackish brown. In all species the nose is large, broad, and naked; the ears are small and rounded; and the hairs are reversed on the nape. The tail is evenly haired except for a fifth or less of the terminal half, which is naked. Females have four mammae and a well-developed pouch that opens forward.

On New Guinea these wallabies generally inhabit lowland rainforests up to 400 meters in elevation. *D. atrata* of Goodenough Island lives in oak forest at elevations of 1,000–1,800 meters in the forested mountains (Flannery 1995). These animals do not appear to be as adapted for hopping as most other wallabies. They are presumed to be mainly nocturnal, but there is some evidence that they move about in daytime in dense forest. Observations of a captive colony revealed a crepuscular pattern of activity (Bourke 1989). The diet consists of roots, leaves, grasses, and fruit. Ganslosser (*in* Grzimek 1990) pointed out a number of unusual behaviors of *Dorcopsis.* When it is sitting or hopping slowly only the end of the tail is in contact with the ground. During mating the male bites the female's neck, a primitive habit also seen in the Didelphidae and Dasyuridae.

Observations of *D. luctuosa* in captivity indicated little agonistic behavior and a tendency to form social groups (Bourke 1989). Females usually give birth to one young at a time (Collins 1973). *D. muelleri* and *D. atrata* may breed throughout the year (Flannery 1995). According to Flannery (1990*b*), naked pouch young of *D. hageni* were found in January and April, a captive male *D. luctuosa* first emerged from the pouch on 22 May 1983 and reached sexual maturity by June 1985, and a specimen of *D. muelleri* was born in captivity on 16 October, first left the pouch on 13 April, and permanently left on 17 May. Another captive *D. muelleri* lived for 7 years and 7 months (Jones 1982).

Natives on Goodenough Island regard *D. atrata* as a valuable food animal. G. G. George (1979) considered it a threatened species, and the IUCN now classifies it as endangered. Thornback and Jenkins (1982) indicated that it is susceptible to hunting and possibly habitat destruction. Flannery (1995) noted that it occupies less than 100 sq km of habitat and is heavily hunted but may still be common.

DIPROTODONTIA; MACROPODIDAE; **Genus DORCOPSULUS**
Matschie, 1916

New Guinean Forest Mountain Wallabies

There are two species (Flannery 1989*b*, 1990*b*):

D. vanheurni, mountains of New Guinea;
D. macleayi, mountains of southeastern Papua New Guinea.

Kirsch and Calaby (1977) suggested that it would be reasonable to put these two species in the genus *Dorcopsis,* and such was done by Ziegler (1977) and Corbet and Hill (1991). Kirsch and Calaby (1977) also stated that *D. van-*

heurni probably is conspecific with *D. macleayi*. Nonetheless, *Dorcopsulus* was accepted as a distinct genus, and *D. vanheurni* as a separate species, by Flannery (1989*b*, 1990*b*) and Groves (*in* Wilson and Reeder 1993).

Head and body length is 315–460 mm, tail length is 225–402 mm, and weight is 1,500–3,400 grams (Flannery 1990*b*). Both species are deep gray-brown above with a darker mark above the hips and light brownish gray below. The chin, lips, and throat are whitish. The pelage is long, thick, soft, and fine. The tail is evenly haired, the terminal quarter to half being bare with a small white tip. Females have four mammae and a well-developed pouch that opens forward. From *Dorcopsis*, *Dorcopsulus* is distinguished by being smaller and more densely furred and in having more of the tail naked. Flannery (1990*b*) noted that with its small size, low-crowned molar teeth, and elongate premolars *D. vanheurni* offers a striking parallel with several members of the Potoroidae in Australia and may fill the same ecological niche.

These wallabies inhabit mountain forests, the known altitudinal ranges being 800–3,100 meters for *D. vanheurni* and 1,000–1,800 meters for *D. macleayi*. Although they are presumed to be nocturnal, Lidicker and Ziegler (1968) observed that *D. vanheurni* apparently is active by day. *D. macleayi* is reported to be fond of the fruit and leaves of *Ficus* and various other trees (Flannery 1990*b*). Two adult female *D. vanheurni* collected by Lidicker and Ziegler (1968) in October were each carrying a single pouch young and did not have visible embryos. A female *D. macleayi* taken in January had two pouch young, and one taken in March had a single pouch young (Flannery 1990*b*).

Lidicker and Ziegler (1968) expressed concern about the ease with which *D. vanheurni* could he caught by small domestic dogs. Flannery (1990*b*) indicated that it was hunted by large-scale drives and burning of the forest. G. G. George (1979) considered *D. macleayi* to be a threatened species, and Thornback and Jenkins (1982) indicated that it is susceptible to hunting and possibly habitat destruction; the IUCN now classifies it as vulnerable.

DIPROTODONTIA; MACROPODIDAE; **Genus DENDROLAGUS**
Schlegel and Miller, 1839

Tree Kangaroos

There are 10 species (Flannery 1989*b*, 1990*b*, 1993, 1995; Flannery, Boeadi, and Szalay 1995; Flannery and Seri 1990; Groves 1982):

D. inustus, northern and western New Guinea, Japen and possibly Waigeo and Salawatti islands;
D. lumholtzi, northeastern Queensland;
D. bennettianus, northeastern Queensland;
D. ursinus, far western New Guinea;
D. matschiei, Huon Peninsula of eastern Papua New Guinea and (perhaps through human introduction) nearby Umboi Island;
D. spadix, south-central Papua New Guinea;
D. goodfellowi, eastern New Guinea;
D. mbaiso, Maokop (Sudirman Range) of west-central New Guinea;
D. dorianus, far western, central, and southeastern New Guinea;
D. scottae, Toricelli Mountains of north-central New Guinea.

Dusky tree kangaroo *(Dendrolagus ursinus)*, photo from New York Zoological Society. Inset: Lumholtz's kangaroo *(D. lumholtzi)*, photo from Denver Museum of Natural History.

Groves (1982) originally considered *D. spadix* and *D. goodfellowi* subspecies of *D. matschiei*, and this arrangement was used by Corbet and Hill (1991). Later, however, Groves (*in* Wilson and Reeder 1993) followed Flannery and Szalay (1982) and Flannery (1989*b*, 1990*b*) in accepting both as full species.

Head and body length is 520–810 mm and tail length is 408–935 mm. Flannery (1990*b*) listed weights of 6.5–14.5 kg, and Flannery and Seri (1990) noted that the maximum recorded weight for a tree kangaroo is 20.0 kg in a wild-caught *D. dorianus*. The pelage usually is fairly long; in some forms it is soft and silky, in others coarse and harsh. Coloration in *D. ursinus* is blackish, brown, or gray above and white or buff below. *D. scottae* is uniformly blackish (Flannery and Seri 1990), *D. spadix* uniformly brownish, and *D. inustus* grizzled in color (Flannery 1990*b*). Both *D. dorianus* and *D. bennettianus* are some shade of brown over most of the body. In *D. lumholtzi* the upper parts are grayish or olive buff, the underparts are white, and the feet are blackish. *D. matschiei* is among the most brilliantly colored of marsupials: its back is red or mahogany brown, its face, belly, and feet are bright yellow, and its tail is mostly yellow. *D. goodfellowi* is similar but also has yellow lines along each side of the spine, and the tail is mostly the same color as the dorsum with some yellow markings (Lidicker and Ziegler 1968). *D. mbaiso* is generally dark above but has distinctly white facial markings and underparts (Flannery, Boeadi, and Szalay 1995).

The forelimbs and hind limbs are of nearly equal proportions. The cushionlike pads on the large feet are covered with roughened skin, and some of the nails are curved. The long, well-furred tail is of nearly uniform thickness and acts as a balancing organ; it is not prehensile but is often used to brace the animal when climbing. The thick fur on the nape and sometimes on the back grows in a reverse direction and apparently acts as a natural water-shedding device as the animals sit with the head lower than the shoulders. Females have a well-developed pouch and four mammae.

Tree kangaroos dwell mainly in mountainous rainforests. Reported altitudinal ranges (in meters) are: *D. inustus,* 100–1,400; *D. bennettianus,* 450–760; *D. ursinus,* 50–2,000; *D. matschiei,* 1,000–3,300; *D. spadix,* 0–800; *D. goodfellowi,* 1,200–2,900; *D. mbaiso,* 3,250–4,200; *D. dorianus,* 600–4,000; and *D. scottae,* 1,200–1,400 (Flannery 1990*b;* Flannery, Boeadi, and Szalay 1995; Flannery and Seri 1990; Ride 1970; Ziegler 1977). They are very agile in trees and reportedly are active both day and night. Often they travel rapidly from tree to tree, leaping as much as 9 meters downward to an adjoining tree. They also jump to the ground from remarkable heights, up to 18 meters or perhaps even more, without injury. They shelter in small groups in trees during the day and spend much of their life there but also frequently descend to the ground. When descending trees they usually back down, unlike the possums (Phalangeridae). On the ground they progress by means of relatively small leaps, leaning well forward to counterbalance the long tail, which is arched upward. Recent studies suggest that *D. dorianus,* the heaviest arboreal marsupial, may actually have become mainly a ground-dweller (Moeller *in* Grzimek 1990) but that *D. lumholtzi* spends only 2 percent of its time on the ground, the remainder being spent in the middle and upper layers of the canopy (Hutchins and Smith 1990). The diet consists mainly of leaves and fruit, obtained either in trees or on the ground. Captives have been observed to eat chickens (Flannery 1990*b*).

There is evidence that related female *D. dorianus* form coalitions that interact in a friendly manner and cooperate in agonistic displays toward unfamiliar males (Ganslosser 1984). Recent field studies of *D. lumholtzi* indicate that it is relatively solitary and nongregarious; females seem to occupy nonoverlapping home ranges of about 1.8 ha., several of which are overlapped by the range of a territorial male (Hutchins and Smith 1990). Groups of female *D. lumholtzi* with a single male can be maintained amicably in captivity, but two males in the presence of females will fight savagely. The species appears to have no definite mating season (P. M. Johnson *in* Strahan 1983). Based on births in captivity, there seems to be no well-defined breeding season in *D. matschiei;* the gestation period is approximately 32 days (Olds and Collins 1973). New Guinea females of various species have been found with pouch young in April, May, September, October, and December (Flannery 1990*b*).

Tree kangaroo *(Dendrolagus mbaiso),* photo by Tim Flannery.

The average litter size is one. A young captive individual was observed first to emerge completely from the pouch at about 305 days and to suckle with only the head in the pouch at about 408 days (Collins 1973). Tree kangaroos seem capable of long life in captivity; the record for *D. ursinus* is 20 years and 2 months (Collins 1973), and specimens of *D. goodfellowi* and *D. matschiei* were still living at 21 years and 23 years and 10 months, respectively (Marvin L. Jones, Zoological Society of San Diego, pers. comm., 1995).

Clearing of rainforest in northeastern Queensland has considerably reduced the range of *D. bennettianus* and *D. lumholtzi,* especially that of the latter species. Although both still appear to be common in protected reserves, there is concern that habitat fragmentation may prevent dispersal and increase inbreeding (Calaby 1971; Hutchins and Smith 1990; Poole 1979). In New Guinea there have been widespread declines of most species mainly because of hunting by people for use as food, and such problems are intensifying with the spread of modern weapons and other technologies into remote areas (Flannery 1990*b;* G. G. George 1979; Thornback and Jenkins 1982). The recently discovered *D. scottae* may number in the low hundreds in a restricted habitat of only 25–40 sq km and is gravely endangered by hunting and disturbance (Flannery and Seri 1990). The IUCN now classifies *D. matschiei, D. goodfellowi,* and *D. scottae* as endangered, *D. mbaiso* and *D. dorianus* as vulnerable, and *D. lumholtzi* and *D. bennetianus* as near threatened. *D. inustus* and *D. ursinus* are on appendix 2 of the CITES.

DIPROTODONTIA; MACROPODIDAE; **Genus SETONIX**
Lesson, 1842

Quokka

The single species, *S. brachyurus,* inhabits southwestern Australia, including Rottnest Island, near Perth, and Bald Island, near Albany (Ride 1970).

Head and body length is about 475–600 mm, tail length is 250–350 mm, and adult weight is 2–5 kg. The hair is short and fairly coarse, and the general coloration is brownish gray, sometimes tinged with rufous. There are no definite markings on the face. The ears are short and rounded. The tail, which is only about twice as long as the head, is sparsely furred and short. Females have four mammae and a well-developed pouch.

On islands the quokka occurs in a variety of habitats with sufficient cover, but on the mainland it seems to be restricted to dense vegetation in swamps amidst dry sclerophyll forest (Ride 1970). An important factor in the quokka's ecology, at least on Rottnest Island, is the diurnal shelter in a thicket or some other shady location where the animal can avoid the heat (Nicholls 1971). An individual returns to the same shelter every day through most of the year but may change sites in May or June. At night the quokka emerges from its shelter to feed. It makes runways and tunnels through dense grass and undergrowth. When moving quickly it hops on its hind legs; when moving slowly it does not use its tail as a third prop for the rear end of the body as do kangaroos and the larger wallabies. Like most macropodids, it is terrestrial, but it can climb to reach twigs up to about 1.5 meters above the ground (Ride 1970). The quokka is herbivorous, feeding on a variety of plants.

Ruminantlike digestion in the Macropodidae was first demonstrated in *Setonix*. The pregastric bacterial digestion in the quokka is much like that in sheep, for most of the 15 or so morphological types of bacteria present in the large stomach region of the quokka are comparable to those in the rumen of sheep. Most of the bacterial fermentation in the stomach of the quokka takes place in the sacculated part. This wallaby seems to occupy a position intermediate to the ruminant and the nonruminant herbivores in its efficient digestion of fiber and the rate of food passage.

During the wet season the feeding area or home range used by a quokka on Rottnest Island covers about 10,000–125,000 sq meters. In the dry season (November–April) this area increases to 20,000–170,000 sq meters (Nicholls 1971). In addition, some individuals move up to 1,800 meters to soaks or fresh-water seepages during the summer since there is almost no free surface water on Rottnest Island at that time. Other individuals do without water (Ride 1970). Population density is 1/0.4 ha. to 1/1.2 ha. (Main and Yadav 1971).

Kitchener (1972) found that on Rottnest Island the quokka population is organized into family groups. Adult males dominate the other members of the family and also form a linear hierarchy among themselves that is usually stable. On hot summer days, however, the adult males may fight intensively for possession of the best shelter sites. Apparently the availability of such sites, rather than food, is

Quokka *(Setonix brachyurus)*, photo from New York Zoological Society.

the main factor in limiting the population. Other than the conflict for shelters there is little evidence of territoriality, and groups of 25–150 individuals may have overlapping home ranges (Nicholls 1971). During the summer some parts of the population concentrate around available fresh water (Ride 1970).

Reproduction in the quokka has been studied in some detail (Collins 1973; Ride 1970; Rose 1978; Shield 1968). Females are polyestrous, with an average estrous cycle of 28 days. They are capable of breeding throughout the year in captivity, but in the wild anestrus occurs from about August through January. On Bald Island most births take place in March or April. Nondelayed gestation is 26–28 days. Litter size is normally one, and in the wild usually only one young can be successfully reared each year. One day after the young is born, however, the female mates again. Embryonic diapause, which was first demonstrated in *Setonix*, then occurs. If the young already in the pouch should die within a period of about 5 months, the embryo resumes development and is born 24–27 days later. If the first young lives, the embryo degenerates when the female enters anestrus. In captivity or under unusually good conditions in the wild the second embryo can resume development even if the first young is successfully raised. The young initially leaves the pouch at about 175–95 days but will return if alarmed or cold and will continue to suckle for three to four more months. Earliest recorded sexual maturity is 389 days for males and 252 days for females. In the wild on Rottnest Island, however, it is unlikely that females give birth until well into their second year. According to Shield (1968), a number of captive females lived for more than 7 years, a marked 10-year-old wild female carried a young, and a wild male lived more than 10 years.

Until the 1930s the quokka was very common in coastal parts of the mainland of southwestern Australia. Subsequently it disappeared, except for a few small colonies on the mainland and the two relatively numerous island populations (Calaby 1971). The Bald Island population is thought to number 200–600 individuals (Main and Yadav 1971). The total number on Rottnest is larger, but Ride (1970) expressed concern for the future of the species there because of the development of the island for recreational purposes. There recently has been some recovery of the mainland population (Kitchener *in* Strahan 1983), but an attempt to reintroduce it to a reserve near Perth was unsuccessful (Short et al. 1992). The IUCN now classifies the quokka as vulnerable, noting that it has declined by at least 50 percent over the last decade and now numbers fewer than 2,500 mature individuals. The species is listed as endangered by the USDI.

DIPROTODONTIA; MACROPODIDAE; **Genus THYLOGALE**
Gray, 1837

Pademelons

Six species are reported to exist (Flannery 1992, 1995; Ride 1970; Ziegler 1977):

- *T. billardierii* (red-bellied pademelon), southeastern South Australia, Victoria, Tasmania, islands of Bass Strait;
- *T. thetis* (red-necked pademelon), eastern Queensland, eastern New South Wales;
- *T. stigmatica* (red-legged pademelon), south-central New Guinea, eastern Queensland, eastern New South Wales;
- *T. brunii* (dusky pademelon), southern New Guinea, Aru and Kei Islands;
- *T. browni*, northern and eastern New Guinea, Bismarck Archipelago;
- *T. calabyi*, southeastern Papua New Guinea.

Hope (1981) described an additional species, *T. christenseni*, from remains at an archeological deposit in western New Guinea and stated that both that species and *T. brunii* survived in that area until less than 5,000 years ago. Flannery (1990*b*) suggested the possibility that both might yet persist there. Maynes (1989) indicated that the presence of *T. brunii* on the Kei Islands and of *T. browni* on Umboi and New Britain in the Bismarck Archipelago might be attrib-

Red-legged pademelons *(Thylogale stigmatica)*, photo by Ernest P. Walker.

utable to introduction by the Melanesians. Flannery (1995) stated that *T. browni* was brought by people to New Ireland in the Bismarck Archipelago about 7,000 years ago. Van Gelder (1977), on the basis of captive hybridization between *T. thetis* and *Macropus rufogriseus*, recommended that *Thylogale* be considered a synonym of *Macropus*.

Head and body length is about 290–670 mm, tail length is 246–510 mm, and weight is 1.8–12.0 kg (Flannery 1990*b*; K. A. Johnson *in* Strahan 1983; Johnson and Rose *in* Strahan 1983; P. M. Johnson *in* Strahan 1983). The fur is soft and thick. *T. thetis* is grizzled gray above, becoming rufous on the shoulders and neck, and often has a light hip stripe. In *T. stigmatica* the upper parts are mixed gray and russet and the yellowish hip stripe is conspicuous. *T. brunii* is gray brown to chocolate brown above with a dark cheek stripe from behind the eye to the corner of the mouth, above which is a white area; it also has a prominent hip stripe. The underparts in these three species are considerably paler than the dorsal parts. The species *T. billardierii* is grayish brown above tinged with olive; a yellowish hip stripe often is evident, and the undersurface is rufous or orange.

This genus is distinguished by features of the incisor teeth; a groove on the third upper incisor is positioned almost at the rear of the crown (Flannery 1990*b*). The comparatively short tail is sparsely haired and thickly rounded. Females have four mammae and a well-developed pouch.

Habitats include rainforest, sclerophyll forest, savannah, thick scrub, and grassland. In New Guinea *T. brunii* has an altitudinal range of 0–4,200 meters (Ziegler 1977), but *T. stigmatica* is found only from sea level to 60 meters (Flannery 1990*b*). Some species are nocturnal or crepuscular, sleeping by day in dense thickets and under bushes. However, K. A. Johnson (1980) found *T. thetis* generally to spend the night grazing in pasture and much of the day traveling widely through the rainforest to seek food and sites for basking.

Pademelons may move about in undergrowth through tunnel-like runways. They are inoffensive and apparently quite curious, as they sometimes will allow a close approach before bounding away. They have a habit, comparable to that of rabbits, of thumping on the ground with their hind feet, presumably as a signal or warning device. In the absence of predators or other disturbing factors they graze in the dusk and early morning on grass near thickets, into which they hop quickly when alarmed. The diet also includes leaves and shoots.

Captive pademelons may be maintained in groups (Collins 1973). *T. billardierii* reportedly is gregarious in the wild. In captivity males of this species were observed to make numerous aggressive displays but not actually to fight. Females also were aggressive toward one another, sometimes fought, and seemed to form a linear dominance hierarchy (Morton and Burton 1973). This species is a seasonal breeder in Tasmania, with most births in April, May, and June (Rose and McCartney 1982). *T. brunii* apparently breeds throughout the year in New Guinea (Flannery 1990*b*). Births in captivity have been recorded in January, February, and July for *T. billardierii* and in September for *T. thetis*. Embryonic diapause is known to occur in these two species. In *T. billardierii* the estrous cycle averages 30 days, nondelayed gestation is 29.6 days, and the period from the end of embryonic diapause to birth is 28.5 days (Rose 1978). Litter size usually is one, but twins have been recorded (Collins 1973). Pouch life of *T. billardierii* is 202 days, weaning occurs about 4 months after the young permanently leave the pouch, and sexual maturity comes at about 14–15 months (Morton and Burton 1973; Rose and McCartney 1982). A specimen of *T. billardierii* lived in captivity for 8 years and 10 months (Collins 1973).

Johnson and Vernes (1994) maintained captive colonies of one adult male and eight adult female *T. stigmatica* and found births to occur throughout the year in captivity. Supplementary studies indicated that young were produced in the wild in Queensland from October to June. A postpartum estrus and mating generally followed birth. The estrous cycle was 29–32 days, gestation was 28–30 days, the pouch was permanently vacated at 184 days, weaning occurred an average of 66 days later, and mean age at sexual maturity was 341 days for females and 466 days for males.

The species *T. billardierii* disappeared from the mainland of Australia during the nineteenth century and also has been exterminated on some of the islands of Bass Strait. It still survives on other of these islands and is common in many parts of Tasmania (Calaby 1971), where it is sometimes considered a pest to agriculture and silviculture and is taken in large numbers for its skin (Ride 1970). Population levels now are thought to be high enough to sustain the regulated hunting and control operations that are carried out (Poole 1978). The distribution of *T. thetis* and *T. stigmatica* has been reduced in Australia by land clearance for agriculture, dairying, and forestry, but these two species still are common in some areas (Calaby 1971; Poole 1979). The status of the New Guinea populations seems to be worse, with widespread disappearances and range fragmentation, apparently caused by excessive hunting by people and their dogs (Flannery 1992). The IUCN now classifies *T. brunii* and *T. browni* as vulnerable and *T. calabyi* as endangered.

DIPROTODONTIA; MACROPODIDAE; **Genus PETROGALE**
Gray, 1837

Rock Wallabies

There are 14 species (Eldridge and Close 1992, 1993; Eldridge, Johnston, and Lowry 1992; Flannery 1989*b*; Kitchener and Sanson 1978; Maynes 1982):

P. xanthopus, eastern South Australia, northwestern New South Wales, southwestern Queensland;
P. persephone, east-central coast of Queensland;
P. rothschildi, Pilbara District and Dampier Archipelago of northwestern Western Australia;
P. lateralis, suitable areas of Western Australia, Northern Territory, South Australia, and western Queensland, including many islands off the western and southern coasts but not Tasmania;
P. penicillata, extreme southeastern Queensland, eastern New South Wales, Victoria;
P. herberti, southeastern Queensland;
P. sharmani, east-central coast of Queensland;
P. mareeba, northeastern Queensland;
P. coenensis, northeastern Cape York Peninsula of Queensland;
P. inornata, east-central Queensland;
P. assimilis, northern Queensland;
P. godmani, eastern Cape York Peninsula of Queensland;
P. brachyotis, northeastern Western Australia, northern Northern Territory, including Groote Eylandt;
P. burbidgei, Kimberley area of northeastern Western Australia.

Petrogale often is considered to include the genus and species *Peradorcas* (see account thereof) *concinna*. *P. pur-*

A. Rock wallaby *(Petrogale penicillata)*, photo by Bob McIntyre through Cheyenne Mountain Zoo. B. Ring-tailed rock wallabies *(P. xanthopus)*, photo from Australian News and Information Bureau.

pureicollis of western Queensland sometimes has been treated as a species separate from *P. lateralis* (Kirsch and Calaby 1977). Much of the recent taxonomic work on *Petrogale* has been based on evaluation of chromosomes, with certain populations that appeared morphologically separable being found to have identical karyotypes and with populations that appeared morphologically indistinguishable being designated as distinct though cryptic species based on karyological differences. Some of these designated species have been reported to form hybrid zones, and other data suggest six various alternatives to the above phylogenetic arrangement (Eldridge and Close 1993; Sharman, Close, and Maynes 1990).

Excluding the very small *P. burbidgei*, the genus is characterized as follows: head and body length about 500–800 mm; tail length, 400–700 mm; adult weight, 3–9 kg; and thick, long, dense fur. In *P. burbidgei*, according to Kitchener and Sanson (1978), head and body length is 290–353 mm, tail length is 252–322 mm, and weight is 0.96–1.43 kg.

The general coloration of the upper parts varies from pale sandy or drab gray to rich dark vinaceous brown; the underparts are paler, usually buffy, yellowish gray, yellowish brown, or white. Several species have stripes, patches, or other striking markings. *P. xanthopus*, the ring-tailed or yellow-footed rock wallaby, is one of the most brightly colored members of the kangaroo family. It is grayish with a white cheek stripe, yellow on the back of the ears, a dark streak from between the ears to the middle of the back, brown patches on the yellow limbs, white underparts, and a tail ringed with brown and pale yellow. In *P. rothschildi* and the subspecies *P. lateralis purpureicollis* the upper back becomes a brilliant purple at certain times of the year. In some other subspecies of *P. lateralis* the tail and armpits are black, there is a white cheek stripe, and the ears have a prominent black patch and whitish margins (Ride 1970). In *P. burbidgei*, according to Kitchener and Sanson (1978), the upper parts are mostly ochraceous tawny and the underparts are ivory yellow.

The tails of rock wallabies are long, cylindrical, bushy, and thickly haired at the tip. They are less thickened at the base than in *Thylogale*, *Macropus*, and *Wallabia* and are used primarily for balancing rather than as props for sitting. The hind foot of *Petrogale* is well padded; the sole is roughly granulated, permitting a secure grip on rock, and edged with a fringe of stiff hair that extends onto the digits. The central hind claws are short, only exceeding the toe pads by 2–3 mm. Females have a well-developed, forward-opening pouch and four mammae.

These wallabies usually inhabit rocky ranges and boulder-strewn outcrops with an associated cover of forest, woodland, heath, or grassland. They are nocturnal, spending their days in rock crevices and caves; occasionally they emerge on warm afternoons for a sunbath. Their agility among the rocks is astonishing; some of their leaps measure up to 4 meters horizontally. They also can scramble up cliff faces and leaning tree trunks with relative ease. The friction of their fur and feet on regular paths, over many generations, imparts a glasslike sheen to limestone. Rock wallabies usually progress in a series of short or long leaps. They travel awkwardly in open country. Their diet consists mainly of grasses, though in dry seasons they can exist for long periods without water by eating the juicy bark and roots of various trees.

The species *P. penicillata* reportedly lives in small groups that have a defended territory, and there is a linear dominance hierarchy among males established through fighting (Russell 1974*b*). Observations of wild *P. assimilis* indicate that a male and a female may form a stable rela-

Rock wallaby (*Petrogale* sp.), photo by Lothar Schlawe.

tionship involving regular social grooming and the sharing of resting sites and other parts of an exclusive home range (Barker 1990). Females are polyestrous (Rose 1978). In a study of captive *P. penicillata* Johnson (1979) found an estrous cycle of 30.2–32.0 days and a gestation period of 30–32 days. There was usually a postpartum mating, and the resultant embryo developed and was born 28–30 days after the premature removal of the original young. Normal pouch life was determined to be 189–227 days, and sexual maturity was attained at 590 days by males and 540 days by females. The following additional life history data were summarized by Collins (1973). Births in captivity have been recorded for *P. xanthopus* during every month and for *P. penicillata* during March, June, August, and September. Gestation in *P. xanthopus* is 31–32 days. Litter size usually is one, but twins occur on occasion. A specimen of *P. penicillata* lived for 14 years and 5 months in captivity.

While rock wallabies have disappeared from some areas and their numbers have declined considerably in the sheep country of southern Australia, they seem to have maintained themselves better than other small macropodids, and several species are common (Calaby 1971; Poole 1979). Portions of their rocky habitat are inaccessible to sheep and rabbits, which compete for forage, though feral goats are a problem in some areas. Perhaps the species most severely affected by European settlement is *P. xanthopus*, classified as endangered by the USDI and generally as near threatened by the IUCN, which not only suffered through habitat alteration but was heavily hunted for its beautiful pelt. It has disappeared in some areas but still occurs in the Flinders Range of South Australia and is locally common in other areas (Calaby 1971; Copley 1983; Gordon, McGreevy, and Lawrie 1978). The subspecies in the Flinders Range, *P. x. xanthopus*, is classified as vulnerable by the IUCN. The population in Queensland is estimated to contain 5,000–10,000 individuals and to be vulnerable to land development (Gordon et al. 1993). The closely related species *P. persephone* also appears to be rare and endangered (Maynes 1982) and is termed endangered by the IUCN. The species *P. sharmani, P. coenensis*, and *P. burbidgei* are designated near threatened by the IUCN.

The most widely distributed species, *P. lateralis* and *P. penicillata*, have been greatly reduced in numbers and range in both southeastern and southwestern parts of the continent. Despised as agricultural pests and valued for their skins, hundreds of thousands were killed in the nineteenth and early twentieth centuries. Remaining populations are continuing to decline because of various factors, including destruction of vegetative cover by introduced rabbits, competition with introduced goats for food and rock shelters, and predation by introduced foxes (D. J. Pearson 1992; Short and Milkovits 1990); both are designated vulnerable by the IUCN. *P. penicillata* was thought to have disappeared in Victoria by 1905 (Harper 1945), but two small colonies are now known to remain (Emison et al. 1978; Wakefield 1971). In the northern half of Australia most species of *Petrogale* seem to be common; the subspecies *P. lateralis purpureicollis* reportedly is thriving in northwestern Queensland (Poole 1979). The conservation of *P. lateralis, P. rothschildi*, and *P. brachyotis* may be assisted by the presence of numerous natural island populations (Calaby 1971; Main and Yadav 1971; Poole 1979). A population of *Petrogale penicillata* became established on the island of Oahu, Hawaii, in 1916 (Lazell, Sutterfield, and Giezentanner 1984; Maynes 1989).

DIPROTODONTIA; MACROPODIDAE; **Genus PERADORCAS**
Thomas, 1904

Little Rock Wallaby

The single species, *P. concinna*, is found in the Kimberley area of northeastern Western Australia and in Arnhem Land in the northern Northern Territory. *Peradorcas* has a history of being switched back and forth by various taxonomists, some regarding it as a distinct genus and others placing it in *Petrogale*. Although the latter procedure was suggested by Poole (1979) based on serological and karyological study and has been followed in most recent systematic treatments (Calaby and Richardson *in* Bannister et al. 1988; Corbet and Hill 1991; Flannery 1989*b*; Groves *in* Wilson and Reeder 1993), *Peradorcas* was retained as a full genus by Ganslosser (*in* Grzimek 1990), Nelson and Goldstone (1986), and Sanson (*in* Strahan 1983) based on morphological and behavioral data. Recent

Little rock wallaby *(Peradorcas concinna)*, photo by G. D. Sanson / National Photographic Index of Australian Wildlife.

chromosomal analyses have indicated that if *Peradorcas* is treated as a separate genus, it should include the evidently closely related *Petrogale brachyotis* and *P. burbidgei* even though the latter two species do not possess continually erupting molar teeth, the character on which *Peradorcas* was originally founded (Eldridge, Johnston, and Lowry 1992).

Head and body length is 290–350 mm, tail length is 220–310 mm, and weight is 1.05–1.70 kg (Sanson *in* Strahan 1983). The fur is short, soft, and silky. The upper back is rusty red to grayish, the rump is reddish to orange, the underparts are white or grayish white, and the tail becomes darker toward the tip. The hind foot is well padded and the sole is roughly granulated. Females have four mammae in the pouch.

Peradorcas resembles *Petrogale burbidgei* but is distinguished by its longer ears and dental characters. Its molar dentition is unique among marsupials in that there are supplementary replacement molars behind the last regular molar. The actual number of molar teeth is not known, but study suggests that as many as nine molars may erupt successively and that there are seldom more than five molars in place at any one time.

According to Sanson (*in* Strahan 1983), the little rock wallaby shelters in sandstone crevices and caves during the day in the dry season and emerges at night to forage for ferns that grow on adjacent blacksoil plains. These ferns are extremely abrasive, and this factor may be associated with the development of the continual molar replacement system found in *Peradorcas*. In the wet season, activity is partly diurnal and the diet includes grasses that grow on the margins of the plains that are not inundated.

Breeding probably occurs throughout the year. In studies of a captive colony Nelson and Goldstone (1986) found *Peradorcas* to have a postpartum estrus and embryonic diapause. The estrous cycle lasted 31–36 days, and gestation about 30 days. Dominant females had shorter cycles than did subordinates. The single young opens its eyes at about 110 days, leaves the pouch at about 160 days, and is independent at about 175 days. Weaning is much more sudden than in *Petrogale*. Sexual maturity is attained in the second year.

Peradorcas inhabits a relatively small area that is subject to some environmental stress. The IUCN includes *P. concinna* in the genus *Petrogale* and designates the species as near threatened.

DIPROTODONTIA; MACROPODIDAE; **Genus LAGORCHESTES**
Gould, 1841

Hare Wallabies

There are four species (Flannery 1989*b;* Ride 1970):

- *L. conspicillatus,* northern Western Australia, including Barrow Island, to northern Queensland;
- *L. leporides,* formerly eastern South Australia, western New South Wales, northwestern Victoria;
- *L. hirsutus,* formerly Western Australia, including islands of Shark Bay, Northern Territory, western South Australia;
- *L. asomatus,* Lake McKay on Western Australia–Northern Territory border.

Head and body length of *L. conspicillatus* is 400–470 mm, tail length is 370–490 mm, and weight is 1.6–4.5 kg; respective figures for *L. hirsutus* are 310–90 mm, 245–380 mm, and 780–1,740 grams (Burbidge and Johnson *in* Strahan 1983). *L. leporides* is usually gray brown above with reddish sides and grayish white underparts; the coloration of *L. hirsutus* is much the same except that long reddish hairs are present on the lower back, imparting a shaggy appearance. *L. conspicillatus* is gray brown or yellowish gray

Hare wallaby (*Lagorchestes* sp.), photo from West Berlin Zoo through Heinz-Georg Klös.

above and whitish or reddish below with reddish patches encircling the eyes and white hip marks. The guard hairs are long and coarse, and all the species have long, soft, thick underfur.

The generic name means "dancing hare." These wallabies are not much larger than hares and resemble them in their movements and, to some extent, in their habits. The nose in *Lagorchestes* is wholly or partially covered by hair; the central hind claw is long and strong and is not hidden by the fur of the foot, and the tail is evenly haired throughout. The members of this genus have 34 teeth, 2 more than other wallabies.

Hare wallabies live in open grassy or spinifex plains with or without shrubs or trees (Ride 1970). They generally are solitary and nocturnal, resting by day in a "hide" or "form" lightly scratched in the ground in the shade of a bush or by a tuft of grass. *L. hirsutus* emits a whistling call when pursued. Jumps of 2–3 meters have been credited to *L. leporides* when pressed, and one individual chased by dogs for 0.4 km doubled back on its track and leaped over the head of a man standing in its path. Hare wallabies are herbivorous, feeding on various grasses and herbs (Collins 1973).

Burbidge and Johnson (*in* Strahan 1983) reported that on Barrow Island *L. conspicillatus* is a selective feeder, browsing mainly on colonizing shrubs and also eating the tips of spinifex leaves; it does not drink even when water is available. It constructs several hides within a home range of about 8–10 ha. Breeding occurs throughout the year, but on Barrow Island there are birth peaks in March and September. Young vacate the pouch at about 150 days, and females become sexually mature at 12 months. Johnson (1993) noted an estrous cycle of 30 days and a gestation period of 29–31 days in this species.

The species *L. asomatus* is known only from a single skull taken in 1932. *L. leporides,* common in South Australia and New South Wales until the mid–nineteenth century, has not been recorded since 1890 (Calaby 1971). Both species are designated extinct by the IUCN. *L. conspicillatus* has declined over much of its range, especially in Western Australia, but still occurs regularly in parts of the Northern Territory and northern Queensland (Calaby 1971; Poole 1979). Its decline seems to be associated with the destruction of vegetative cover by sheep, competition for habitat with introduced rabbits, and predation by introduced cats (Ingleby 1991*b*). A population of about 10,000 individuals is present on Barrow Island off the west coast of Western Australia (Short and Turner 1991). *L. conspicillatus* is designated generally as near threatened by the IUCN, and the subspecies *L. c. conspicillatus,* of Barrow Island, is classified as vulnerable. *L. hirsutus* disappeared from most of its mainland range long ago, but about 20 years ago the presence of two small colonies each containing only 6–10 animals was confirmed in the Tanami Desert Sanctuary in the Northern Territory. One of these colonies has since increased to about 20 individuals and has been studied in some detail (Lundie-Jenkins, Corbett, and Phillips 1993); the other was destroyed, apparently by introduced foxes. *L. hirsutus* is moderately abundant and well protected on Bernier and Dorre islands in Shark Bay. Populations there are known to fluctuate, but recent surveys carried out by Short and Turner (1992) during a drought and thus probably when numbers were minimal indicated the presence of 2,600 individuals on Bernier Island and 1,700 on Dorre. *L. hirsutus* is listed as endangered by the USDI and is on appendix 1 of the CITES. The IUCN classifies the island subspecies *L. hirsutus bernieri* and *L. h. dorreae* as vulnerable, the formerly widespread mainland subspecies *L. h. hirsutus* as extinct, and the surviving unnamed subspecies in the central desert as critically endangered.

Ride (1970) suggested that an important factor in the decline of hare wallabies on the mainland may have been alteration of grassland habitat through trampling and grazing by sheep and cattle. The aborigines of Australia avidly hunted hare wallabies for food but actually may have benefited the animals. These people regularly set winter fires in order to clear areas for easier hunting and thereby produced a mosaic of different regenerating vegetative stages. This process not only provided food for *Lagorchestes* but also prevented the buildup of brush and the devastation of the habitat by lightning-caused fires during the summer. The decline of hare wallabies coincided with the removal of the aborigines from large areas and the reduction of winter fires. The two recently discovered mainland colonies of *L. hirsutus* were in localities where regular winter burning is still practiced (Australian National Parks and Wildlife Service 1978; Bolton and Latz 1978; Ingleby 1991*b*).

DIPROTODONTIA; MACROPODIDAE; **Genus ONYCHOGALEA**
Gray, 1841

Nail-tailed Wallabies

There are three species (Calaby 1971; Gordon and Lawrie 1980; Ride 1970):

O. unguifera, northern Western Australia to northeastern Queensland;
O. lunata, formerly southern Western Australia, southwestern Northern Territory, South Australia, probably southwestern New South Wales;
O. fraenata, formerly inland parts of eastern Queensland, New South Wales, and northern Victoria.

In the two living species, *O. unguifera* and *O. fraenata,* head and body length is 430–700 mm, tail length is 360–730 mm, and weight is 4–9 kg; *O. lunata* was smaller (Burbidge *in* Strahan 1983; Gordon *in* Strahan 1983). The fur is soft, thick, and silky. In *O. unguifera* the upper parts are fawn-colored. The general color in the other two species is gray. In *O. fraenata,* the bridled wallaby, the white shoulder stripes run from beneath the ears along the back of the neck and down around the posterior part of each shoulder to the white undersurface of the body; the center of the neck is black or gray. In *O. lunata* the white shoulder stripes form a crescent around the posterior region of each shoulder and do not extend onto the neck, which is dark rufous. In all three species the underparts are white and white hip stripes, often indistinct, are present.

Nail-tailed wallabies inhabit woodland, savannah, brushland, or steppe. They are mainly nocturnal, and although they occasionally move about in daylight, they spend most of the day in a shallow nest scratched out beneath a tussock of grass or a bush. They seem to depend on thick vegetation for cover. When disturbed, *O. unguifera* sometimes emits a quickly repeated "u-u-u" and then flees. All species are remarkably swift when startled from their shelter and often escape by running into thick brush or a tree hollow. Unlike other wallabies, they carry their arms outward almost at a right angle to the body. When hopping they move their arms in a rotary motion, which has led to the common name "organ grinders." The diet is somewhat different from that of other wallabies, for it seems to consist mainly of the roots of various species of coarse grass and other herbaceous vegetation.

Nail-tailed wallaby *(Onychogalea fraenata)*, photo by S. Carruthers / National Photographic Index of Australian Wildlife.

Nail-tailed wallabies are shy and usually solitary. Females are said to give birth to a single young, usually in early May. A specimen of *O. fraenata* was maintained in captivity for 7 years and 4 months (Jones 1982).

Both *O. fraenata* and *O. lunata* are listed as endangered by the USDI and are on appendix 1 of the CITES. The IUCN designates the former as endangered but the latter as extinct. Both have suffered drastic declines largely through habitat alteration by farming and grazing and perhaps because of predation by introduced foxes and dogs. Newsome (1971*a*) suggested that removal of the thickets in which these wallabies sheltered through livestock grazing and deliberate clearing left them homeless, insecure, and greatly vulnerable to predation. Still reasonably common in some parts of its range until the 1930s, *O. lunata* now is very rare, if not extinct (Thornback and Jenkins 1982). The latest reliable records are from central Australia in the early 1960s (Poole 1979; Ride 1970). *O. fraenata* was common in parts of Queensland and New South Wales in the mid–nineteenth century but subsequently became rare. There were no confirmed sightings from 1937 to 1973, and the species was considered possibly extinct. In the latter year, however, a small population was discovered in an area of about 100 sq km in central Queensland. About half of the area has been purchased by the state government to form a reservation for the species (Gordon and Lawrie 1980; Poole 1979). The third species of the genus, *O. unguifera*, is still widespread in northern Australia and reportedly is common in several localities (Calaby 1971). It may have survived through an association with wetter habitats than those used by the other species of *Onychogalea* and also by many other now endangered or extinct Australian marsupials. It thus would be less dependent on the core areas of rich vegetation that were needed by the others during droughts but were degraded by livestock, introduced animals, and other problems caused by humans (Ingleby 1991*a*).

DIPROTODONTIA; MACROPODIDAE; **Genus WALLABIA**
Trouessart, 1905

Swamp Wallaby

The single species, *W. bicolor*, occurs in eastern Queensland, eastern New South Wales, Victoria, and southeastern South Australia (Ride 1970). Some of the species here assigned to the genus *Macropus* often have been put in *Wallabia*. Calaby (1966), however, argued that because of its karyology, reproductive physiology, behavior, and dental morphology *W. bicolor* should be placed in a monotypic genus. Calaby and Richardson (*in* Bannister et al. 1988), Flannery (1989*b*), and most other recent systematic accounts have followed this arrangement, though Kirsch (1977*b*), on the basis of serological investigation, thought that *W. bicolor* probably could be included in *Macropus*. Van Gelder (1977) recommended that *Wallabia* be considered a synonym of *Macropus* since hybridization in captivity had occurred between *W. bicolor* and *M. agilis*.

Head and body length is 665–847 mm, tail length is 640–862 mm, and weight is 10.3–15.4 kg in females and 12.3–20.5 kg in males (Merchant *in* Strahan 1983). The fur is long, thick, and coarse. In the north the back and head are reddish brown and the belly is orange; in the south the general coloration is brownish or grayish black with grayish sides and sometimes brownish red underparts. The paws, toes, and terminal part of the tail are usually black. There may be a distinct light-colored stripe extending from the upper lip to the ear. Females have 4 mammae and a well-developed, forward-opening pouch.

Swamp wallaby *(Wallabia bicolor)*, photo from Zoological Society of San Diego.

Swamp wallaby *(Wallabia bicolor)*, photo by Joan M. Dixon.

Despite its common name, *W. bicolor* is not restricted to swamps. It does inhabit moist thickets in gullies and even mangroves but also is found in open forest in upland areas as long as there are patches of dense cover. It is nocturnal and hops heavily with the body well bent over and the head held low (Ride 1970). A study in Victoria found this species to occupy elongated home ranges, with long axes measuring up to 600 meters, and to be mainly a browser (Edwards and Ealey 1975). In Queensland the diet includes pasture, brush, and agricultural crops (Kirkpatrick 1970*b*).

The swamp wallaby usually is solitary, but unrelated animals may gather at an attractive food source (Kirkpatrick 1970*b*). There does not appear to be any territorial defense (Edwards and Ealey 1975). Individuals of both sexes maintain small, overlapping home ranges of about 16 ha. (Troy and Coulson 1993). Females are polyestrous, with an estrous cycle averaging 31 (29–34) days (Kaufmann 1974). Births in captivity have been recorded from January to May and from October to December (Collins 1973), but there appears to be no sharply demarcated breeding season in the wild. Under good conditions females can breed continuously and give birth about every eight months (Edwards and Ealey 1975; Kirkpatrick 1970*b*). Average undelayed gestation is 36.8 (35–38) days, and litter size usually is one, but twins have been reported (Collins 1973; Kaufmann 1974). This is the only marsupial with a gestation period longer than its estrous cycle. Thus, there is a prepartum estrus and mating during the last 3–7 days of pregnancy. Presumably, after this mating there is a near-term embryo in one uterus and a segmenting egg in the other. After the first embryo is born, its suckling induces diapause in the other embryo. Should the first young die or be removed from the pouch, the second embryo will resume development and another birth will occur in about 30 days. Otherwise the second embryo will remain dormant until late in the pouch life of the first young and then will resume development in time to be born when the first young vacates the pouch at an age of about 250 days (Collins 1973; Kaufmann 1974; Russell 1974*a*). Sexual maturity is attained by both sexes at about 15 months, and maximum life span in the wild may be 15 years (Kirkpatrick 1970*b*). A captive individual lived 12 years and 5 months (Collins 1973).

Although the swamp wallaby has declined in numbers and distribution because of clearing of its habitat, it still is widespread and common (Calaby 1971). Approximately 1,500 skins are marketed each year in Queensland, and a larger number of animals are shot as agricultural pests, but such killing is not considered a threat to the survival of the species (Kirkpatrick 1970*b*). The swamp wallaby was introduced on Kawau Island, New Zealand, about 1870 and still occurs there (Maynes 1977*a*).

DIPROTODONTIA; MACROPODIDAE; **Genus MACROPUS**
Shaw, 1790

Wallabies, Wallaroos, and Kangaroos

There are 3 subgenera and 14 species (Caughley 1984; Collins 1973; Dawson and Flannery 1985; Flannery 1989*b*, 1995; Kirsch and Calaby 1977; Kirsch and Poole 1972; Richardson and Sharman 1976; Ride 1970):

subgenus *Notamacropus* Dawson and Flannery, 1985

M. irma (western brush wallaby), southwestern Western Australia;

M. greyi (toolache wallaby), formerly southeastern South Australia, Victoria;
M. parma (parma wallaby), eastern New South Wales;
M. dorsalis (black-striped wallaby), southeastern Queensland, eastern New South Wales;
M. agilis (agile wallaby), southern New Guinea and (possibly through introduction) several islands to the east, northeastern Western Australia, northern Northern Territory, northern and eastern Queensland;
M. rufogriseus (red-necked wallaby), coastal areas from eastern Queensland to southeastern South Australia, Tasmania, islands in Bass Strait;
M. eugenii (tammar wallaby), southwestern Western Australia, southern South Australia, several offshore islands;
M. parryi (whiptail wallaby), eastern Queensland, northeastern New South Wales;

subgenus *Macropus* Shaw, 1790

M. giganteus (eastern gray kangaroo), eastern and central Queensland, New South Wales, Victoria, extreme southeastern South Australia, Tasmania;
M. fuliginosus (western gray kangaroo), southern Western Australia, southern South Australia, western and central New South Wales, southern Queensland, western Victoria, Kangaroo Island;

subgenus *Osphranter* Gould, 1842

M. bernardus (black wallaroo), Arnhem Land of northern Northern Territory;
M. robustus (wallaroo, euro), Australia except Tasmania;
M. antilopinus (antilopine wallaroo), northeastern Western Australia to northern Queensland;
M. rufus (red kangaroo), Australia except extreme north, east coast, and extreme southwest.

Until recently the genus *Macropus* usually was restricted to the last six species listed above, with the remaining eight species being placed in the genus *Wallabia*. The latter genus is now considered to include only *W. bicolor*, though Kirsch (1977*b*) and Van Gelder (1977) suggested that even that species might belong in *Macropus*. The red kangaroo (*M. rufus*) often has been put in the separate genus or subgenus *Megaleia* Gistel, 1848.

The kangaroos and wallaroos (the last six species listed above) include the largest living marsupials. Measurements (from Grzimek 1975) for *M. giganteus* and *M. fuliginosus* are: head and body length for males, 1,050–1,400 mm, and for females, 850–1,200 mm; tail length for males, 950–1,000 mm, and for females, approximately 750 mm. For *M. robustus* and *M. antilopinus* measurements are: head and body length for males, 1,000–1,400 mm, and for females, 750–1,000 mm; tail length for males, 800–900 mm, and for females, 600–700 mm. Measurements for *M. rufus* are: head and body length for males, 1,300–1,600 mm, and for females, 850–1,050 mm; tail length for males, 1,000–1,200 mm, and for females, 650–850 mm. Weight is 20–90 kg but seldom exceeds about 55 kg in males and 30 kg in females. When standing in a normal, plantigrade position, male gray and red kangaroos usually are about 1.5 meters tall and sometimes reach nearly 1.8 meters. Reports of kangaroos 2.1 meters (7 feet) tall seem unfounded (W. E. Poole, Commonwealth Scientific and Industrial Research Organization, Division of Wildlife Research, pers. comm., 1977). When standing on their hind toes, however, as when in an aggressive position, males may reach or slightly exceed 2.1 meters. Wallaroos are shorter on average than kangaroos, but they are more heavily built. Male *M. antilopinus*, the largest wallaroo, may weigh just as much as male gray and red kangaroos (Russell 1974*a*). *M. bernardus*, the smallest wallaroo, is only two-thirds as large as *M. robustus* and *M. antilopinus* (Richardson and Sharman 1976). The fur of kangaroos and wallaroos is generally thick and coarse, being especially long and shaggy in the wallaroos.

Red-necked wallabies *(Macropus rufogriseus)*, photo by Bernhard Grzimek.

Eastern gray kangaroos *(Macropus giganteus)*, photo by Lothar Schlawe.

The muzzle is completely hairless in wallaroos, partly haired in the red kangaroo, and fully haired in the gray kangaroos. General coloration is as follows (Ride 1970; Russell 1974*a*): *M. giganteus,* silvery gray; *M. fuliginosus,* light gray brown to chocolate; *M. robustus,* red brown to very dark blue gray (looking black); *M. antilopinus,* reddish tan or bluish gray; *M. bernardus,* males dark sooty brown (looking black), females paler; *M. rufus,* males usually rich reddish brown, females usually bluish gray, but in a few areas males are blue and females red.

Wallabies of the genus *Macropus* (the first eight species listed above) are generally smaller than kangaroos and wallaroos. Overall, head and body length is about 400–1,050 mm, tail length is 330–750 mm, and weight is 2.5–24 kg. Most species average about 700–800 mm in head and body length, 600–700 mm in tail length, 20 kg in weight for males, and 12 kg in weight for females. There are, however, two especially small species, *M. eugenii* and *M. parma.* A series of mature specimens of *M. parma* from New South Wales measured as follows: head and body length, 424–527 mm; tail length, 416–544 mm; and weight, 2.6–5.9 kg (Maynes 1977*b*). The sexes were approximately the same size. In coloration *M. parma* is rich brown with a dark dorsal stripe extending from neck to shoulders and has white underparts and throat; *M. eugenii* is grayish brown, often with reddish shoulders; *M. agilis* has a yellowish brown color, a white cheek stripe, and a fairly distinct whitish hip stripe; *M. rufogriseus* has a fawny gray body with red nape and shoulders; *M. dorsalis* has a dark brown spinal stripe and a distinct white hip stripe; *M. parryi* is pale gray with a marked white face stripe and a very long, slender tail; *M. irma* also has a distinct white face stripe, black and white ears, and a crest of black hair on the tail; *M. greyi* is colored like *M. irma,* but its tail is crested with pale hair.

The enlarged hindquarters in *Macropus* are powerfully muscled, and the tapered tail acts as a balance and rudder when the animal is leaping and as a third leg when it is sitting. The tail is strong enough to support the weight of the entire animal. Females have a well-developed, forward-opening pouch and four mammae.

Habitat varies widely in this genus. Most wallabies rely on dense vegetation for cover but usually move into open forest or savannah to feed. *M. parma* and *M. dorsalis* generally live in moist forest. The gray kangaroos, *M. giganteus* and *M. fuliginosus,* are found in forests and woodland and also seem to require heavy cover. *M. robustus* and *M. bernardus* occur mainly in mountains or rough country. *M. antilopinus* and *M. rufus* dwell on grass-covered plains or savannah but, again, depend on places with denser vegetation for shelter. *M. robustus* sometimes shelters in caves. All species are primarily crepuscular or nocturnal, feeding from late afternoon to early morning and resting during the day, but they sometimes move about in daylight. The most diurnal species is *M. parryi* and the most nocturnal in habit are *M. parma* and *M. rufogriseus* (Kaufmann 1974; Maynes 1977*b;* Russell 1974*a*).

When moving slowly, as in grazing or browsing, these animals exhibit an unusual, "five-footed" gait, balancing on their tail and forearms while swinging their hind legs forward, then bringing their arms and tail upward. They are remarkably developed in the leaping mode of progression. At a slow pace the leaps of wallaroos and kangaroos usually measure 1.2–1.9 meters, and at increased speeds they may leap 9 or more meters. One gray kangaroo jumped nearly 13.5 meters on a flat. Normally they do not jump higher than 1.5 meters. Speeds of about 48 km/hr are probably attained for short distances when these animals are pressed in relatively open country.

All species of *Macropus* are herbivorous, and most are mainly grazers. In some areas well over 90 percent of the

Tammar wallaby *(Macropus eugenii)*, photo by L. F. Schick / National Photographic Index of Australian Wildlife.

food eaten by kangaroos and wallaroos is grass (Russell 1974*a*). Some species can go for long periods without water—two to three months have been recorded for *M. robustus.* Where they shelter in cool caves they can exist indefinitely without water, even when outside temperatures exceed 45° C. Some species dig in the ground for water or eat succulent roots. In certain areas where there is almost no fresh water *M. eugenii* obtains what it needs from salty plant juices and even is able to drink sea water (Ride 1970).

Caughley, Sinclair, and Wilson (1977) estimated that in 496,000 sq km of New South Wales, overall population densities were 3.18/sq km for the combined populations of *M. giganteus* and *M. fuliginosus* and 4.18/sq km for *M. rufus.* In the Pilbara District of northwestern Western Australia the density of *M. robustus* averaged 9.3/sq km but in some favorable areas was as high as 40/sq km. Kaufmann (1974) found a density of 1/2 ha. to 1/4 ha. for *M. parryi.* Other densities, as summarized by Main and Yadav (1971),

Parma wallaby *(Macropus parma)*, photo by Wolfgang Dressen.

Agile wallaby *(Macropus agilis)*, photo from San Diego Zoological Society.

are: 1/1.6 ha. for *M. eugenii*, 1/10.3 ha. for *M. robustus*, and 1/24 ha. to 1/89 ha. for *M. rufus*. The red kangaroo sometimes becomes relatively mobile, covering considerable distances to seek food and readily shifting its range in response to environmental conditions. One marked individual moved 338 km from the point of release (Bailey 1992). In contrast, *M. giganteus* and *M. robustus* are sedentary, seemingly preferring to remain in a restricted home range even when food and water become scarce. In such a way of life the euro is assisted by its ability to withstand dehydration better than the kangaroos, to survive on food containing less nitrogen, and to minimize water loss by sheltering in caves during the daytime (Russell 1974*a*). In a semiarid part of New South Wales, Croft (1991*a*, 1991*b*) found *M. robustus* to utilize weekly home ranges of 116–283 ha. in summer and winter, and over a period of years no individual moved more than 7 km. In the same locality, *M. rufus* used weekly ranges of 259–516 ha., but during a prolonged dry spell there was a general movement to a more favorable area 20–30 km away.

Social structure is variable in *Macropus* and does not seem fully related to the size, taxonomy, or ecology of the animals. The small *M. parma*, the medium-sized *M. rufogriseus* and *M. agilis*, and the large *M. robustus* are often solitary but may aggregate temporarily in the vicinity of favored resources (food, water, shelter); the medium- and large-sized species *M. dorsalis, M. parryi, M. giganteus, M. fuliginosus, M. antilopinus*, and *M. rufus* occur in organized groups (often called "mobs"); and the small *M. eugenii* is intermediate in social activity (Croft 1989; Russell 1974*a*). With respect to *M. parma*, Maynes (1977*b*) made sightings of 52 solitary animals, 14 pairs, and 5 groups of 3. McEvoy (1970) reported that the only associations of *M. rufogriseus* are of females and offspring, which separate not more than a month after the young leave the pouch, and mated pairs, which stay together for only 24 hours. In *M. robustus* the young may remain with its mother for several months after pouch life ends, and a few males may follow a near-estrous female, but other observed groups represent only chance aggregations at some favorable site (Kirkpatrick 1968). *M. agilis* sometimes has been referred to as gregarious, but several studies have indicated that it is most frequently seen alone, in temporary aggregations in the vicinity of resources, or in consorting units (Dressen 1993). It also may be found in small groups made up mainly of females sharing the same resting and feeding areas. Such groups are not necessarily related, however, as it is thought that young animals quickly separate from the females (Kirkpatrick 1970*a*).

The gray and red kangaroos and the antilopine wallaroo usually occur in small groups of 2 to at least 10 individuals that appear to be more than just random, temporary assemblies (Russell 1974*a*, 1974*b*). Such groups are in a different category from the large but unorganized aggregations of kangaroos that sometimes collect at a favorable feeding site. Newsome (1971*b*), for example, observed a gathering of 1,500 *M. rufus* in central Australia and stated that aggregations of 50–200 were common during periods of drought.

A male *M. rufus* may gain temporary control of several females, along with their young, but there usually is no permanent association between adults of opposite sexes (Russell 1974*b*, 1979). As in other species of *Macropus*, there appears to be no territorial defense, but the male may fight fiercely with other males that challenge his possession of the females, and there may also be agonistic behavior relative to competition for food and resting sites. Croft (1980) stated that in such situations male *M. rufus* might suddenly rear up and begin hitting each other from an upright position (boxing) and also kick with the hind feet. Collins (1973) reported that captive *M. rufus*, as well as other species of *Macropus*, could be kept in groups but that the presence of more than one adult male would lead to trouble.

The most social of the larger species of *Macropus* may be *M. giganteus*. Mobs of more than 10 animals comprising mainly adult females and young are often observed. Large males are thought to join the groups only when females are in estrus (Kirkpatrick 1967). Males form dominance hierarchies through aggressive interaction to determine access to females, and females also form them (Grant 1973; Russell 1974*b*). Kaufmann (1975) studied two mobs each comprising 20–25 animals, which were divided into subgroups averaging 3.7 individuals. He found a persistent group structure but no territorial defense or permanent association between males and females.

The most gregarious species in the genus are *M. dorsalis* and *M. parryi*. The latter, known as the "whiptail wallaby," was studied intensively in northeastern Queensland by Kaufmann (1974). It is perhaps the most social of all marsupials. In the study area were three loosely organized but discrete mobs each having a year-round membership of 30–50 individuals. The animals in any one mob did not keep together in a single group but usually split into varying subgroups of fewer than 10 individuals. All the animals in a particular mob, however, used the same home range, measuring 71–110 ha., which had little overlap with the home ranges of other mobs. Members of adjacent mobs mingled peaceably in the zones of overlap but kept mostly to their own undefended ranges. Within each mob, males established a dominance hierarchy through ritualized bouts of pawing, which did not cause injury. Larger males were dominant over smaller ones. The hierarchy functioned only to determine access to estrous females and ensured that fathering of offspring was limited to higher-ranking adult males. Courtship involved wild chasing of females by males. As they approached maturity, some subadult males left their natal mobs and joined other mobs, but no females were observed to do this.

Extensive information now is available on reproduction in *Macropus;* however, only a brief compilation can be provided here. A detailed summary also was given by Hume et al. (1989). The females of all species that have been investigated are polyestrous, with lengthy reproductive seasons but relatively brief intervals of receptivity lasting perhaps only a few hours (Kaufmann 1974). Calculated average lengths of estrous cycles (in days) are: *M. eugenii,* 28.4; *M. parma,* 41.8; *M. agilis,* 30.6; *M. rufogriseus,* 32.4; *M. parryi,* 42.2; *M. giganteus,* 45.6; *M. fuliginosus,* 34.9; *M. robustus,* 32.8; *M. rufus,* 34.3 (Kaufmann 1974; Poole and Catling 1974; Rose 1978). A capability to breed throughout the year has been demonstrated in *M. parma, M. agilis, M. rufogriseus, M. parryi, M. giganteus, M. fuliginosus, M. robustus,* and *M. rufus* (Collins 1973; Kaufmann 1974; Maynes 1977*b;* Merchant 1976; Poole 1975; Russell 1974*a*). Births in captive *M. dorsalis* have been recorded in March, September, and October (Collins 1973). Most other species are not well known in this regard. *M. eugenii,* however, is an exception to the general pattern. This species apparently has a rigid breeding season extending from January to June or August, with most births occurring between mid-January and mid-February; females can produce only one young per year. The season is the same in a natural population on Kangaroo Island off South Australia, in an introduced population on Kawau Island off northern New Zealand, and in captivity in Australia. The season is reversed when animals are taken to the Northern Hemisphere (Maynes 1977*a*). A restricted breeding season also is found in populations of *M. rufogriseus* on Tasmania and islands in Bass Strait (Poole 1979).

Most of those species of *Macropus* capable of continuous breeding in captivity or under favorable natural conditions may nonetheless be dependent on environmental factors. Hence, the introduced population of *M. parma* on Kawau Island bred throughout 1970 and 1971 but only from March to July in 1972, when less food was available (Maynes 1977*a*). The red kangaroo *(M. rufus)* and the euro *(M. robustus)* are opportunistic breeders, with both sexes remaining potentially fertile throughout the year. Young are produced continuously when there is adequate rainfall and forage, but breeding may cease altogether during prolonged drought (Newsome 1975; Tyndale-Biscoe 1973). In the wild the gray kangaroos *(M. giganteus* and *M. fuliginosus)* breed mainly from September to March, a period that follows winter rainfall and coincides with the time of maximum growth of vegetation. In New South Wales most births occur from November to January (Poole 1973, 1975, 1976).

Nondelayed, average gestation periods (in days) are: *M. eugenii,* 28.3; *M. parma,* 34.5; *M. agilis,* 29.2; *M. rufogriseus,* 29.6; *M. parryi,* 36.3; *M. giganteus,* 36.4; *M. fuliginosus,* 30.6; *M. robustus,* 32.3; *M. rufus,* 33.0 (Kaufmann 1974; Maynes 1973; Poole 1975; Rose 1978). Litter size usually is one, but twins have been recorded in *M. eugenii, M. rufogriseus* (also one set of triplets), *M. rufus* (Collins 1973), and *M. giganteus* (Poole 1975).

Embryonic diapause occurs in most species of *Macropus.* In these species, unlike in most mammals, conception alone does not affect the estrous cycle, and the next period of receptivity and mating come at the same time as they would have if the female had not become pregnant (Russell 1974*a*). After the second mating, estrus finally is suppressed through the suckling stimulus of the newly born first young. The embryo resulting from the second mating develops only to the blastocyst stage and then becomes quiescent until the first young is approaching the end of pouch life or perishes or is experimentally removed. Subsequently the second embryo resumes development, and birth occurs at a time when the pouch is vacant.

Whiptail wallaby *(Macropus parryi)*, photo by I. R. McCann / National Photographic Index of Australian Wildlife.

Embryonic diapause is perhaps best known in *M. rufus* (Russell 1974*a;* Tyndale-Biscoe 1973). A female of this species mates within two days after giving birth. The resultant embryo develops into a blastocyst of approximately 85 cells and then becomes dormant, provided the original young still is suckling in the pouch. When the pouch young is about 204 days old, or at any time sooner if the pouch young dies or is removed, the embryo resumes development. In 31 days, and within a day after the first young permanently leaves the pouch, a second birth occurs. Immediately thereafter the female again becomes receptive, and another mating can take place. If the pouch young is lost before it is 204 days old the female becomes receptive 33 days after such loss. As a result of this process, a female red kangaroo can produce one young approximately every 240 days as long as favorable conditions hold. Under such conditions most adult females examined in the field are found to have one quiescent embryo, one pouch young, and one accompanying young outside of the pouch. When food supplies are low a pouch young may die from inadequate lactation, but the diapausing embryo will then resume growth and may reach term by the time conditions are better. If there has been no environmental improve-

Western gray kangaroo *(Macropus fuliginosus)* with young, photo by Mary Eleanor Browning.

ment, the second young also may die, but another fertilization would have taken place and a third embryo would be on its way. In severe droughts, which are common in much of Australia, females become anestrous and breeding ceases.

The reproductive patterns of three other species are known to resemble closely those of *M. rufus* in that there is continuous breeding under suitable conditions, a postpartum estrus before the pouch young is two days old, and embryonic diapause. These species, the average period from premature loss of pouch young to next birth, and the age of pouch young when the second young normally is born are: *M. agilis,* 26.5 days, 7 months; *M. rufogriseus,* 26.7 days, 9 months; and *M. robustus,* 30.8 days, 8–9 months (Kaufmann 1974; Kirkpatrick 1968, 1970*a;* McEvoy 1970).

In still another three species the pattern is somewhat different. *M. eugenii* is not a continuous breeder but does have a postpartum estrus. If the pouch young is lost while the breeding season is still in progress, the next birth occurs in 27 days (Rose 1978). If the first pouch young survives, the diapausing embryo is retained through the anestrous period of the female and resumes development at the beginning of the next breeding season, about 11 months after fertilization (Russell 1974*a*). In *M. parma* some females have an estrus 4–13 days after giving birth, but most do not become receptive until the pouch young is 45–105 days old, and a few do not enter estrus at all while carrying a pouch young. If fertilization does occur when a pouch young is present, embryonic diapause occurs and removal of the pouch young is followed by birth in 30.5–32.0 days. If the pouch young survives, it vacates the pouch when it is about 212 days old, and the next young is born 6–11 days later (Maynes 1973). In *M. parryi* there is no postpartum estrus, and mating usually does not occur again until the pouch young is 150–210 days old. Embryonic diapause then may follow. If the pouch young dies, the next young is born about a month later; otherwise the next young is born just after the first young vacates the pouch at 9 months (Kaufmann 1974).

Embryonic diapause also is known to occur in *M. irma* (Collins 1973), *M. dorsalis* (Poole and Catling 1974:300), and *M. giganteus.* In the latter species, at least, this phenomenon appears rare, having been found in only about 5 percent of wild females with pouch young that were shot for study (Clark and Poole 1967; Poole 1973), and in only seven captive females during 10 years of observation (Poole and Catling 1974). In the latter animals diapause followed matings that took place 160–209 days after the first young was born. In both *M. giganteus* and *M. fuliginosus* (in which embryonic diapause has not been confirmed) females with pouch young may become receptive as early as 150 days after the first young is born. Conception, however, does not generally occur until there will be time before birth for the first young to vacate the pouch, which it does at an average age of about 320 days. The interval between successive births is approximately one year in both species (Poole 1977). If the pouch young dies, the female returns to estrus within an average of 10.92 days in *M. giganteus* and 8.25 days in *M. fuliginosus* (Poole and Catling 1974).

The red kangaroo, largest of marsupials, weighs only 0.75 grams at birth (Sharman and Pilton 1964). At that time it has a large tongue and well-developed nostrils, forelimbs, and digits but is embryonic in other external features. Like other newborn marsupials, it scrambles from the birth canal to the pouch without the assistance of the mother and grasps one of the mammae. It first releases the nipple at about 70 days, first protrudes its head from the pouch at 150 days, temporarily emerges at 190 days, and permanently vacates the pouch at 235 days. It then continues to suckle by placing its head in the pouch, and it is finally weaned at about 1 year.

Despite differences in size, most other species of *Macropus* do not take much less time to develop than *M. rufus*, and some take longer. For example, age (in days) at first departure from the pouch, permanent emergence, and weaning, respectively, for three species are: *M. parma*, 175, 207–18, 290–320 (Maynes 1973); *M. agilis*, 176–211, 207–37, 365 (Merchant 1976); and *M. parryi*, 240, 275, 450 (Kaufmann 1974).

Another large species, *M. robustus*, has a pouch life of about the same length as that of *M. rufus* (Collins 1973), but in the gray kangaroos development is longer, average age at first emergence from the pouch being 283.9 days in *M. giganteus* and 298.4 days in *M. fuliginosus*. Respective figures for final departure are 319.2 days and 323.1 days. Lactation in both species usually exceeds 18 months, longer than in any other marsupials (Poole 1975).

Sexual maturity also takes fairly long in gray kangaroos (Poole 1973; Poole and Catling 1974). Average age for captive males was 29.0 months in *M. fuliginosus* and 42.5 months in *M. giganteus*. A few wild males mated as early as 20 months, whereas some did not begin until 72 months. Average age for first mating in captive females was about 22 months in *M. fuliginosus* and 21 months in *M. giganteus*. Wild females generally became sexually mature between 20 and 36 months. In *M. rufus* females usually mature at 15–20 months, and males at about 20–24 months, but the period is lengthened under adverse environmental conditions. In the rigorous habitat of northwestern Australia *M. robustus* has not been observed to breed before 25 months, but captive males may breed at 22 months and females may breed at about 18 months. Sexual maturity in *M. parma* was attained at 11.5–16.0 months in captivity, as early as 12 months in the wild in Queensland, and as early as 19 months, but usually not until 2 years, on Kawau Island, New Zealand (Maynes 1973, 1977*a*, 1977*b*). In *M. agilis* males mature at 14 months and females at 12 months (Merchant 1976). In *M. parryi* maturity occurs at 18–29 months, but because of social factors males are prevented from mating until they are 2–3 years old (Kaufmann 1974).

Although most individuals probably do not even live to maturity, wallabies, kangaroos, and wallaroos have a potentially great life span. Some notable longevities in captivity are: *M. eugenii*, 9 years and 10 months; *M. agilis*, 10 years and 2 months; *M. rufogriseus*, 15 years and 2 months; *M. dorsalis*, 12 years and 5 months; *M. parryi*, 9 years and 8 months; *M. giganteus*, 24 years; *M. robustus*, 19 years and 7 months; *M. antilopinus*, 15 years and 11 months; and *M. rufus*, 16 years and 4 months (Collins 1973; Jones 1982; Kirkpatrick 1965). Some outstanding estimated ages for wild individuals are: *M. parma*, 7 years; *M. rufogriseus*, 18.6 years; *M. giganteus*, 19.9 years; *M. robustus*, 18.6 years; and *M. rufus*, 27 years (Bailey 1992; Kirkpatrick 1965; Maynes 1977*b*).

The effect of people on *Macropus* has varied considerably. One species, *M. greyi*, is designated as extinct by the IUCN. It was still common in southeastern Australia in the first years of the twentieth century but subsequently declined rapidly through competition from livestock, bounty and sport hunting, and killing for its beautiful pelt. The last wild individuals were recorded in 1924, though a few captives survived until about 1937 (Calaby 1971; Harper 1945).

Two additional species were once thought to be extinct. *M. bernardus*, which occupies a restricted area in the Northern Territory, had not been collected since 1914 but was rediscovered in 1969 (Goodwin and Goodwin 1973; Parker 1971). It now is considered to be maintaining its original numbers and distribution (Calaby 1971) but is designated as near threatened by the IUCN. *M. parma*, of the

Wallaroo *(Macropus robustus)*, photo from San Diego Zoological Society.

Antilopine wallaroo *(Macropus antilopinus)*, photo by I. R. McCann / National Photographic Index of Australian Wildlife.

coastal forests of New South Wales, seemed to have disappeared through clearing of its habitat, with the last known specimens taken in 1932. During the 1870s, however, the species, along with several other kinds of wallabies, had been introduced on Kawau Island, New Zealand. There a relatively dense population built up that was unknown to science until 1965. Then in 1967 *M. parma* again was found in its original range in New South Wales. Later investigation showed that the species still occurred along several hundred kilometers of the coast but at low densities (Maynes 1974, 1977*b*; Poole 1979; Ride 1970). *M. parma* now is classified as near threatened by the IUCN and as endangered by the USDI. A closely related species, *M. eugenii*, also was introduced to Kawau Island in the 1870s and still occurs there. It was common on the mainland of Australia until about 1920 but subsequently has nearly vanished in South Australia and become restricted to a few isolated colonies in Western Australia because of habitat loss and competition with domestic sheep. A number of natural island populations also were extirpated or reduced, but the species still occurs in satisfactory numbers on Kangaroo Island, South Australia, and on several islands off the coast of Western Australia (Calaby 1971; Harper 1945; Poole 1979; Poole, Wood, and Simms 1991). The IUCN designates *M. eugenii* generally as near threatened but notes that the subspecies *M. e. eugenii*, of Nuyt's Archipelago off South Australia, is extinct in the wild.

The five other species of wallaby in the genus *Macropus* have fared better in modern times. Although Harper (1945) expressed concern about heavy sport and skin hunting of *M. parryi* and *M. irma*, and even though these and other species have lost habitat in recent years, all five still are widely distributed and relatively common (Calaby 1971; Poole 1979). In certain areas suitable habitat actually has increased through scattered opening of the forest for agriculture (Kaufmann 1974; Ride 1970). *M. irma* is designated as near threatened by the IUCN.

In Tasmania *M. rufogriseus* probably is more common now than when settlement began (Calaby 1971). It is considered a pest because of its damage to regenerating eucalyptus forests (Ride 1970). It and other species are sometimes shot in large numbers for alleged destruction of crops and pasture. In addition, several species are still important in the skin trade, though hunting is now regulated in order to ensure that overall populations are not adversely affected. In 1976, 45,259 *M. parryi* were killed for commercial purposes in Queensland (Poole 1978). *M. dorsalis* was introduced on Kawau Island, New Zealand, in the 1870s and still occurs there. *M. rufogriseus* was brought to the South Island of New Zealand at about the same time, eventually increased to an estimated 750,000 individuals, and then was reduced by control in the 1960s to only about 3,500 (Wodzicki and Flux 1971).

Surprisingly, perhaps, the effects of European settlement thus far have seemed less detrimental to the larger wallaroos and kangaroos than to many smaller species of marsupials. Gray kangaroos, however, have disappeared from many densely settled localities, and there is concern for some remaining populations, such as that of *M. giganteus* in southeastern South Australia (Poole 1977). The subspecies *M. giganteus tasmaniensis* of Tasmania was greatly reduced in numbers and distribution by the early twentieth century through excessive sport and commercial hunting and today occurs in just a few small parts of its original range (Barker and Caughley 1990; Harper 1945). It is now listed as endangered by the USDI and as near threatened by the IUCN. *M. fuliginosus fuliginosus* of Kangaroo Island off South Australia, recently confirmed as a highly distinctive subspecies (Poole, Carpenter, and Simms 1990) and also designated as near threatened by the IUCN, may be subject to loss of habitat as human activity increases in its limited range (Poole 1976, 1979). Otherwise the gray kangaroos, as well as the red kangaroo *(M. rufus)* and the two larger wallaroos *(M. robustus* and *M. antilop-*

Red kangaroos *(Macropus rufus)*, photos by Lothar Schlawe.

inus), still occur over most of their original ranges and generally are not thought to be in any immediate jeopardy (Calaby 1971; Poole 1979). Indeed, there is evidence that at least some of these species increased in numbers following settlement. They seem to have been aided by establishment of artificial water holes for livestock, as well as by the grazing of stock, which crops the long, dry grass avoided by the kangaroos and causes the sprouting of soft, green shoots. Favorable grazing conditions for the kangaroos thus were created over vast areas (Newsome 1971*a;* Poole 1979; Russell 1974*a*). The subspecies *M. robustus isabellinus,* restricted to Barrow Island off Western Australia, is classified as vulnerable by the IUCN.

According to Newsome (1975), livestock and introduced rabbits may yet cause the collapse of the red kangaroo. He postulated that simultaneous competition with other herbivores (sheep, cattle, or rabbits), especially during drought, could eventually bring about conditions under which *M. rufus* would become rare. He also observed that in northwestern Australia *M. rufus* initially increased in numbers following introduction of sheep but then lost in competition with the sheep and *M. robustus.* The latter species had invaded the plains because overgrazing by sheep had eliminated the luxuriant grass cover and allowed the spread of highland vegetation favored by *M. robustus.* Edwards (1989) indicated that the ecological relationship between kangaroos and sheep is complex and not fully understood but that competition does occur under certain circumstances.

For the present, at least, kangaroos are relatively common in Australia. According to Grigg et al. (1985), surveys from 1980 to 1982 indicated that there were then about 8,351,000 red, 1,774,000 western gray, and 8,978,000 eastern gray kangaroos in Australia. Mainly because of a severe drought, by 1984 numbers had fallen to 6,330,000 red, 1,162,000 western gray, and 5,791,000 eastern gray kangaroos. New surveys reported by Fletcher et al. (1990) suggested that populations had recovered to approximately their 1980 levels, but no overall estimate was calculated. The USDI (1995) used a total estimate of 26,200,000 red and gray kangaroos for 1992 but cited other estimates of barely half that number and noted that there subsequently had been a renewed decline in association with another drought. In any case, the total number of these three species plus that of the wallaroos probably approximates or exceeds the number of humans (about 18 million) in Australia. No other terrestrial region in the world of comparative size still has more large wild mammals than people.

The major question, however, is not whether there are plenty of kangaroos but whether populations can sustain present hunting levels, especially considering the additional factors of human habitat modification and drought. Kangaroos have long been heavily hunted, first by the aborigines and then by European settlers for meat and skins, and later by stockmen because of alleged damage to fences and pasture. During the late nineteenth and early twentieth centuries millions of kangaroos were killed for bounty payments alone.

Subsequently an industry developed based on the harvesting and processing of about 300,000–400,000 skins per year (Poole 1978). In the late 1950s, improved refrigerating equipment allowed a great increase in the take of kangaroos for use as pet food. Such hunting, along with continued killing for skins and pest control, was not well regulated and coincided with severe droughts in parts of Australia. Kangaroo numbers dropped alarmingly over large areas (Ride 1970; Russell 1974*a*).

The total recorded harvest of *M. giganteus, M. robustus,* and *M. rufus* in Queensland amounted to at least 200,000 animals annually from 1954 to 1983 and was over 1 million in several of those years. In 1983 the kill was 502,161 gray kangaroos, 62,100 wallaroos, 263,346 red kangaroos, and about 21,000 of the smaller wallabies (Poole 1984). In New South Wales the legal take in 1975 was 60,300 gray kangaroos and 48,100 red kangaroos, or 3.8 percent and 2.3 percent, respectively, of the estimated populations in the state (Caughley, Sinclair, and Wilson 1977). In 1992 the total commercial kill of the three species was 2,676,000, or about 10 percent of the total population estimated by the USDI (1995).

Much controversy centers on how many kangaroos can be safely harvested and whether commercial utilization jeopardizes the species or helps by giving them a value that discourages their mass destruction as pests (Australian National Parks and Wildlife Service 1988; Poole 1984). During the 1970s the Australian state governments increased their control of kangaroo hunting. From 1973 to 1975 the Australian federal government prohibited the general export of kangaroo products but subsequently ended the ban as the states established acceptable management programs (Poole 1978). Most exports of kangaroo skins had been going to the United States, but in 1974 the USDI listed *M. giganteus, M. fuliginosus,* and *M. rufus* as threatened. Accompanying regulations prohibited the importation of these species and products thereof until there was satisfactory certification that wild populations would not be adversely affected. Such certification subsequently was received, and importation was opened in 1981. The USDI proposed removing the three species from threatened status in 1983, but in 1984, following drastic population declines attributed to a severe drought in Australia, the proposal was withdrawn. Gaski (1988) reported that the number of skins imported to the United States averaged more than 1 million annually in 1963–66, was 158,254 in 1981, and was about 50,000 in 1986. Many conservationists continue to oppose commercial importation and do not consider kangaroo populations to be safely managed in Australia. Over such opposition, and despite the role of its regulations in improving such management, the USDI did proceed to cancel the regulations and remove the three species from threatened status in 1995.

DIPROTODONTIA; **Family BURRAMYIDAE**

Pygmy Possums

This family of two Recent genera and five species is found in Australia, Tasmania, and New Guinea. The members of the Burramyidae sometimes have been placed in the family Phalangeridae but now are considered to represent a distinct group (Kirsch 1977*b*). The sequence of genera presented here follows that of Kirsch and Calaby (1977). The genera *Distoechurus* and *Acrobates* also were placed in the Burramyidae when that family was first resurrected in the 1970s, but recently they were moved to a new family, the Acrobatidae (Aplin and Archer 1987; Archer 1984). Nonetheless, the karyotype of *Distoechurus* has been found to be very similar to that of the burramyids (Westerman, Sinclair, and Woolley 1984).

Head and body length is 60–120 mm and tail length is 70–175 mm. The pelage is soft. Digital structure is much like that of the Phalangeridae, but neither genus has opposable digits on the forefoot. The dental formula is: (i 3/2, c 1/0, pm 2–3/3, m 3–4/3–4) × 2 = 34–40. According to Archer (1984), the Burramyidae are characterized by reduction of the first upper and second lower premolars, a bicuspid tip on

"Dormouse" possum *(Cercartetus lepidus)*, photo by Peter Crowcroft.

the third lower premolar, and a well-developed posteromesial expansion of the tympanic wing of the alisphenoid. Both genera have prehensile tails and are good climbers.

The known geological range of this family is middle Miocene to Recent in Australasia, though a possible burramyid was reported from beds of late Oligocene age in Tasmania (Marshall 1984).

DIPROTODONTIA; BURRAMYIDAE; **Genus CERCARTETUS**
Gloger, 1841

"Dormouse" Possums

There are four species (Aitken 1977*a;* Barritt 1978; Dixon 1978; Kirsch and Calaby 1977; Ride 1970; Wakefield 1970*a*):

C. concinnus, southwestern Western Australia, southern South Australia, western Victoria, southwestern New South Wales;
C. nanus, southeastern South Australia to southeastern Queensland, Tasmania;
C. lepidus, southeastern South Australia, western Victoria, Kangaroo Island, Tasmania;
C. caudatus, New Guinea, northeastern Queensland.

C. lepidus long was known only from Tasmania, except for fossil and subfossil remains on the mainland. In 1964, however, it was discovered alive on Kangaroo Island, and in the 1970s specimens were collected in southeastern South Australia and western Victoria. The species *C. lepidus* and *C. caudatus* sometimes have been placed in a separate genus, *Eudromicia* Mjöberg, 1916.

Head and body length is 70–120 mm, tail length is 70–175 mm, and adult weight is about 15–40 grams. In *C. concinnus* the upper parts are reddish brown and the underparts and feet are white; *C. nanus* is grayish or fawn-colored above and slaty below; and in *C. lepidus* and *C. caudatus* the upper parts vary from dull rufous and rich brown to bright pale fawn and the underparts are white to yellowish. *C. caudatus* has well-marked, broad black bands passing from the nose through the eyes, not quite reaching the ears. The pelage is dense and soft. The prehensile, cylindrical tail is well furred at the base and scantily haired for the remainder of its length. With the approach of winter *C. nanus* becomes very fat and its tail becomes greatly enlarged, especially at the base, so that stored food will be available during periods of torpor.

The common name, "dormouse" possum, reflects the superficial resemblance of this marsupial to the dormouse *(Glis glis)* so well known in Europe. In addition, both genera are nocturnal and undergo hibernation in cold weather. "Dormouse" possums have large, thin, almost naked ears. The forefoot looks somewhat like a human hand, and the great toe of the hind foot is thumblike and widely opposable. The claws are small, but the pad of each digit is expanded into two lobes. The pouch is well defined in females. The normal number of mammae is four except in *C. concinnus,* which has six (Collins 1973). *C. nanus* usually has four functional and two nonfunctional teats (Ward 1990*a*).

These tiny, arboreal possums dwell in forest, heath, or shrubland. They construct a small, dome-shaped nest of grass or bark in a hollow limb, hollow stump, crevice in a tree trunk, or thick clump of vegetation. They sometimes reside in abandoned birds' nests. *C. nanus* usually makes a nest of soft bark and may travel as much as 500 meters to

"Dormouse" possum *(Cercartetus nanus)*, photo by Stanley Breeden. Inset: underside of right foot and tip of toe *(C. concinnus)*, photo from *The Mammals of South Australia*, F. Wood Jones.

secure the desired kind of bark. Possums of this genus hang by the tail when reaching from one branch to another and also skillfully use the tail when climbing or descending. They are nocturnal, leaping and running freely at night. They generally sleep soundly during the day but sometimes become active in cloudy weather. The species *C. nanus* and *C. lepidus* have been found to undergo alternate periods of activity and dormancy throughout the year in Tasmania. Neither species experienced prolonged hibernation; the longest period of torpor was 6 days for *C. lepidus* and 12 days for *C. nanus*. Body temperature during dormancy was about equal to air temperature (Hickman and Hickman 1960). Experimental inducement of deep hibernation in *Cercartetus nanus* for periods of up to 35 days has been achieved by Geiser (1993). All species seem to be omnivorous, feeding on leaves, nectar, fruit, nuts, insects and their larvae, spiders, scorpions, and small lizards. One captive specimen of *C. nanus*, however, was a strict vegetarian, refusing to eat meat or insects, and lived a record 8 years (Perrers 1965).

These possums have been maintained in pairs in captivity, and a group of three males and one female was found hibernating together, but generally *Cercartetus* seems to be solitary in the wild (Collins 1973). *C. concinnus* and *C. nanus* may breed throughout the year in Victoria, but most births of the latter species occur from November to March. After the birth of a litter females return to estrus and mate again; there probably is a subsequent period of embryonic diapause, with the young being born soon after the previous litter is weaned. Most females produce two litters per year, but some produce three (Ward 1990*a*, 1990*c*). Minimum gestation in *C. concinnus* is 51 days (Hayssen, Van Tienhoven, and Van Tienhoven 1993). In New Guinea a wide range of birth dates has been recorded for *C. caudatus*, but there is a breeding recession during the dry season from mid-May to early August (Dwyer 1977). Captive *C. caudatus* have bred twice yearly, in January and February and from late August to early November; litters have contained one to four young, which emerged from the pouch at 34 days of age, became independent at 92 days, and first bred at 15 months (Atherton and Haffenden 1982). Average litter size is five in *C. concinnus*, with a maximum of six. Ward (1990*a*) recorded litter sizes of four to six in *C. nanus*, though modal size was four; pouch life is 30 days, the young become independent immediately after weaning and reach sexual maturity at as early as 4.5–5.0 months. Available records from Tasmania indicate that litters of about four *C. lepidus* are born between September and January and that they approach adult size and become independent at 3 months (Green *in* Strahan 1983). *C. nanus* appears to do well in captivity; in addition to the specimen mentioned above that lived 8 years, a number of individuals have survived for 4–6 years (Collins 1973); maximum longevity in the wild is at least 4 years (Ward 1990*a*).

The subspecies *C. caudatus macrurus*, of northeastern Queensland, is designated as near threatened by the IUCN.

DIPROTODONTIA; BURRAMYIDAE; **Genus**
BURRAMYS
Broom, 1895

Mountain Pygmy Possum

The single species, *B. parvus*, now inhabits the mountains of eastern Victoria and extreme southeastern New South Wales. Until recently it was known only from fossilized cranial and dental material found at the Wombeyan Caves, near Goulburn in New South Wales. The remains are from the late Pleistocene and are estimated to be about 15,000 years old. There was controversy regarding the specimens, with some scientists thinking that the fossils represented a kind of miniature kangaroo. In the early 1950s, however, the material was examined by W. D. L. Ride, who showed that *Burramys* actually was a small possum related to *Cercartetus*. Additional fossils were located in the late 1950s. Then in August 1966 a live possum of an unknown kind was found in a ski hut on Mount Hotham, Victoria. Upon examination it was identified as *B. parvus*. In the words of Ride (1970:16): "The dream dreamed by every paleontologist had come true. The dry bones of the fossil had come together and were covered with sinews, flesh and skin." Subsequently, from 1970 to 1972, 3 more specimens were taken at Mount Hotham; 1 was trapped in the Falls Creek area of the Bogong High Plains, Victoria; and 19 were collected in the Kosciusko National Park, New South Wales (Dimpel and Calaby 1972; Dixon 1971). More recently many hundreds have been live-trapped in the course of field investigations (Mansergh and Scotts 1990).

Head and body length is 101–30 mm, tail length is 131–60 mm, and weight is 30–60 grams (Calaby *in* Strahan 1983). The upper parts are brownish gray, and the underparts are paler. The body fur continues onto the tail for about 1 cm, and for the rest of its length the prehensile tail is almost naked. The premolar teeth are unusually large. There is a distinct pouch with two pairs of mammae (Dixon 1971).

All specimens have been taken in mountainous areas at altitudes of about 1,500–1,800 meters. The species apparently is restricted to dense patches of shrubs associated with snow gum trees and large boulders in the subalpine to alpine zones (Dimpel and Calaby 1972). The climate in the Kosciusko National Park is severe, with an average annual precipitation of about 125–200 cm, much of it winter snow. Calaby, Dimpel, and Cowan (1971) suggested that *Burramys*, unlike most possums, is not arboreal but mainly terrestrial and adapted to climbing shrubs. They also thought that the animal probably lives in holes among or under rocks. When released under controlled conditions *Burramys* ran rapidly on the ground and over rocks in a ratlike manner and did not hesitate to dive into holes. It climbed shrubs rapidly by grasping the stems with its forefeet and hind feet. Dixon (1971) reported that a captive was well adapted for climbing and jumping but made only occasional use of its prehensile tail.

A number of recent investigations have substantially increased our knowledge of this newly discovered genus (Fleming 1985; Kerle 1984*a*, 1984*b*; Mansergh 1984*a*, 1984*b*; Mansergh and Scotts 1990). *Burramys* is strictly nocturnal. In captivity it has entered prolonged bouts of deep torpor lasting up to 20 days (Geiser and Broome 1991), and it apparently undergoes winter hibernation in the wild. The diet consists mainly of seeds, fruits, worms, and arthropods. *Burramys* is the only marsupial known to establish caches of durable foods; such storage would facilitate winter survival. Nonbreeding captives are tolerant of one another, thus suggesting that huddling, another winter survival strategy, occurs in the wild. Breeding females, however, defend their nests. The pygmy possum population in 6 ha. of suitable habitat on Mount Higginbotham, Victoria, has been estimated to contain 84–122 breeding females. Densities in other areas are much lower.

Females are polyestrous, with a mean estrous cycle of 20.3 days. The short alpine summer requires a brief gestation period and rapid development of young. In the wild, births occur in mid- and late November (spring), though in captivity they have been recorded in several other months.

Mountain pygmy possum *(Burramys parvus)*, photo from Australian News and Information Bureau.

Limited evidence indicates that gestation lasts 13–16 days. Litters contain up to eight young, though only four actually survive to be carried in the pouch. They leave the pouch after 3 weeks, open their eyes at 5–6 weeks, are weaned at 8–9 weeks, and attain adult weight at 3–4 months. Females invariably breed in the first spring after they are born, but some males do not. Females have been known to live for at least 11 years in the wild, but males only up to 4 years.

The mountain pygmy possum is listed as endangered by the IUCN and USDI. Its restricted range may be jeopardized by habitat destruction and other environmental modifications associated with the development of ski resorts. The largest amount of remaining habitat, about 8 sq km in Mount Kosciusko National Park, New South Wales, is estimated to contain 500 adult *Burramys*. Another population, of much greater density but limited to a 6-ha. area on Mount Higginbotham, Victoria, contains 400 individuals. The remaining tracts of good habitat are all small and are separated from one another by large, unsuitable areas (Caughley 1986; Mansergh 1984*a*).

DIPROTODONTIA; **Family PSEUDOCHEIRIDAE**

Ring-tailed and Greater Gliding Possums

This family of 6 Recent genera and 16 species inhabits Australia, Tasmania, New Guinea, and certain nearby islands. The members of the Pseudocheiridae long were regarded as components of the family Phalangeridae. Kirsch (1977*b*), however, explained that they differ in serology, karyology, and other characteristics, and he placed them in the subfamily Pseudocheirinae of the family Petauridae. On the basis of additional serological studies, Baverstock (1984) elevated the Pseudocheirinae to familial rank but indicated that this group is closely related to the Petauridae. This view has been supported by Aplin and Archer (1987), Baverstock, Birrell, and Krieg (1987), and most other recent authorities. Szalay (1994), however, regarded the group only as a tribe within the family Petauridae. The sequence of genera presented herein attempts to express the affinities discussed by Flannery and Schouten (1994) but does not precisely follow their order of presentation.

New Guinean ring-tailed possum *(Pseudochirulus forbesi)*, photo by Pavel German.

Head and body length is 163–480 mm and tail length is 170–550 mm. The pelage may be woolly or silky. According to McKay (1989), the dentition of the Pseudocheiridae is characterized as follows: three upper incisors on each side of the jaw, only the first pair prominent, and also a reduced upper canine; three upper premolars, the third elongated and bearing two or three cusps; four strongly selenodont upper molars, with crescent-shaped ridges connecting the cusps; the large, procumbent first lower incisor compressed and bladelike; a second, vestigial lower incisor may be present in some species but lost in others; three large lower premolars; and four lower molars, all approximately equal in size. The skull is characterized by a robust zygomatic arch and posterior palatal vacuities. The forefoot has the first two digits at least partly opposable to the other three, an adaptation to grasping small branches. The tail is prehensile and may have a ventral friction pad of naked calloused skin or only a small naked area at the tip. In the digestive tract the caecum is greatly enlarged to act as a fermentation chamber.

All members of the Pseudocheiridae are arboreal. One genus has attained a gliding ability through development of a membrane uniting the front and hind limbs. This genus, *Petauroides,* has an analogous phylogenetic position to that of *Petaurus* in the family Petauridae (Kirsch 1977*b*).

The geological range of the Pseudocheiridae is Miocene to Recent in Australasia (McKay 1989). Details on Miocene pseudocheirid fossils and their significance to the phylogeny of the family were presented by Woodburne, Tedford, and Archer (1987).

DIPROTODONTIA; PSEUDOCHEIRIDAE; **Genus PSEUDOCHIRULUS**
Matschie, 1915

New Guinean and Queensland Ring-tailed Possums

There are seven species (Flannery 1995; Flannery and Schouten 1994; Kirsch and Calaby 1977; Musser and Sommer 1992; Ride 1970; Ziegler 1977):

P. canescens, New Guinea and Japen and Salawatti islands to the west;
P. mayeri, Central Range of New Guinea;
P. caroli, west-central New Guinea;
P. schlegeli, known only by the type from the Vogelkop Peninsula of western New Guinea and by one other specimen probably from the same area;
P. forbesi, New Guinea;
P. herbertensis, from Kuranda to Ingham in northeastern Queensland;
P. cinereus, from Mount Lewis to Thornton Peak in northeastern Queensland.

These species long were placed in the genus *Pseudocheirus* together with *P. peregrinus.* However, Flannery and Schouten (1994), interpreting the molecular analyses of Baverstock (1984) and Baverstock et al. (1990), treated *Pseudochirulus* as a separate genus with a content as given above. Further studies suggest that *P. forbesi* might comprise several species (Musser and Sommer 1992).

Head and body length is about 167–368 mm, tail length is 151–395 mm, and weight is 105–1,450 grams (Flannery and Schouten 1994; Winter *in* Strahan 1983). In most species the fur is dense, soft, and woolly. The upper parts of the body are gray or brown, often very dark, and the underparts are white, yellowish, or almost as dark as the back. Some species have dark and light markings on the head, a median stripe on the back, or stripes around the thighs. The distal end of the tail is usually bare for some length on the undersurface and only sparsely haired on the upper surface. In most species the prehensile end of the tapered tail is usually curled into a ring, hence the common name of these possums. The first two digits of the forefoot are opposable to the other three digits. Females have four mammae, but normally only two are functional.

All species are arboreal forest dwellers. Their habits are not well known but are thought to be comparable to those of *Pseudocheirus.* Some species are known to build nests like those of *Pseudocheirus* or to shelter in hollow trees. The diet probably is largely folivorous (leaf-eating) and also includes fruit. *P. herbertensis* appears to be solitary except when males are attracted to estrous females. Mating

New Guinean ring-tailed possum *(Pseudochirulus mayeri)*, photo by Tim Flannery.

may occur throughout the year but seems to peak in April and May. Usually two young are reared. They emerge from the pouch after four to five months and may then be carried on the mother's back for a short period; thereafter they are left in the nest and make increasingly longer forays alone (Winter *in* Strahan 1983). *P. forbesi* also appears to breed year-round but usually has only a single young (Flannery 1990*b*). *P. herbertensis* may be in jeopardy because of logging of its tropical forest habitat (Winter 1984*a*). That species, as well as *P. cinereus* of the same region, is designated as near threatened by the IUCN.

DIPROTODONTIA; PSEUDOCHEIRIDAE; **Genus PSEUDOCHEIRUS**
Ogilby, 1837

Common Ring-tailed Possum

Flannery and Schouten (1994) recognized a single species, *P. peregrinus,* occurring along the eastern coast of Australia from the Cape York Peninsula of northeastern Queensland to southeastern South Australia, in southwestern Western Australia, and on Kangaroo Island and Tasmania. Kennedy (1992) regarded the isolated subspecies of *P. peregrinus* in Western Australia as a separate species, *P. occidentalis,* and it is possible that other populations also warrant specific status. The genera *Pseudochirulus, Pseudochirops, Petropseudes,* and *Hemibelideus* (see accounts thereof) sometimes have been included in *Pseudocheirus.*

Head and body length is 287–353 mm, tail length is 287–360 mm, and weight is 700–1,110 grams (Flannery and Schouten 1994; McKay *in* Strahan 1983). The fur is short, dense, and soft. The coloration is highly variable, ranging from predominantly gray to rich red, and the underparts are paler. There are always distinctive white ear tufts and a white tail tip. The tail is tapered, with the fur progressively shorter distally and a long, bare friction pad on the undersurface of the distal portion. This prehensile end of the tail is commonly curled into a ring. The first two digits of the forefoot are opposable to the other three digits. Females have four functional mammae.

The common ring-tailed possum is scansorial and nocturnal. It may be found in rainforest, sclerophyll forest, woodland, or brush. It generally shelters by day in a large, dome-shaped nest of interwoven leaves, bark, and ferns located in the branches of a shrub or tree, or it may rest in a tree hollow, which it lines with leafy material. The height of the nest varies from only a few centimeters above the ground in heavy cover to about 25 meters in mistletoe clusters (Collins 1973). In the wild this possum seems to be strictly herbivorous, feeding on a variety of leaves, fruits, flowers, bark, and sap.

The movements of this possum are generally slow, and its quiet, retiring manner may make it seem uninteresting to human observers. When it moves from one limb to another it usually keeps hold of the old resting place with its tail until it has grasped the new branch with its forefeet. It sleeps with its head beneath its body and between its hind feet. Although it is agile and graceful, its usual reaction to a sudden encounter is to remain motionless with a vacant stare. If the intruder remains, the possum may slowly creep away. On the ground it travels at fair speed with a waddling gait.

The scarce western subspecies, *P. p. occidentalis,* has been

Ring-tailed possum *(Pseudocheirus peregrinus)*, photo by G. Weber / National Photographic Index of Australian Wildlife.

reported to occur at densities of about 0.1–4.5/ha. and to have an average home range of 1.0–2.5 ha. (Jones, How, and Kitchener 1994). Density of the eastern populations commonly is 12–16/ha., up to 19/ha. in favorable habitat, and home range averages 0.37 ha. In areas where density is high individuals are very aggressive toward one another. Otherwise this genus is semisocial and even gregarious, with reports in the east of up to three adults per nest, up to eight individuals constructing their nests in close proximity, and groups of males and females remaining together all year (Collins 1973; How 1978; How et al. 1984; McKay *in* Strahan 1983; Ride 1970). *Pseudocheirus* is polyestrous, with an estrous cycle of about 28 days. Breeding season varies: late April–December in Victoria, early winter and early summer near Sydney, and late summer and autumn in northern Queensland. Females sometimes rear two litters in a year. In the east litter size is one to three, with a mode of two. The young leave the pouch at about 18 weeks, are weaned at 6–7 months, and disperse at 8–12 months. Females reach sexual maturity at about 1 year (Collins 1973; How 1978; How et al. 1984; Ride 1970; Tyndale-Biscoe 1973). The western subspecies has birth peaks in April–June and September–November, there usually is only a single young, and pouch life is about 104 days (Jones, How, and Kitchener 1994). Life span in the wild is four to five years, and the record captive longevity is eight years (Collins 1973; How 1978).

In Tasmania, where presumably the cold winters result in desirable pelts, *P. peregrinus* is an important fur bearer; approximately 7.5 million were taken for this purpose from 1923 to 1955 (Ride 1970). In parts of Western Australia and Victoria *P. peregrinus* has become uncommon or restricted in distribution because of loss of habitat to agricultural development (Emison et al. 1978; Winter 1979). The subspecies *P. peregrinus occidentalis* of Western Australia has lost 50–90 percent of its habitat (Kennedy 1992) and is now classified as vulnerable by the IUCN.

DIPROTODONTIA; PSEUDOCHEIRIDAE; **Genus PSEUDOCHIROPS**
Matschie, 1915

Coppery and Silvery Ring-tailed Possums

There are five species (Flannery 1990*b;* Flannery and Schouten 1994; Kirsch and Calaby 1977; McKay *in* Bannister et al. 1988; Ride 1970; Ziegler 1977):

P. cupreus, Central Range of New Guinea;
P. albertisii, northern and western New Guinea and nearby Japen Island;
P. coronatus, Arfak Mountains of far western New Guinea;
P. corinnae, Central Range of New Guinea;
P. archeri, northeastern Queensland.

Coppery ring-tailed possum *(Pseudochirops cupreus)*, photo by Pavel German.

Kirsch and Calaby (1977) and most other authorities once considered *Pseudochirops* to be a subgenus of *Pseudocheirus* and also to include the species *P. dahli*. More recent serological and morphological studies have suggested that *Pseudochirops* should be elevated to generic rank (Baverstock 1984) and that *P. dahli* should be placed in the separate genus *Petropseudes* (see account thereof). Corbet and Hill (1991) restricted *Pseudochirops* to the single species *P. archeri*, but Groves (*in* Wilson and Reeder 1993) accepted the general arrangement given above. *P. coronatus* usually was considered a synonym of *P. albertisii* until it was restored to specific rank by Flannery and Schouten (1994).

Head and body length is 289–410 mm, tail length is 258–371 mm, and weight is 640–2,250 grams (Flannery 1990*b;* Flannery and Schouten 1994; Winter *in* Strahan 1983). The upper parts are coppery or silvery green, the underparts are paler, and there are dark dorsal stripes and pale facial markings on some of the species. The tail is prehensile and has distally bare areas similar to those of *Pseudocheirus.* McKay (1989) characterized *Pseudochirops* as follows: fur short, dense, and fine; tail shorter than head and body, tapering rapidly from a thickly furred base to sparsely furred tip, friction pad long; pupil a vertical slit. Flannery (1990*b*) added that the molar teeth of *Pseudochirops* are larger and more complex than those of *Pseudocheirus.*

The following natural history information was taken from Flannery (1990*b*) and Winter (*in* Strahan 1983). The New Guinea species are found mainly in undisturbed montane forests at elevations of 1,000–4,000 meters and may be declining in response to human habitat disruption. *P. archeri,* of Australia, lives in upland rainforest and seems to be able to survive in areas that have been partially logged over. Both *P. archeri* and *P. corinnae* are unusual in that they sleep on exposed branches rather than in a nest. In contrast, *P. cupreus* nests both in tree hollows and in burrows under tree roots. All species presumably are nocturnal but sometimes move about by day. All are primarily folivorous but may also eat fruit. *P. archeri* is solitary and breeds mainly in the latter half of the year. *P. cupreus* breeds throughout the year. A single young is normally produced.

The IUCN classifies *P. albertisii* and *P. corinnae* of New Guinea as vulnerable; both are thought to be undergoing a substantial decline. *P. archeri,* of Australia, is designated near threatened.

Rock possum *(Petropseudes dahli)*, photo by Pavel German.

DIPROTODONTIA; PSEUDOCHEIRIDAE; **Genus PETROPSEUDES**
Thomas, 1923

Rock Possum

The single species, *P. dahli,* is found in northeastern Western Australia, northwestern Northern Territory, and northwestern Queensland (Flannery and Schouten 1994). Although *Petropseudes* long was considered a full genus or subgenus, Kirsch and Calaby (1977) treated it as a synonym of *Pseudochirops.* More recent systematic accounts have restored it to generic rank (Groves *in* Wilson and Reeder 1993; McKay *in* Bannister et al. 1988). Part of the following information was taken from Nelson and Kerle (*in* Strahan 1983).

Head and body length is 325–450 mm, tail length is 200–275 mm, and weight is 1,280–2,000 grams. The fur is long and woolly. The upper parts are grayish, more or less suffused with rufous, the rump is brighter rufous, and the underparts are lighter gray. The short but prehensile tail is rufous gray and lacks a white tip; it is thickly furred at the base, but the tip and the terminal two-thirds of the undersurface are almost naked. Unlike in the other pseudocheirids, the first two digits of the forefoot are not opposable to the other three digits. The claws are reduced and blunt. Females have a well-developed, anteriorly opening pouch and two mammae.

This possum inhabits rocky outcrops on savannahs; it is nocturnal, spending the day in caves or crevices among the rocks and usually selecting the darkest recesses. It is not known to make a nest, and when at rest it often lies squeezed flat in a crevice or sits in a bent-over position in a corner. When the animal is in the latter position the tail is curved forward and upward, but when the animal is walking the tail is held straight out. *Petropseudes* is a good climber and enters trees at night to feed on fruits, flowers, and leaves. It generally associates in pairs, a male and a female, sometimes accompanied by a young, but aggregations of up to nine have been observed. Breeding apparently occurs throughout the year, and there normally is a single offspring.

Rock possum *(Petropseudes dahli)*, photo by Pavel German.

DIPROTODONTIA; PSEUDOCHEIRIDAE; **Genus HEMIBELIDEUS**
Collett, 1884

Brush-tipped Ring-tailed Possum

The single species, *H. lemuroides*, is found only in a small highland area in northeastern Queensland (Winter *in* Strahan 1983). *Hemibelideus* sometimes is considered a subgenus of *Pseudocheirus*, but new serological and morphological studies suggest that the two are generically distinct (Baverstock 1984; Johnson-Murray 1987). Chromosomal studies point to possible close affinity between *Hemibelideus* and *Petauroides* (McQuade 1984).

Head and body length is 313–52 mm, tail length is 335–73 mm, and weight is 810–1,270 grams (Winter *in* Strahan 1983). The upper parts vary in color from light fawn to dark blackish gray, and the underparts are yellowish gray. The head is dark brown with a tinge of red, and the limbs are dark brown with black near the ends. The ears project only a short distance beyond the thick fur of the head. The pelage is soft and woolly, even on the feet. The prehensile tail is black, bushy, thickly covered with fur for the whole length on the upper surface, and slightly tapering; the naked underside of the tip is very short.

Hemibelideus possesses suggestions of gliding membranes, for it has small folds of skin (less than 25 mm in width) along the side of the body. It has been observed making long jumps from tree to tree and from tree to ground. Some zoologists think that this genus bears characters transitional between the other ring-tailed possums and the greater gliding possum *(Petauroides)*. It is said to be very agile and active in the trees—some of its jumps almost have the appearance of true glides. It seems to use its furry tail as a rudder during these jumps.

Brush-tipped ring-tailed possum *(Hemibelideus lemuroides)*, photo from Queensland Museum.

According to Winter (*in* Strahan 1983), this possum is found in rainforest at elevations above about 450 meters. It is strictly arboreal and nocturnal, spends the day in a tree hollow, and emerges just after dark to forage. Leaps from branch to branch often cover 2–3 meters. The diet consists mostly of leaves. *Hemibelideus* is frequently seen in groups of two individuals (mother and young or male and female) and in family groups of three. Up to three may share a den, and feeding aggregations of as many as eight have been found in a single tree. Females have two mammae in the pouch, but usually only one young is reared. Young have been recorded in the pouch from August to November and riding on the mother's back from October to April. Winter (1984) cautioned that *Hemibelideus* may be in jeopardy because of logging of its tropical forest habitat. Laurance (1990) found that the genus could be relatively abundant on large tracts of primary forest but that it declined by more than 97 percent when such areas were fragmented into small areas, thereby preventing normal movement through trees. The IUCN now designates *Hemibelideus* as near threatened.

DIPROTODONTIA; PSEUDOCHEIRIDAE; **Genus PETAUROIDES**
Thomas, 1888

Greater Gliding Possum

The single species, *P. volans*, occurs in the coastal region from eastern Queensland south to southern Victoria (Ride 1970). The use of the name *Petauroides* in place of the former *Schoinobates* Lesson, 1842, was explained by McKay (1982).

Head and body length is 300–480 mm and tail length is 450–550 mm. Adult weight is 0.9–1.5 kg (Flannery and Schouten 1994; How 1978). The fur is soft and silky. Coloration is variable, ranging from black to smoky gray or creamy white on the upper parts and sooty gray, grayish white, or pure white on the underparts. All-white and white-headed individuals are common. The long tail is prehensile and evenly furred except the underpart of the tip, which is naked. Females have a well-developed pouch and two mammae.

This attractive marsupial glides with the use of a patagium, consisting of a fold of skin extending from the elbow to the leg. The position of the arms of *Petauroides* when gliding is entirely different from that of gliding squirrels and of the lesser gliding possums *(Petaurus):* they are bent at the elbows, so that the forearms are directed in toward the head and the hands almost meet on the front part of the chest. Grzimek and Ganslosser (*in* Grzimek 1990) noted that the gliding membrane apparently has a secondary use as a blanket: the animal wraps it around its body as a protection against loss of heat.

According to Wakefield (1970*b*), the attributes of the greater gliding possum have been confused with those of the largest of the lesser gliding possums, *Petaurus australis.* Whereas the latter species can glide up to about 114 meters, is highly maneuverable in the air, and calls loudly while gliding, *Petauroides* is a sedentary, slow-moving, silent animal of "minor gliding ability." McKay (*in* Strahan 1983), however, wrote that the glides of *Petauroides* may cover a horizontal distance of up to 100 meters and involve changes in direction of as much as 90 degrees.

The habitat is sclerophyll forest and tall woodland (Ride 1970). Although this possum apparently travels across open ground at times, it spends nearly its entire life in trees. It is nocturnal, sheltering by day in hollows high up in trees. Some individuals make a nest of stripped bark or leaves in their den, but often no material is added. There is some indication that this possum transports nesting material in its rolled-up tail. The musty eucalyptus smell of this animal permeates the shelter. The diet is very specialized, consisting mainly of the leaves, bark, and bud debris of certain species of *Eucalyptus* (Marples 1973).

Population densities of 0.24–0.83/ha. have been reported. Captive pairs or groups can be maintained in a sufficiently large enclosure. Wild adult males utilize home ranges that are separate from those of other males but overlap those of females. The ranges of females overlap to some extent, but the females tend to avoid one another. Home range size averaged about 1.5 ha. in one area and 2.6 ha. in another. Some males are monogamous and others are bigamous. There is some interaction between the sexes all year, but contact peaks just prior to and during the mating season. Females are polyestrous and have one litter per season. In New South Wales mating occurs from March to May and births occur from April to June. In Victoria the young usually are born in July and August. Litters contain a single young. It first releases the nipple at about 6 weeks but spends up to 6 months in the pouch and then another 4 months as a dependent nestling. It may sometimes be carried on the back of the mother. There evidently is no paternal care. Full independence comes at about 10–13 months, and sexual maturity is attained in the second year of life. Longevity may be 15 years in the wild (Collins 1973; Henry 1984; How 1978; Kerle and Borsboom 1984; Smith 1969; Tyndale-Biscoe and Smith 1969).

Kennedy (1992) regarded *Petauroides* as "potentially vulnerable" because of the continued fragmentation and disturbance of the coastal forests on which it depends. McKay (*in* Strahan 1983) observed that the conservation of *Petauroides* is utterly dependent on the maintenance of sufficient amounts of old-growth forest and that its abundance in undisturbed areas contrasts strongly with its absence from pine plantations and its scarcity in regenerated forest that lacks old trees with suitable hollows for nesting.

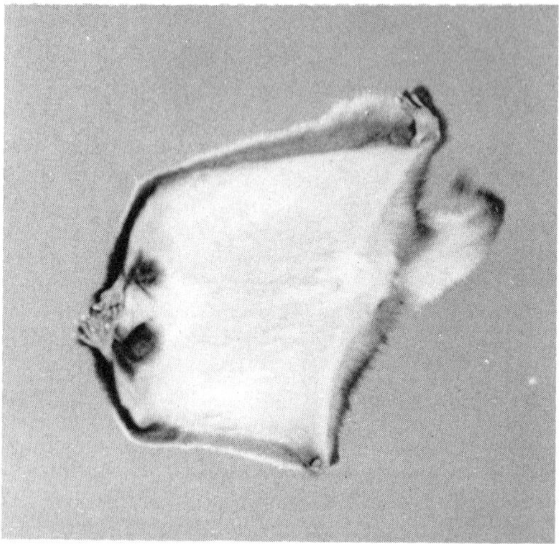

Greater gliding possum *(Petauroides volans)*, photos by Alan Root through Bernhard Grzimek.

DIPROTODONTIA; **Family PETAURIDAE**

Gliding and Striped Possums

This family of 3 Recent genera and 10 species inhabits Australia, Tasmania, New Guinea, and certain nearby islands. The members of the Petauridae sometimes have been placed in the family Phalangeridae. Kirsch (1977*b*), however, explained that the Petauridae differ from the Phalangeridae in serology, karyology, and other characteristics. He regarded the Pseudocheiridae (see account thereof) as one subfamily of the Petauridae and also recognized two other subfamilies: Petaurinae, for the genera *Gymnobelideus* and *Petaurus,* and Dactylopsilinae, for the genus *Dactylopsila.* Flannery and Schouten (1994) maintained the latter two subfamilies, though Edwards and Westerman (1992) found DNA analysis to indicate that *Gymnobelideus* actually is more closely related to *Dactylopsila* than to *Petaurus.* On the basis of serological studies, Baverstock (1984) and Baverstock, Birrell, and Krieg (1987) suggested that the genera *Distoechurus* and *Acrobates* have affinity to the Petauridae, but these genera now are placed in a new family, the Acrobatidae (see account thereof).

Head and body length is 120–320 mm and tail length is 150–480 mm. The pelage may be woolly or silky. All members of the Petauridae are arboreal. One genus, *Petaurus,* has attained a gliding ability through development of a membrane uniting the front and hind limbs. According to McKay (1989), the dentition of the Petauridae is characterized as follows: three upper incisors on each side of the jaw, the first pair normally longer than the succeeding pairs and projecting anteriorly; the upper canine small in the Dactylopsilinae but larger and laterally compressed in the Petaurinae; three single-cusped upper premolars, the first and second laterally compressed and the third conical; four bunodont upper molars, decreasing in size from first to last, each having four low pointed cusps; the procumbent first lower incisor slightly curved in the Petaurinae and strongly curved and greatly enlarged in the Dactylopsilinae; a second, vestigial lower incisor; three small lower premolars; and four lower molars, decreasing in size from first to fourth. The skull is characterized by a slender and rounded zygomatic arch and a palate without posterior vacuities. The forefoot shows no specialization for climbing other than enlarged claws; in the Petaurinae the second to fifth digits are all subequal in length, but in the Dactylopsilinae the fourth digit is considerably elongated as an adaptation for feeding. The tail is semiprehensile; in the Petaurinae it is entirely furred, but in the Dactylopsilinae there is a small ventral naked patch at the tip. The caecum of the digestive tract is large in the Petaurinae but very small in the Dactylopsilinae.

The geological range of the Petauridae is Miocene to Recent in Australasia (Kirsch 1977*a*).

Striped possum: Left, *Dactylopsila trivirgata* young, photo by John Warham; Right, *D. palpator,* photo by Howard Hughes through Australian Museum, Sydney.

DIPROTODONTIA; PETAURIDAE; **Genus PETAURUS**
Shaw and Nodder, 1791

Lesser Gliding Possums

There apparently are five species (Colgan and Flannery 1992; Flannery 1990*b;* Flannery and Schouten 1994; McKay *in* Bannister et al. 1988; Ride 1970; Smith 1973; Suckling *in* Strahan 1983; Ziegler 1981):

P. breviceps (sugar glider), Halmahera Islands in the North Moluccas, New Guinea and many nearby islands, Bismarck Archipelago, Louisiade Archipelago, northern and eastern Australia;
P. norfolcensis (squirrel glider), eastern Queensland, eastern New South Wales, Victoria, southeastern South Australia;
P. gracilis, northeastern Queensland;
P. abidi, northwestern coastal Papua New Guinea;
P. australis (fluffy glider), coastal parts of Queensland, New South Wales, and Victoria.

Flannery (1990*b*) indicated that the forms *biacensis* and *tafa* of New Guinea may be species distinct from *P. breviceps.* Flannery and Schouten (1994) treated *P. biacensis,* which is found on Biak and Supiori islands off northwestern New Guinea, as a separate species but treated *tafa* as a subspecies of *P. breviceps.* Those authorities also indicated that a population in the D'Entrecasteaux Islands off southeastern New Guinea may be an undescribed species related to *P. breviceps.*

Head and body length is 120–320 mm and tail length is 150–480 mm. Weights are 79–160 grams in *P. breviceps,* 200–260 grams in *P. norfolcensis* (Suckling *in* Strahan 1983), and 228–332 grams in *P. abidi* (Flannery 1990*b*). The larger *P. australis* weighs 435–710 grams (Henry and Craig 1984). The fur is fine and silky. In *P. breviceps, P. norfolcensis, P. gracilis,* and *P. abidi* the upper parts are generally grayish and the underparts are paler. A dark dorsal stripe runs from the nose to the rump, and there are stripes on each side of the face from the nose through the eye to the ear. In *P. australis* the upper parts usually are dusky brown, markings are less conspicuous, the feet are black, and the underparts are orange yellow.

Lesser gliding possums resemble flying squirrels *(Glaucomys)* in form and in having a large gliding membrane, but the tail of the former is furred all around and not as flattened as that of *Glaucomys.* As in *Glaucomys,* but not the greater gliding possum *(Petauroides),* the gliding membrane extends all the way from the outer side of the forefoot to the ankle and is opened by spreading the limbs straight out. Females have a well-developed pouch during the breeding season, and the number of mammae usually is four but occasionally is two. The pouch morphology of *P. australis* is unique among marsupials in having two compartments separated by a well-furred septum (Craig 1986).

All species inhabit wooded areas, preferably open forest, and are arboreal and largely nocturnal. They generally shelter by day in a leaf nest in a tree hollow. *P. breviceps* usually collects leaves by hanging by its hind feet, passing the leaves via the forefeet to the hind feet and then to the tail, which then coils around the nest material. The tail cannot be used for gliding when it is employed in this manner, so the possum transports its load of leaves by running along branches to its hollow.

These possums are extremely active. *P. breviceps* can glide up to about 45 meters and has been observed to leap at and catch moths in flight, and *P. australis* can glide at

A. Lesser gliding possum *(Petaurus norfolcensis)*, photo by Ernest P. Walker. B. *Petaurus* sp., photo by Bernhard Grzimek.

least 114 meters; vertical and lateral angle can be changed in the air (Ride 1970; Smith 1973). All species are omnivorous, feeding on sap, blossoms, nectar, insects and their larvae, arachnids, and small vertebrates (Collins 1973; Smith 1973; Wakefield 1970*b*). *P. australis* probably eats mostly nectar and arboreal arthropods, but it also removes bark from various eucalyptus trees to get the sugary sap. It makes characteristic V-shaped cuts in the bark that channel the sap to its mouth, placed at the bottom of the V. In a study of *P. australis* in coastal New South Wales, Goldingay (1990) found that individuals devoted 90 percent of the time outside of their dens to foraging, 70 percent of that time feeding on *Eucalyptus* nectar, and probably assisted in the cross-pollination of the eucalyptus trees.

In a study in remnant natural vegetation in Victoria, Suckling (1984) found the average population density of *P. breviceps* to vary from 2.9/ha. in summer to 6.1/ha. in autumn. Average home range was about 0.5 ha., though males extended their range slightly during the breeding season. In a study of the same species in New South Wales, Quin et al. (1992) found home ranges nearly 10 times as large. The species of *Petaurus* are social to some extent and produce a variety of sounds. *P. breviceps* has an alarm call resembling the yapping of a small dog and a high-pitched cry of anger (Smith 1973). While gliding, pairs of *P. australis* communicate by loud calls that are audible to humans at a distance of several hundred meters (Wakefield 1970*b*). The loud vocalizations of that species may also serve to define territories of a group (Goldingay 1994). Collins (1973) reported that captive *P. australis* can be maintained in pairs and that *P. breviceps* and *P. norfolcensis* can be kept in groups but that established groups might attack newly introduced individuals. Fleming (1980) reported that individuals of *P. breviceps* huddle together to conserve energy during cold weather and that they may simultaneously enter daily torpor when winter food supplies are low.

In the wild *P. breviceps* nests in groups of up to seven adult males and females and their young (Ride 1970; Smith 1973; Suckling 1984). All members probably are related and descended from an original colonizing pair. Groups appear mutually exclusive, territorial, and agonistic toward one another. One or two dominant, usually older males are responsible for most territorial maintenance, aggression against intruders, and fathering of young. There are a few lone adults, and the young generally leave their natal group at 10–12 months. This species has a complex chemical communication system based on scents produced by frontal, sternal, and urogenital glands of males and by pouch and urogenital glands of females. Each animal of a group has its own characteristic smell, which identifies it to other individuals and which is passively spread around the group's territory. In addition, a dominant male actively marks the other members of the group with his scent.

Studies of *P. australis* (Craig 1985, 1986; Goldingay 1992; Goldingay and Kavanagh 1990, 1993; Henry and Craig 1984) show that the species is thinly distributed. Most estimates of population density range from about 1/25 ha. to 1/6 ha. A population is divided into pairs or small groups of adults, with or without dependent young, that occupy largely separate home ranges of 30–100 ha. These areas include a variety of habitats, especially eucalyptus forest, favored for denning and feeding. The pairs are stable, monogamous units; one unit existed throughout a 41-month study period. Polygynous units consisting of one adult male and several adult females also may be found, especially where food resources are abundant. Breeding occurs mainly from about June to December in Victoria and New South Wales, but at one site in the latter state females were observed to give birth predominantly from February to April, presumably in association with availability of food resources. Breeding is year-round in Queensland. Almost always a single young per pair is raised in a year, though there is a record of twins. Both parents provide care for the young. Dispersal and independence has been observed at various ages from 9 to 24 months.

The species *P. breviceps* is polyestrous, with an estrous cycle averaging about 29 days. A female may produce a second litter during a breeding season if the first is lost or weaned (Smith 1971, 1973). In captivity there seems to be no definite breeding season for *P. breviceps* and *P. norfolcensis* (Collins 1973). The same is true for *P. breviceps* in the wild in New Guinea and Arnhem Land in the Northern Territory, but in southeastern Australia the young of this species are born only from June to November (Flannery 1990*b*; Smith 1973). Gestation is about 16 days in *P. breviceps* and slightly less than 3 weeks in *P. norfolcensis;* litter size usually is one or two in these species but may be as large as three in *P. breviceps* (Collins 1973; Smith 1973). The young of *P. breviceps* weigh about 0.19 grams at birth, first release the nipple at about 40 days, first leave the pouch at about 70 days, first leave the nest at 111 days, and are independent shortly thereafter (Collins 1973; Smith 1973). Sexual maturity in this species appears to come late in the first year for females and early in the second year for males. Lesser gliding possums may have a long life span in captivity, the recorded maximums being 14 years in *P. breviceps,* 11 years and 11 months in *P. norfolcensis,* and 10 years in *P. australis* (Collins 1973; Jones 1982). Wild *P. australis* are known to have lived at least 6 years (Goldingay and Kavanagh 1990).

The species *P. breviceps* is comparatively common in Australia. It was introduced into Tasmania in 1835 and has since spread over the island (Smith 1973). The considerably larger *P. australis,* however, appears to have become rare in some areas. It is a wide-ranging species and depends on maintenance of substantial tracts of woodland (Emison et al. 1975, 1978; Mackowski 1986). Both *P. australis* and *P. norfolcensis,* the latter also being rare and confronted with habitat fragmentation (Kennedy 1992), are designated as near threatened by the IUCN.

P. abidi and *P. gracilis,* which occur in very restricted areas of threatened habitat, are classified by the IUCN as vulnerable and endangered, respectively. *P. gracilis* apparently numbers fewer than 2,500 mature individuals and is continuing to decline.

DIPROTODONTIA; PETAURIDAE; **Genus GYMNOBELIDEUS**
McCoy, 1867

Leadbeater's Possum

The single species, *G. leadbeateri,* occurs in eastern Victoria and possibly in southeastern New South Wales (Ride 1970). It had been known only from six specimens collected between 1867 and 1909 but was rediscovered in 1961 (Lindenmayer and Dixon 1992).

Head and body length in four specimens is 152–68 mm and tail length is 190–99 mm. Weight is 120–65 grams (Smith 1984*a*). The sexes are equal in size. The fur is soft but neither as long nor as silky as in *Petaurus.* The upper parts are gray or brownish gray with a dark middorsal stripe extending from the forehead to the base of the tail. The fur of the ventral surface and the inner surface of the limbs is a dull creamy yellow at the tips and light gray beneath. The markings about the ears and the eyes are very similar to those of the sugar glider, *Petaurus breviceps.*

In many respects this possum closely resembles the sugar glider, but it lacks the gliding membrane. Unlike in other possums, the tail is flattened laterally, narrow at the base, and bushes out evenly to the tip. It is not prehensile to any marked degree and appears to be used for balance when the animal is climbing or jumping. The digits are very wide or spatulate at the tip and bear short, strong claws. *Gymnobelideus* is very active and appears to rely on the grip of the toe pads rather than the claws when climbing. The well-developed pouch of the female contains four mammae.

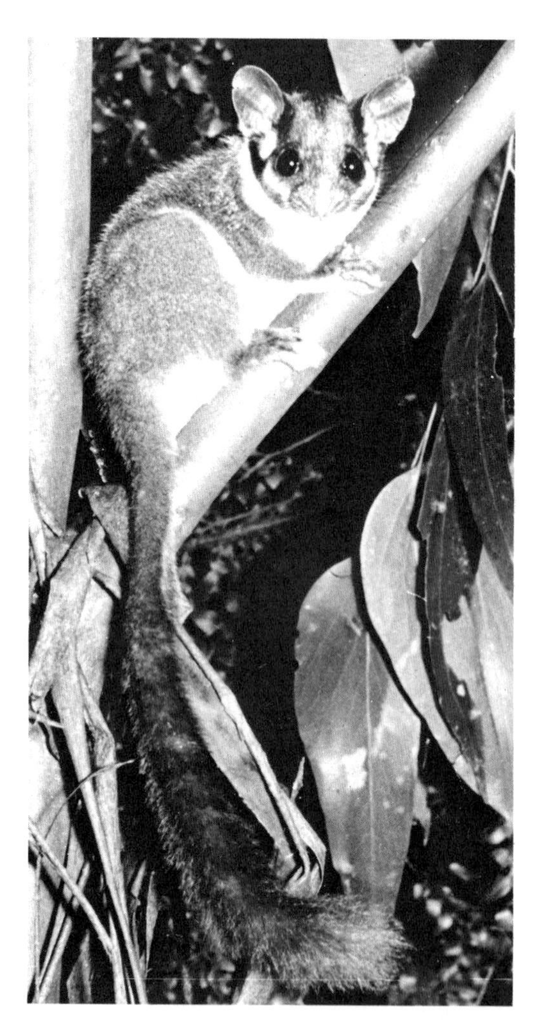

Leadbeater's possum *(Gymnobelideus leadbeateri)*, photos by Norman A. Wakefield, of Fisheries and Wildlife, Victoria, through J. McNally.

This possum is found in dense, wet sclerophyll forest at altitudes of up to about 1,200 meters. It is arboreal and nocturnal and constructs a nest of loosely matted bark about 10–30 meters up in a large hollow tree. The diet consists mainly of plant and insect exudates and also includes a variety of arthropods. A notch may be gnawed into a tree in order to obtain gums (Smith 1984*b*).

According to Smith (1984*a*), populations have a density of 1.6–2.9/ha. and are grouped into colonies of one to eight individuals. Each colony occupies a den tree centered in a defended territory of 1–2 ha. There are prolonged territorial disputes that involve chasing and grappling. Colonies usually consist of a single adult female, her mate, one to two unrelated adult males, and one or more generations of offspring. Aggression between adult females is prevalent, but males appear to move freely between colonies. Births occur in all months except January and February, with peaks from April to June (autumn) and October to December (spring). Females are polyestrous and can give birth within 30 days of losing a litter. Gestation probably lasts 15–17 days. Litters usually contain one to two young. They remain in the pouch for about 90 days and then spend another 5–40 days in the nest. Female offspring disperse at 7–14 months, and males at 11–26 months. Since the young females are excluded from established colonies, they suffer a high rate of mortality. A captive female, however, was reproductively active until the age of 9 years. A wild male was still alive at 7.5 years.

The first four known specimens of this species apparently came from the Bass River Valley and Koo-wee-rup Swamp, which are lowland areas southeast of Melbourne. Each of these was taken long before it was actually described, and little or no information was available regarding the living population that it represented. This lack of information, plus the clearing of the involved habitat, led to the assumption that the species was extinct. In 1931, however, it was learned that another specimen had been taken in 1909 in a highland area farther inland. Initial searches for a population in this area failed, but in 1961 individuals were observed in the Cumberland Valley of eastern Victoria. Subsequent investigation has shown that the species occurs in numerous localities over a large area of mountain forest and that it apparently increased in numbers and range following severe brush fires in 1939 (Ride 1970).

Recently, however, forest clearing has destroyed a substantial area of habitat. If additional timbering is carried out as planned, the old trees that the species depends upon for nesting could be eliminated over almost its entire known range. There is some question about the extent of future logging and the degree to which the needs of the species will be considered relative to such activity (Thornback and Jenkins 1982). The species and its remnant habitat also are closely tied to a narrow set of climatic conditions that could be severely affected by the "greenhouse effect" (Lindenmayer et al. 1991). The population is estimated to number 5,000 individuals and is expected to undergo further decline and fragmentation (Kennedy 1992). Leadbeater's possum now is classified as endangered by the IUCN and USDI.

DIPROTODONTIA; PETAURIDAE; **Genus DACTYLOPSILA**
Gray, 1858

Striped Possums

There are two subgenera and four species (Collins 1973; Flannery 1990*b;* Kirsch and Calaby 1977; Ride 1970):

subgenus *Dactylopsila* Gray, 1858

D. trivirgata, New Guinea and certain nearby islands, northeastern Queensland;
D. tatei, Fergusson Island off southeastern Papua New Guinea;
D. megalura, central and western New Guinea;

subgenus *Dactylonax* Thomas, 1910

D. palpator, central and eastern New Guinea.

Dactylonax was considered a full genus by McKay (*in* Bannister et al. 1988). Ziegler (1977) did not consider *D. tatei* and *D. megalura* to be specifically distinct from *D. trivirgata.* However, the above arrangement was followed by Corbet and Hill (1991) and Groves (*in* Wilson and Reeder 1993).

Head and body length is 170–320 mm and tail length is 165–400 mm. *D. trivirgata* weighs 246–470 grams; a male *D. palpator* weighed 550 grams, a female 320 grams (Flannery 1990*b;* Van Dyck *in* Strahan 1983). In the subgenus *Dactylopsila* the fur is thick, close, woolly, and rather harsh; in *Dactylonax* it is silky and dense. All species have three parallel, dark stripes on the back; in the subgenus *Dactylopsila* these stripes are black on a basal color of white or gray, and in *Dactylonax* they are smoky brown on a basal color of grayish tawny. In *D. trivirgata* and *D. megalura* there is a black chin spot. The bushy tail is well haired except at the undersurface of the tip and is mostly dark-colored except that the tip is usually whitish. Females have two mammae and a well-developed pouch. Flannery (1990*b*) noted that the pouch of *D. palpator* is divided into two subunits, each with a single nipple.

This genus is characterized by large first incisor teeth and a slender, elongated fourth digit with a hooked nail on the forefoot. Both features are more pronounced in the subgenus *Dactylonax* and are shared by the aye aye *(Daubentonia),* a primate found in Madagascar. The claw of the fourth front digit of *Dactylonax* is much smaller than those of the other digits.

Striped possums live in rainforest or sclerophyll forest and are nocturnal and arboreal. The members of the subgenus *Dactylopsila,* at least, are superb climbers that seldom if ever descend to the ground. They shelter by day in dry leaf nests in hollows, where they lie curled into a ball, flat and not rolled up in a sitting position like most possums. The subgenus *Dactylonax,* however, seems less adept at climbing, may spend considerable time hunting insects in rotting logs on the forest floor, and nests on the ground among tree roots as well as in tree hollows (Flannery 1990*b*). The extremely unpleasant and penetrating odor of striped possums is of glandular origin, but it cannot be ejected in "skunklike" manner. Nonetheless, the odor, together with the black and white markings, provide a striking case of convergence with North American skunks (Flannery and Schouten 1994). When angered, these marsupials utter a loud and prolonged "throaty gurgling shriek."

Striped possum *(Dactylopsila tatei),* photo by Pavel German.

A captive striped possum cleaned itself elaborately upon emerging at night and then began to hunt for food with long-legged striding movements, so that it seemed to flow rather than jump from branch to branch. Striped possums sniff loudly around likely food sources, use their incisors to tear wood and gnaw out wood-boring grubs, and use their long finger to hook grubs from deep and narrow holes. They also tap their forefeet rapidly on loose bark, presumably to disturb insects beneath it. Their natural diet consists mainly of insects but also includes fruits and leaves. A captive occasionally attacked and consumed mice. Smith (1982) reported that *D. trivirgata* evidently feeds mainly on ants and other social insects, which it obtains by breaking into nests with its incisors.

Flannery (1990*b*) reported that *D. trivirgata* and *D. palpator* have been found in small female groups but that males of the latter species have always been found alone. Females with single young have been collected in New Guinea throughout much of the year. According to Van Dyck (*in* Strahan 1983), mating in *D. trivirgata* may occur in Australia from February to August, and up to two young are born. One specimen of *D. trivirgata* lived in captivity for 9 years and 7 months (Marvin L. Jones, Zoological Society of San Diego, pers. comm., 1995).

Both *D. tatei* and *D. megalura* are found in restricted areas of vulnerable habitat (Kennedy 1992). They are classified by the IUCN as endangered and vulnerable, respectively. *D. tatei* long was known only by nine specimens collected in 1935 and could not be located by several expeditions from the 1950s to the 1980s, but another specimen was taken in 1992 (Flannery and Schouten 1994).

DIPROTODONTIA; **Family TARSIPEDIDAE; Genus TARSIPES**
Gray, 1842

Honey Possum

The single known genus and species, *Tarsipes rostratus*, is found in the southwestern part of Western Australia (Ride 1970). This genus once was usually placed in the family Phalangeridae. More recently, it came to be recognized morphologically and serologically as one of the most divergent of Australian marsupials and was put in its own superfamily (Kirsch 1977*b*; Kirsch and Calaby 1977). Later serological and morphological studies suggested that this superfamily also should include the genera *Acrobates* and *Distoechurus* (Aplin and Archer 1987; Baverstock et al. 1987). The use of the name *T. rostratus* in place of the previously used *T. spenserae* was explained by Mahoney (1981).

Head and body length is 65–90 mm, tail length is 70–105 mm, and weight is 7–11 grams for males and 8–16 grams for females (Russell and Renfree 1989). The head is pale brown, and the body is grayish brown with three dark stripes along the back. The central stripe, which is almost black, extends from the head to the base of the tail; the two outer stripes are fainter and do not reach the tail. The underparts are pale yellowish or white, the limbs are pale rufous, and the feet are white. Some individuals are more grayish, but the stripes are always present. The fur is rather coarse, short, and close. The long whiplike tail is almost hairless and has a naked prehensile undertip.

This possum can be distinguished from the other small Australian marsupials by its coloration and its extremely long snout, which is about two-thirds of the length of the rest of the head. The long tongue, bristled at the tip, can be extended about 25 mm beyond the nose. Ridges on the hard palate scrape honey and pollen off the tongue when it is withdrawn through a channel formed by flanges on the upper and lower lips, assisted by the long, slender lower incisors. The pair of procumbent lower incisors are the only well-developed teeth; the at most 20 other teeth are reduced to small pegs, reflecting the soft diet of the animal. All the digits except the united second and third of the hind feet, which are equipped with functional claws, are expanded at the tip and bear a short nail much like those of the tarsier *(Tarsius)*, hence the generic name *Tarsipes* for the honey possum. Females have a well-developed pouch and four mammae.

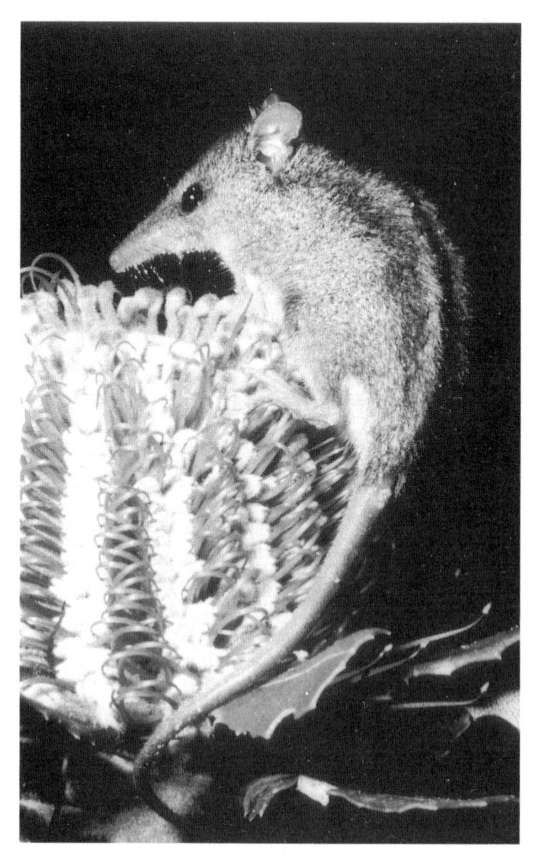

Honey possum *(Tarsipes rostratus)*, photo by P. A. Woolley and D. Walsh.

The honey possum dwells on tree and shrub heaths and is an active, nimble climber. It moves short distances to trees and shrubs that are producing its favorite flowers, and several animals may assemble in one desirable place. It often hangs upside down, especially when feeding on flowers. Sometimes it shelters in deserted birds' nests, though it often constructs its own refuge. It has three peaks of activity in each 24-hour period: between 0600 and 0800, 1700 and 1900, and 2330 and 0130 (Vose 1973). Nocturnal movements are concerned primarily with feeding. The diet consists almost exclusively of nectar and pollen, there being little evidence that insects are a significant component (Russell and Renfree 1989). Like some bats and hummingbirds, the honey possum is well adapted for probing into flowers and licking up its food. Nearly every floret in a favorite flower is explored.

Individuals usually occupy permanent, overlapping home ranges of about 1 ha. *Tarsipes* is very gregarious in the laboratory, huddling together in groups of two or more. Animals are often caught in a torpid state during cold or wet weather, and presumably huddling helps to minimize body heat loss. Large adult females interact very little with one another and are dominant to males; when accompanied by young, they may exclude other individuals from their range. Females with pouch young have been found throughout the year but predominantly in early autumn

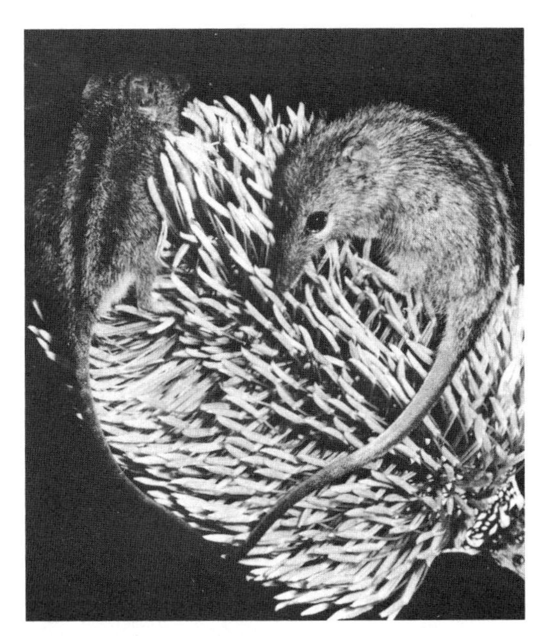

Honey possum *(Tarsipes rostratus)*, photo by M. B. Renfree.

(February–March), winter (May–July), and spring (September–October), when pollen and nectar are most abundant. Few young seem to be born in the summer months of December and January. Females are polyestrous, and some breed at least twice a year. Embryonic diapause such as in the Macropodidae also occurs in *Tarsipes.* Dormant embryos, carried in the uterus while the female is lactating, resume development after the young have left the pouch. The active gestation period appears to be 21–28 days. Litter size is usually two or three, but from one to four young have been recorded. With a weight of under 5 milligrams each, the young are the smallest known mammalian neonates. Lactation lasts about 10 weeks; for the the first 8 weeks the young remain in the pouch and for the last 2 weeks they are left in a shelter while the mother forages. They may then ride about on her back for another 1–2 weeks. Adult size is attained at about 8 months, and females may breed in their first year of life. Following the spring breeding peak there is a drastic population decline and most individuals apparently do not survive more than a year (Renfree 1980; Renfree, Russell, and Wooller 1984; Renfree and Wooller *in* Strahan 1983; Russell and Renfree 1989).

Vose (1973) cautioned that the existence of the honey possum in the wild is directly dependent on a continued supply of blossoms from eucalyptus, *Banksia* and *Callistemon.* Large-scale urbanization and habitat destruction in southwestern Australia is affecting the feeding of this species. Kennedy (1992) indicated that it already had lost up to 50 percent of its habitat and designated it "potentially vulnerable."

The geological range of the family Tarsipedidae is late Pleistocene to Recent in Australia (Marshall 1984).

DIPROTODONTIA; **Family ACROBATIDAE**

Feather-tailed Possums

This family of two genera and two species occurs only in New Guinea and eastern Australia. The two genera, *Distoechurus* and *Acrobates,* once were usually placed in the family Phalangeridae, but serological data led to their transfer to the Burramyidae (Kirsch 1977*b*). For a time they were considered to have affinity to the Petauridae, but new serological and morphological studies indicated a closer relationship to the Tarsipedidae and, in any case, that they warranted inclusion in an entirely separate family (Aplin and Archer 1987; Archer 1984; Baverstock, Birrell, and Krieg 1987). The latter arrangement was followed by Groves (*in* Wilson and Reeder 1993), though Corbet and Hill (1991) continued to include the two genera in the Burramyidae, and Szalay (1994) considered them to constitute only a tribe of the Petauridae.

Head and body length is 60–120 mm and tail length is 65–155 mm. The distichous or "pen" tail is distinctive in having paired lateral fringes of long stiff hairs that give a featherlike appearance. The genus *Acrobates* has a narrow patagium, or gliding membrane, that extends from the forelimbs to the hind limbs. The dental formula is: (i 3/2, c 1/0, pm 3/3, m 4/4) × 2 = 40. Aplin and Archer (1987) provided a detailed technical diagnosis of the family based in part on the structures of both the external and the internal ear. The external ear is uniquely complex, with a prominent anterior helical process, paired antitragal processes, and a well-defined bursa. In the internal ear the auditory bulla lacks an alisphenoid component and instead is formed from the petrosal, squamosal, and ectotympanic bones; the bulla is underrun medially and posteriorly by secondary tympanic processes from basi- and exoccipital bones; the primary tympanic cavity is complexly compartmentalized by numerous septa. Other diagnostic characters include a distinct lingual eminence at the rear of the tongue, a long upper canine that projects well below the level of the upper incisor teeth, greatly enlarged posterior palatal vacuities, and an intestinal caecum moderately to very elongate. The Burramyidae and Petauridae lack these and other key features, but the Tarsipedidae show some similarity in the structure of the internal ear and associated basicranial components.

In a study of the organogenesis and fetal membranes of *Distoechurus,* Hughes et al. (1987) found that the bilaminar yolk sac is totally invasive and that the maternal endometrial glands exhibit total degeneration. These two features are without parallel in marsupials so far studied and may provide further evidence of the distinctiveness of the Acrobatidae.

DIPROTODONTIA; ACROBATIDAE; **Genus DISTOECHURUS**
Peters, 1874

Feather-tailed Possum

The single species, *D. pennatus,* is found in suitable habitat throughout New Guinea (Ziegler 1977).

Head and body length is 103–32 mm, tail length is 126–55 mm, and weight is 38–62 grams (Flannery 1990*b*). The general body coloration is dull buff, light brown, slightly darker and more olivaceous, or slightly darker and grayer. In contrast to the dull, plain body, the head is strikingly ornamented. The face is white, with two broad well-defined dark brown or black bands that pass from the sides of the muzzle through the eyes to the top of the head just between the ears, and there is a conspicuous black patch just below each ear. The basal part of the tail is well furred, and the remainder is nearly naked but fringed laterally with long, relatively stiff hairs. The pelage is soft, thick, and woolly.

Feather-tailed possum *(Distoechurus pennatus)*, photo by P. A. Woolley and D. Walsh.

Gliding membranes are not present. The claws are sharp and curved; the terminal pads of the digits are not expanded. The eyes are large, and the ears are small and naked. The tip of the tail is prehensile. Females have one medially placed teat in a well-developed pouch, which opens anteriorly. Two lateral teats have also been reported.

This possum is primarily an inhabitant of young regrowth and disturbed forest but also is found in gardens and may occur in primary rainforest. The elevational range is sea level to 1,900 meters. It sometimes nests in tree hollows but probably also nests in leafy vegetation. It is active, arboreal, and nocturnal and has been reported to eat blossoms, fruit, insects, and other invertebrates. Most individuals seem to nest alone, but an adult male and an adult female with a single pouch young were collected together in a leafy nest in a small tree in January. Other females that were lactating or that had pouch young have been collected from June to October. Individuals have been kept in captivity for as long as 20 months (Collins 1973; Flannery 1990*b;* Woolley 1982*a;* Ziegler 1977).

DIPROTODONTIA; ACROBATIDAE; **Genus ACROBATES**
Desmarest, 1817

Pygmy Gliding Possum, or Feathertail Glider

The single species, *A. pygmaeus,* occurs along the eastern coast of Australia, from the extreme northern tip of the Cape York Peninsula to southeastern South Australia (Ride 1970). Another named species, *A. pulchellus,* was based on a single specimen from a small island to the north of New Guinea and is usually considered to represent an introduced pet *A. pygmaeus* (Groves *in* Wilson and Reeder 1993). However, Flannery (1990*b*) suggested that *Acrobates* probably does occur in New Guinea but has not yet been discovered because the most likely habitat has not been well explored.

Head and body length is 65–80 mm, tail length is 70–80 mm, and weight is 10–14 grams (Russell *in* Strahan 1983). The pelage is soft and silky. The upper parts are grayish brown, and the underparts and inner sides of the limbs are white. The ears are sparsely haired and bear tufts of hair at their base. The tail is light brown throughout, the ventral surface being somewhat paler than the dorsal surface.

This beautiful, delicate creature is the smallest marsupial capable of gliding flight and is indeed, as the name *Acrobates pygmaeus* indicates, a pygmy acrobat. The gliding membrane is a very narrow fold of skin, fringed with long hairs along its margin, that extends from each limb along the sides of the body. The structure of the tail is such that it provides an additional plane: it is flattened by a fringe of hairs along the sides, a characteristic that has given *Acrobates* the common name "feathertail glider." The fringe of hairs spreads from each side to a total width of 8 mm. The tip of the tail is without hair below, and the tail is somewhat prehensile. The tips of the fingers and toes have expanded, deeply scored or striated pads, which assist the animal in clinging to surfaces. The fourth digit on all four limbs is the longest, and the claws of the digits are sharp. The well-developed pouch encloses four mammae and is lined with yellow hair.

This active, agile marsupial inhabits sclerophyll forest and woodland and is much like a flying squirrel *(Glaucomys)* in its actions. It can glide from one tree to another as well as travel short distances, from limb to limb, by air. Glides of up to 20 meters can be achieved (Flannery and Schouten 1994). Because of its small size and nocturnal habits, *Acrobates* often is overlooked. Some individuals are caught by domestic cats or are discovered when trees are felled. This animal makes small, spherical nests of dry leaves in hollow branches and knotholes of trees at a height of 15 meters or more. During cold days it may become torpid in the nest. Jones and Geiser (1992) experimentally induced deep torpor, during which body temperature fell to 2° C, for periods of up to 5.5 days. Insects, including larvae,

Pygmy gliding possum *(Acrobates pygmaeus)*, photo by Stanley Breeden.

probably form the main part of the diet. Nectar and other plant products also are eaten.

The feathertail glider does not appear to be territorial, shows remarkable conspecific tolerance, and has an extended breeding season (Collins 1973; Fleming and Frey 1984; Ward 1990*b*; Ward and Renfree 1988*a*, 1988*b*). Family units consisting of one or both parents plus the offspring of one or two litters are present for much of the year, and such units freely share their nests with other individuals. The resulting aggregations may be the basis of reports that families contain up to 16 individuals. Artificial nest boxes have yielded up to 14 animals, though groups more commonly have 2–5 individuals. In Victoria adult males are usually solitary from January to April and then form pairs with the females from May to August. Births occur there from July to February, with a peak from August to November. Females undergo a postpartum estrus and a period of embryonic diapause and normally produce two litters during the season. A female may carry as many as 4 young in her pouch, but litter size averages 2.5 at weaning. When they are well furred, the young ride on the mother's back. The young are left in the nest at 60 days and weaned at 95–100 days. A second litter may then be born immediately. The young from the first litter remain with the mother while she raises the second litter. Sexual maturity is attained at 8 months in females and by 12–18 months in males. According to Jones (1982), a captive specimen lived 7 years and 2 months.

World Distribution of Marsupials

For maximum usefulness, it has been necessary to devise the simplest practicable outline of the approximate distribution of the genera in the sequence used in the text. The tabulation should be regarded as an index guide to the marsupials or to geographic regions. At the same time it gives a good overall picture of the general distribution of marsupial orders.

The major geographic distribution of the genera of Recent marsupials that appears in the tabulation is designed to show their natural distribution at present or within comparatively recent times. Most of the animals occupy only a portion of the geographic region that appears at the head of the column. Some are limited to the tropical regions, others to temperate zones. Also, many restricted ranges cannot be designated either by letters to show the general area or by footnotes because of limited space on the tabulation. *It therefore should not be assumed that a mark indicating that an animal occurs within a geographic region implies that it inhabits that entire area.* For more detailed outlines of the ranges of the respective genera, it is necessary to consult the generic texts.

Explanation of Geographic Column Headings

Europe and Asia constitute a single landmass, but this landmass comprises widely different types of zoogeographic areas created by high mountain ranges, plateaus, latitudes, and prevailing winds.

Most islands are included with the major landmasses nearby unless otherwise specified, although in many instances some of the marsupials indicated for the continental mass do not occur on the islands.

With Europe are included the British Isles and other adjacent islands, including those in the Arctic.

With Asia are included the Japanese Islands, Taiwan, Hainan, Sri Lanka, and other adjacent islands, including those in the Arctic.

With North America are included Mexico and Central America south to Panama, adjacent islands, the Aleutian chain, the islands in the arctic region, and Greenland but not the West Indies.

With South America are included Trinidad, the Netherlands Antilles, and other small adjacent islands but not the Falkland and Galapagos Islands unless named in footnotes.

With Africa are included only Zanzibar Island and small islands close to the continent but not the Cape Verde or Canary Islands.

The island groups treated separately are:

Southeastern Asian islands, in which are included the Andamans, the Nicobars, the Mentawais, Sumatra, Java, the Lesser Sundas, Borneo, Sulawesi, the Moluccas, and the many other adjacent small islands;

New Guinea and small adjacent islands;

the Australian region, in which are included Australia, Tasmania, and adjacent small islands;

the Philippine Islands and small adjacent islands;

the West Indies;

Madagascar and small adjacent islands.

Footnotes indicate the major easily definable deviations from the distribution indicated in the tables.

Symbols

■	The marsupials occur on or adjacent to the land or in the water area.
N	Northern portion
S	Southern portion
E	Eastern portion
W	Western portion
Ne	Northeastern portion
Se	Southeastern portion
Sw	Southwestern portion
Nw	Northwestern portion
C	Central portion

Examples: N,C = northern and central; Nc = north-central. Numerals refer to footnotes indicating clearly defined limited ranges within the general area.

Genera of Recent Mammals	page	North America	West Indies	South America	Madagascar	Africa	Europe	Asia	Southeast Asia Islands	Philippine Islands	New Guinea	Australian Region	Antarctic Region	Arctic Region	Atlantic Ocean	Indian Ocean	Pacific Ocean
DIDELPHIMORPHIA MARMOSIDAE																	
Gracilinanus	70			■													
Marmosops	71	■S		■													
Marmosa	72	■S		■													
Micoureus	73	■S		■													
Thylamys	74			■													
Lestodelphys	74			■S													
Metachirus	75	■S		■													
Monodelphis	76	■S		■													
DIDELPHIMORPHIA CALUROMYIDAE																	
Caluromys	78	■S		■													
Caluromysiops	79			■Nw													
DIDELPHIMORPHIA GLIRONIIDAE																	
Glironia	80			■Wc													
DIDELPHIMORPHIA DIDELPHIDAE																	
Philander	81	■S		■													
Didelphis	82	■		■													
Chironectes	84	■S		■													
Lutreolina	85			■													
PAUCITUBERCULATA CAENOLESTIDAE																	
Caenolestes	87			■Nw													
Rhyncholestes	88			■1													
MICROBIOTHERIA MICROBIOTHERIIDAE																	
Dromiciops	90			■Sw													
DASYUROMORPHIA DASYURIDAE																	
Murexia	93										■						
Phascolosorex	94										■						
Neophascogale	95										■						
Phascogale	96											■					
Antechinus	98										■	■2					
Planigale	100										■S	■N,E					
Ningaui	101											■					
Sminthopsis	102										■S	■2					
Antechinomys	105											■					
Parantechinus	106											■W					
Dasykaluta	107											■Nw					
Pseudantechinus	107											■					
Myoictis	108										■						
Dasyuroides	110											■Nc					
Dasycercus	111											■					
Dasyurus	112										■	■2					
Sarcophilus	115											■3					
DASYUROMORPHIA MYRMECOBIIDAE																	
Myrmecobius	116											■S					
DASYUROMORPHIA THYLACINIDAE																	
Thylacinus	118											■3					

1. Chile only. 2. And Tasmania. 3. Tasmania only.

Genera of Recent Mammals	page	North America	West Indies	South America	Madagascar	Africa	Europe	Asia	Southeast Asia Islands	Philippine Islands	New Guinea	Australian Region	Antarctic Region	Arctic Region	Atlantic Ocean	Indian Ocean	Pacific Ocean
PERAMELEMORPHIA PERAMELIDAE																	
Macrotis	122											■C,S					
Chaeropus	123											■					
Isoodon	124										■S	■1					
Perameles	126											■1					
PERAMELEMORPHIA PERORYCTIDAE																	
Echymipera	128										■2	■Ne					
Rhynchomeles	129								■3								
Microperoryctes	130										■						
Peroryctes	132										■						
NOTORYCTEMORPHIA NOTORYCTIDAE																	
Notoryctes	133											■					
DIPROTODONTIA PHASCOLARCTIDAE																	
Phascolarctos	135											■E					
DIPROTODONTIA VOMBATIDAE																	
Vombatus	138											■E,1					
Lasiorhinus	139											■E					
DIPROTODONTIA PHALANGERIDAE																	
Ailurops	141								■4								
Strigocuscus	142								■5								
Trichosurus	143											■1					
Wyulda	145											■Nw					
Spilocuscus	146										■2	■Ne					
Phalanger	147								■6		■7	■Ne					
DIPROTODONTIA POTOROIDAE																	
Hypsiprymnodon	149											■Ne					
Potorous	150											■S,1					
Bettongia	151											■1					
Caloprymnus	153											■C					
Aepyprymnus	153											■E					
DIPROTODONTIA MACROPODIDAE																	
Lagostrophus	156											■W					
Dorcopsis	157										■						
Dorcopsulus	158										■						
Dendrolagus	159										■	■Ne					
Setonix	161											■Sw					
Thylogale	162										■2	■E,1					
Petrogale	163											■					
Peradorcas	165											■N					
Lagorchestes	166											■					
Onychogalea	167											■					
Wallabia	168											■E					
Macropus	169										■S	■1					
DIPROTODONTIA BURRAMYIDAE																	
Cercartetus	180										■	■					
Burramys	181											■Se					

1. And Tasmania. 2. And Bismarck Archipelago. 3. Seram only. 4. Sulawesi and Talaud Islands only. 5. Sulawesi only. 6. Moluccas, Seram, and Timor. 7. And Bismarck Archipelago and Solomon Islands.

Genera of Recent Mammals	page	North America	West Indies	South America	Madagascar	Africa	Europe	Asia	Southeast Asia Islands	Philippine Islands	New Guinea	Australian Region	Antarctic Region	Arctic Region	Atlantic Ocean	Indian Ocean	Pacific Ocean
DIPROTODONTIA PSEUDOCHEIRIDAE																	
Pseudochirulus	183										■	■Ne					
Pseudocheirus	184											■1					
Pseudochirops	185										■	■Ne					
Petropseudes	186											■N					
Hemibelideus	187											■Ne					
Petauroides	187											■E					
DIPROTODONTIA PETAURIDAE																	
Petaurus	190								■2		■3	■					
Gymnobelideus	191											■Se					
Dactylopsila	193										■	■Ne					
DIPROTODONTIA TARSIPEDIDAE																	
Tarsipes	194											■Sw					
DIPROTODONTIA ACROBATIDAE																	
Distoechurus	195										■						
Acrobates	196											■E					

1. And Tasmania. 2. Moluccas only. 3. And Bismarck and Louisiade Archipelagoes.

Appendix

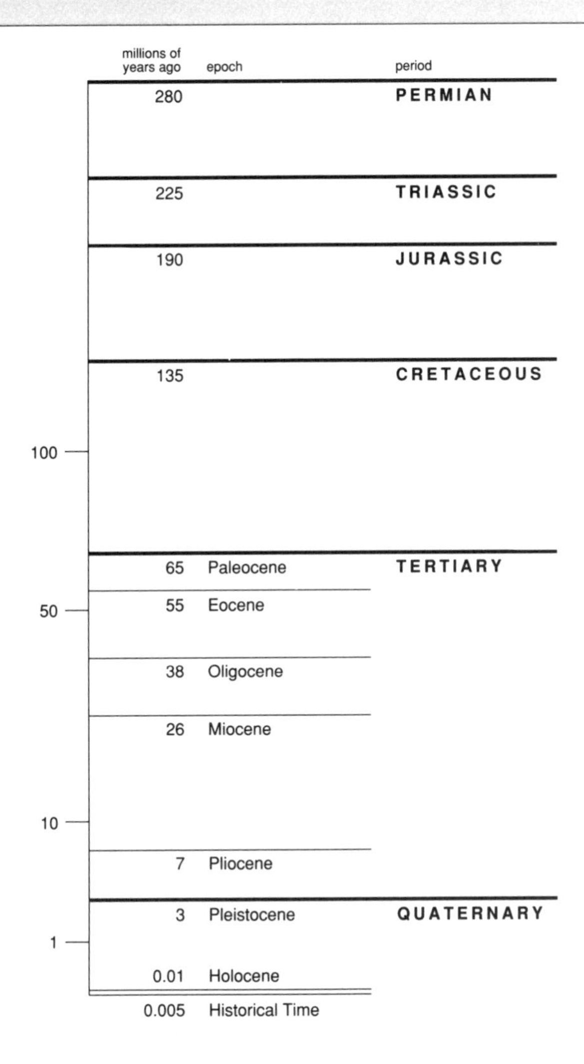
GEOLOGICAL TIME
millions of years ago
epoch
period
280
PERMIAN
225
TRIASSIC
190
JURASSIC
135
CRETACEOUS
100
65
Paleocene
TERTIARY
50
55
Eocene
38
Oligocene
26
Miocene
10
7
Pliocene
3
Pleistocene
QUATERNARY
1
0.01
Holocene
0.005
Historical Time

LENGTH

scales for comparison of metric and U.S. units of measurement

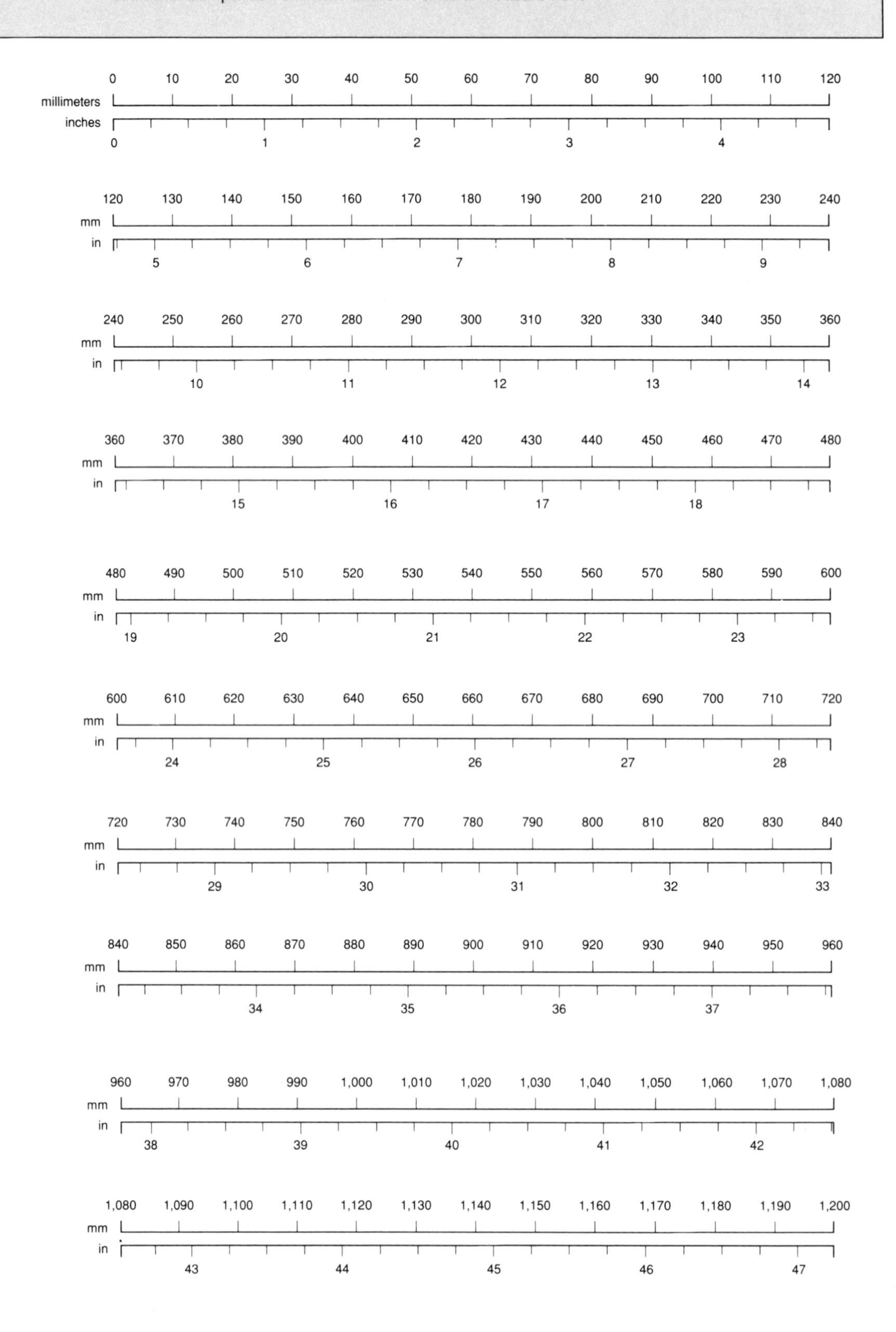

LENGTH

scales for comparison of metric and U.S. units of measurement

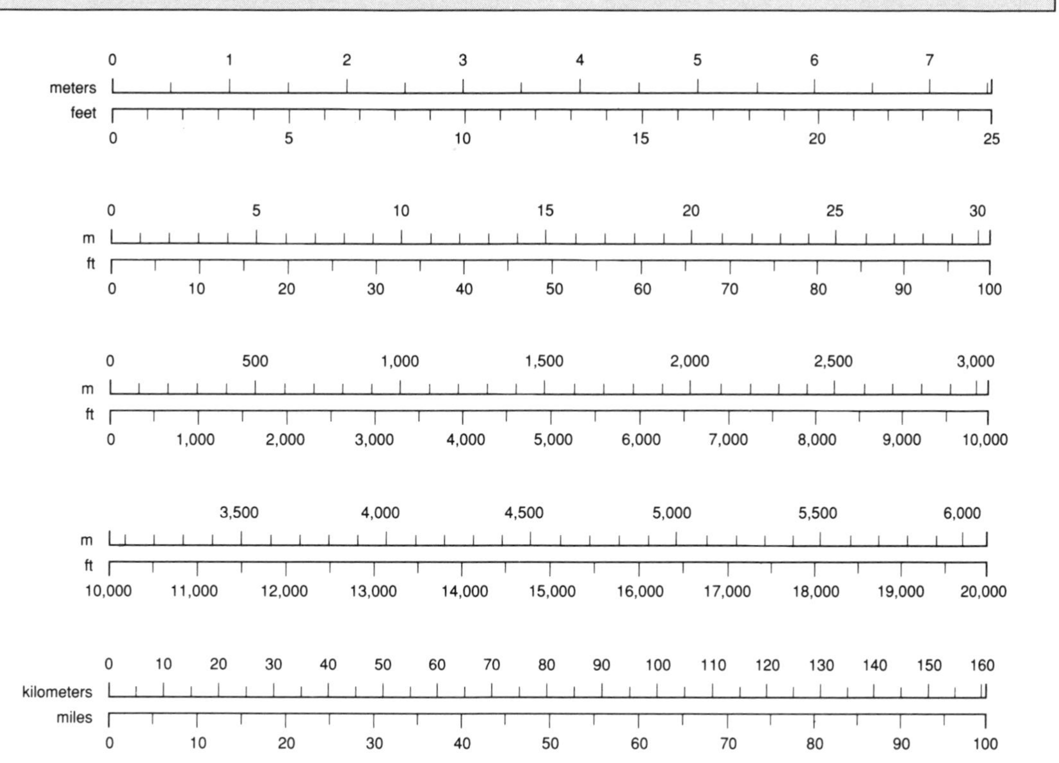

CONVERSION TABLES

	U.S. to Metric		Metric to U.S.	
	to convert	*multiply by*	*to convert*	*multiply by*
LENGTH	in. to mm.	25.4	mm. to in.	0.039
	in. to cm.	2.54	cm. to in.	0.394
	ft. to m.	0.305	m. to ft.	3.281
	yd. to m.	0.914	m. to yd.	1.094
	mi. to km.	1.609	km. to mi.	0.621
AREA	sq. in. to sq. cm.	6.452	sq. cm. to sq. in.	0.155
	sq. ft. to sq. m.	0.093	sq. m. to sq. ft.	10.764
	sq. yd. to sq. m.	0.836	sq. m. to sq. yd.	1.196
	sq. mi. to ha.	258.999	ha. to sq. mi.	0.004
VOLUME	cu. in. to cc.	16.387	cc. to cu. in.	0.061
	cu. ft. to cu. m.	0.028	cu. m. to cu. ft.	35.315
	cu. yd. to cu. m.	0.765	cu. m. to cu. yd.	1.308
CAPACITY (liquid)	fl. oz. to liter	0.03	liter to fl. oz.	33.815
	qt. to liter	0.946	liter to qt.	1.057
	gal. to liter	3.785	liter to gal.	0.264
MASS (weight)	oz. avdp. to g.	28.35	g. to oz. avdp.	0.035
	lb. avdp. to kg.	0.454	kg. to lb. avdp.	2.205
	ton to t.	0.907	t. to ton	1.102
	l. t. to t.	1.016	t. to l. t.	0.984

Abbreviations

avdp.	avoirdupois
cc.	cubic centimeter(s)
cm.	centimeter(s)
cu.	cubic
ft.	foot, feet
g.	gram(s)
gal.	gallon(s)
ha.	hectare(s)
in.	inch(es)
kg.	kilogram(s)
lb.	pound(s)
l. t.	long ton(s)
m.	meter(s)
mi.	mile(s)
mm.	millimeter(s)
oz.	ounce(s)
qt.	quart(s)
sq.	square
t.	metric ton(s)
yd.	yard(s)

WEIGHT

scales for comparison of metric and U.S. units of measurement

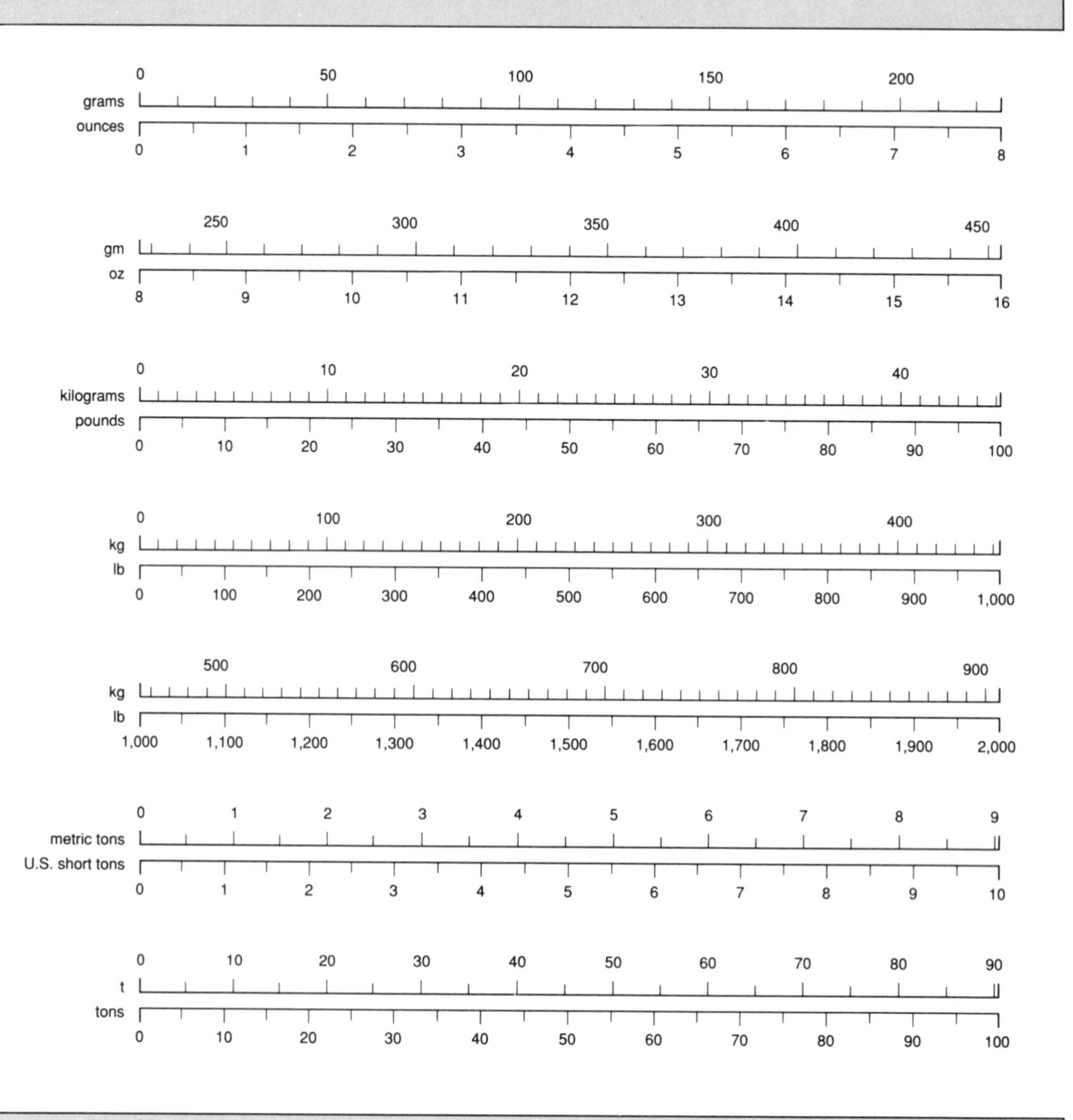

TEMPERATURE

scales for comparison of metric and U.S. units of measurement

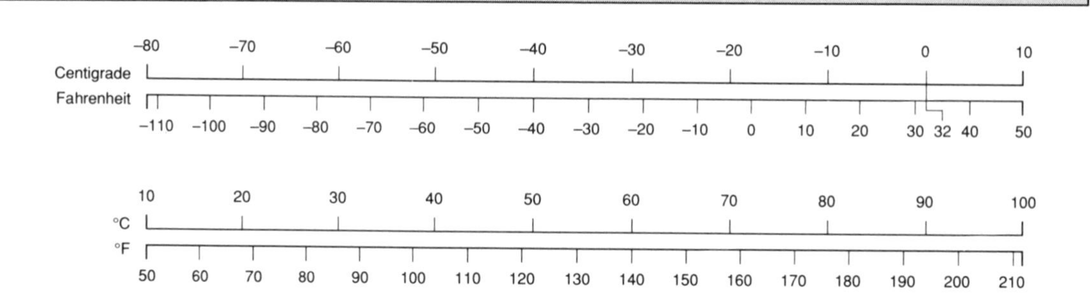

AREA

HECTARE

10,000.0 square meters

107,639.1 square feet

ACRE

4,046.86 square meters

43,560.0 square feet

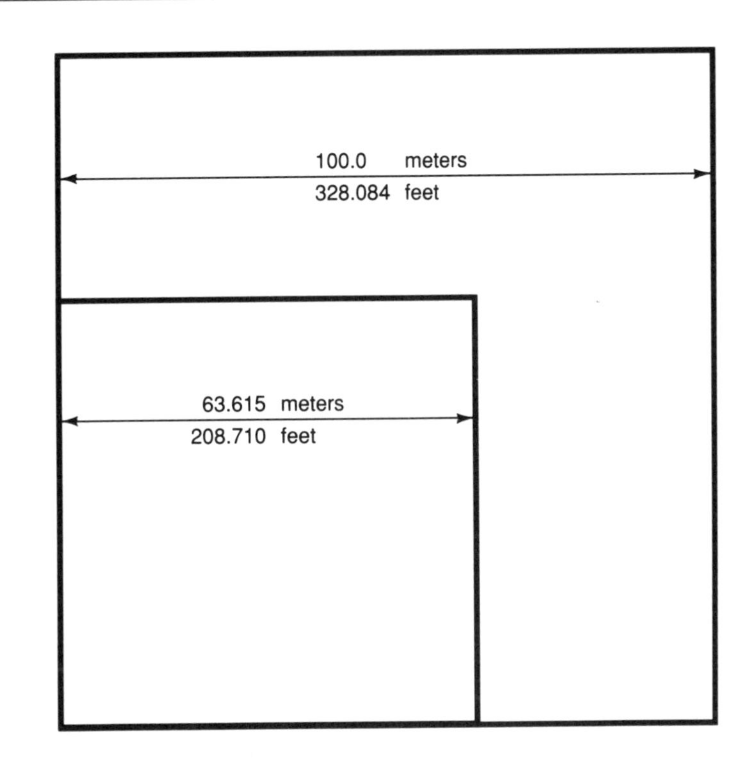

Literature Cited

A

Aitken, P. F. 1971*a*. The distribution of the hairy-nosed wombat [*Lasiorhinus latifrons* (Owen)]. Part I: Yorke Peninsula, Eyre Peninsula, the Gawler Ranges and Lake Harris. S. Austral. Nat. 45:93–104.

———. 1971*b*. *Planigale tenuirostris* Troughton, the narrow-nosed planigale, an addition to the mammal fauna of South Australia. S. Austral. Nat. 46:18.

———. 1971*c*. Rediscovery of the large desert sminthopsis (*Sminthopsis psammophilus* Spencer) on Eyre Peninsula, South Australia. Victorian Nat. 88:103–11.

———. 1972. *Planigale gilesi* (Marsupialia, Dasyuridae); a new species from the interior of southeastern Australia. Rec. S. Austral. Mus. 16:1–14.

———. 1977*a*. The little pygmy possum [*Cercartetus lepidus* (Thomas)] found living on the Australian mainland. S. Austral. Nat. 51:63–66.

———. 1977*b*. Rediscovery of swamp antechinus in South Australia after 37 years. S. Austral. Nat. 52:28–30.

———. 1979. The status of endangered Australian wombats, bandicoots, and the marsupial mole. *In* Tyler (1979), 61–65.

Alonso-Mejía, A., and R. A. Medellín. 1992. *Marmosa mexicana*. Mammalian Species, no. 421, 4 pp.

Anderson, S. 1982. *Monodelphis kunsi*. Mammalian Species, no. 190, 3 pp.

Anderson, T. J. C., A. J. Berry, J. N. Amos, and J. M. Cook. 1988. Spool-and-line tracking of the New Guinea bandicoot, *Echymipera kalubu* (Marsupialia, Peramelidae). J. Mamm. 69:114–20.

Aplin, K. P. 1987. Basicranial anatomy of the Early Miocene diprotodontian *Wynyardia bassiana* (Marsupialia: Wynyardiidae) and its implications for wynyardiid phylogeny and classification. *In* Archer (1987*a*), 369–91.

Aplin, K. P., and M. Archer. 1987. Recent advances in marsupial systematics with a new syncretic classification. *In* Archer (1987*a*), xv–lxxii.

Aplin, K. P., and P. A. Woolley. 1993. Notes on the distribution and reproduction of the Papuan bandicoot *Microperoryctes papuensis* (Peroryctidae, Peramelemorphia). Sci. New Guinea 19:109–12.

Archer, M. 1974*a*. New information about the Quaternary distribution of the thylacine (Marsupialia, Thylacinidae) in Australia. J. Roy. Soc. W. Austral. 57:43–50.

———. 1974*b*. Regurgitation or mercyism in the western native cat, *Dasyurus geoffroii*, and the red-tailed wambenger, *Phascogale calura* (Marsupialia, Dasyuridae). J. Mamm. 55:448–52.

———. 1975. *Ningaui*, a new genus of tiny dasyurids (Marsupialia) and two species, *N. timealeyi* and *N. ridei*, from arid western Australia. Mem. Queensland Mus. 17:237–49.

———. 1976. Revision of the marsupial genus *Planigale* Troughton (Dasyuridae). Mem. Queensland Mus. 17:341–65.

———. 1977. Revision of the dasyurid marsupial genus *Antechinomys* Krefft. Mem. Queensland Mus. 18:17–29.

———. 1979*a*. The status of Australian dasyurids, thylacinids, and myrmecobiids. *In* Tyler (1979), 29–43.

———. 1979*b*. Two new species of *Sminthopsis* Thomas (Dasyuridae: Marsupialia) from northern Australia, *S. butleri* and *S. douglasi*. Austral. Zool. 20:327–45.

———. 1981. Results of the Archbold Expeditions. No. 104. Systematic revision of the marsupial dasyurid genus *Sminthopsis* Thomas. Bull. Amer. Mus. Nat. Hist. 168: 61–224.

———. 1982*a*. Review of the dasyurid (Marsupialia) fossil record, integration of data bearing on phylogenetic interpretation, and suprageneric classification. *In* Archer (1982*a*), 397–443.

———. 1982*b*. A review of the Miocene thylacinids (Thylacinidae, Marsupialia), the phylogenetic position of the Thylacinidae and the problem of apiorisms in character analysis. *In* Archer (1982*a*), 445–76.

———. 1984. The Australian marsupial radiation. *In* Archer and Clayton (1984), 633–808.

Archer, M., and J. A. W. Kirsch. 1977. The case for the Thylacomyidae and Myrmecobiidae Gill, 1872, or why are marsupial families so extended? Proc. Linnean Soc. New South Wales 102:18–25.

Armstrong, D. M. 1972. Distribution of mammals in Colorado. Monogr. Mus. Nat. Hist. Univ. Kansas, no. 3, x + 415 pp.

Armstrong, D. M., and J. K. Jones, Jr. 1971. Mammals from the Mexican state of Sinaloa. I. Marsupialia, Insectivora, Edentata, Lagomorpha. J. Mamm. 52:747–57.

Arundel, J. H., I. K. Barker, and I. Beveridge. 1977. Diseases of marsupials. *In* Stonehouse and Gilmore (1977), 141–54.

Ashley, E., D. Lunney, J. Robertshaw, and R. Harden. 1990. Distribution and status of bandicoots in New South Wales. *In* Seebeck et al. (1990), 43–50.

Aslin, H. J. 1974. The behaviour of *Dasyuroides byrnei* (Marsupialia) in captivity. Z. Tierpsychol. 35:187–208.

———. 1975. Reproduction in *Antechinus maculatus* Gould (Dasyuridae). Austral. Wildl. Res. 2:77–80.

———. 1977. New records of *Sminthopsis ooldea* Troughton from South Australia. S. Austral. Nat. 52:9–12.

———. 1980. Biology of a laboratory colony of *Dasyuroides byrnei* (Marsupialia: Dasyuridae). Austral. Zool. 20:457–71.

———. 1983. Reproduction in *Sminthopsis ooldea* (Marsupialia: Dasyuridae). Austral. Mamm. 6:93–95.

Atherton, R. G., and A. T. Haffenden. 1982. Observations on the reproduction and growth of the long-tailed pygmy possum, *Cercartetus caudatus* (Marsupialia: Burramyidae), in captivity. Austral. Mamm. 5:253–59.

Atramentowicz, M. 1992. Optimal litter size: does it cost more to raise a large litter in *Caluromys philander?* Can. J. Zool. 70: 1511–15.

Australian National Parks and Wildlife Service. 1978. Australian endangered species. Mammals, nos. 1–23.

———. 1988. Kangaroos in Australia: conservation status and management. Austral. Natl. Parks Wildl. Serv. Occas. Pap., no. 14, 45 pp.

B

Bailey, P. 1992. A red kangaroo, *Macropus rufus,* recovered 25 years after marking in north-western New South Wales. Austral. Mamm. 15:141.

Banfield, A. W. F. 1974. The mammals of Canada. Univ. Toronto Press, xxv + 438 pp.

Bannister, J. L., J. H. Calaby, L. J. Dawson, J. K. Ling, J. A. Mahoney, G. M. McKay, B. J. Richardson, W. D. L. Ride, and D. W. Walton. 1988. Zoological catalogue of Australia. Volume 5. Mammalia. Australian Government Publ. Serv., Canberra, x + 274 pp.

Barker, I. K., I. Beveridge, A. J. Bradley, and A. K. Lee. 1978. Observations on spontaneous stress-related mortality among males of the dasyurid marsupial *Antechinus stuartii* Macleay. Austral. J. Zool. 26:435–47.

Barker, R. D., and G. Caughley. 1990. Distribution and abundance of kangaroos (Marsupialia: Macropodidae) at the time of European contact: Tasmania. Austral. Mamm. 13:157–66.

Barker, S. C. 1990. Behaviour and social organisation of the allied rock-wallaby *Petrogale assimilis,* Ramsay, 1877 (Marsupialia: Macropodoidea). Austral. Wildl. Res. 17: 301–11.

Barkley, L. J., and J. O. Whitaker, Jr. 1984. Confirmation of *Caenolestes* in Peru with information on diet. J. Mamm. 65:328–30.

Barnett, A. 1991. Records of the grey-bellied shrew opossum, *Caenolestes caniventer* and Tate's shrew-opossum, *Caenolestes tatei* (Caenolestidae, Marsupialia), from Ecuadorian montane forests. Mammalia 55:443–45.

Barritt, M. K. 1978. Two further specimens of the little pigmy possum [*Cercartetus lepidus* (Thomas)] from the Australian mainland. S. Austral. Nat. 53:12–13.

Baverstock, P. R. 1984. The molecular relationships of Australian possums and gliders. *In* Smith and Hume (1984), 1–8.

Baverstock, P. R., M. Adams, and M. Archer. 1984. Electrophoretic resolution of species boundaries in the *Sminthopsis murina* complex (Dasyuridae). Austral. J. Zool. 32:823–32.

Baverstock, P. R., M. Adams, M. Archer, N. L. McKenzie, and R. How. 1983. An electrophoretic and chromosomal study of the dasyurid marsupial genus *Ningaui* Archer. Austral. J. Zool. 31:381–92.

Baverstock, P. R., M. Adams, T. Reardon, and C. H. S. Watts. 1987. Electrophoretic resolution of species boundaries in Australian Microchiroptera. III. The Nycticeiini—*Scotorepens* and *Scoteanax* (Chiroptera: Vespertilionidae). Austral. J. Biol. Sci. 40: 417–33.

Baverstock, P. R., M. Archer, M. Adams, and B. J. Richardson. 1982. Genetic relationships among 32 species of Australian dasyurid marsupials. *In* Archer (1982*a*), 641–50.

Baverstock, P. R., J. Birrell, and M. Krieg. 1987. Albumin immunologic relationships among Australian possums: a progress report. *In* Archer (1987*a*), 229–34.

Baverstock, P. R., T. Flannery, K. Aplin, J. Birrell, and M. Krieg. 1990. Albumin immunologic relationships of the bandicoots (Perameloidea: Marsupialia)—a preliminary report. *In* Seebeck et al. (1990), 13–18.

Baverstock, P. R., B. J. Richardson, J. Birrell, and M. Krieg. 1989. Albumin relationships of the Macropodidae (Marsupialia). Syst. Zool. 38:38–50.

Baynes, A. 1982. Dasyurids (Marsupialia) in late Quaternary communities in southwestern Australia. *In* Archer (1982*a*), 503–10.

———. 1984. Native mammal remains from Wilgie Mia aboriginal ochre mine: evidence on the pre-European fauna of the western arid zone. Rec. W. Austral. Mus. 11:297–310.

Bolton, B. L., and P. K. Latz. 1978. The western hare-wallaby, *Lagorchestes hirsutus* (Gould) (Macropodidae), in the Tanami Desert. Austral. Wildl. Res. 5:285–93.

Bourke, D. W. 1989. Observations on the behaviour of the grey dorcopsis wallaby, *Dorcopsis luctuosa* (Marsupialia: Macropodidae), in captivity. *In* Grigg, Jarman, and Hume (1989), 633–40.

Brosset, A. 1989. Camouflage chez le yapock *Chironectes minimus.* Rev. Ecol. 44:279–81.

Broughton, S. K., and C. R. Dickman. 1991. The effect of supplementary food on home range of the southern brown bandicoot, *Isoodon obesulus* (Marsupialia: Peramelidae). Austral. J. Ecol. 16:71–78.

Brown, R. 1973. Has the thylacine really vanished? Animals 15:416–19.

Bryant, S. L. 1989. Growth, development, and breeding pattern of the long-nosed potoroo, *Potorous tridactylus* (Kerr, 1792) in Tasmania. *In* Grigg, Jarman, and Hume (1989), 449–56.

Bublitz, J. 1987. Untersuchungen zur Systematik der Rezenten Caenolestidae Trouessart, 1898. Bonner Zool. Monogr., no. 23, 96 pp.

Bucher, J. E., and H. I. Fritz. 1977. Behavior and maintenance of the woolly opossum *(Caluromys)* in captivity. Lab. Anim. Sci. 27:1007–12.

Buchmann, O. L. K., and E. R. Guiler. 1974. Locomotion in the potoroo. J. Mamm. 55: 203–6.

———. 1977. Behavior and ecology of the Tasmanian devil, *Sarcophilus harrisii. In* Stonehouse and Gilmore (1977), 155–68.

C

Cabrera, A. 1957. Cat logo de los mamiferos de América del Sur. Rev. Mus. Argentino Cien. Nat. "Bernardo Rivadavia," 4:1–732.

Calaby, J. H. 1966. Mammals of the upper Richmond and Clarence rivers, New South Wales. CSIRO Div. Wildl. Res. Tech. Pap., no. 10, 55 pp.

———. 1971. The current status of Australian Macropodidae. Austral. Zool. 16:17–31.

Calaby, J. H., L. K. Corbett, G. B. Sharman, and P. G. Johnston. 1974. The chromosomes and systematic position of the marsupial mole, *Notoryctes typhlops.* Austral. J. Biol. Sci. 27:529–32.

Calaby, J. H., H. Dimpel, and I. M. Cowan. 1971. The mountain pygmy possum, *Burramys parvus* Broom (Marsupialia) in the Kosciusko National Park, New South Wales. CSIRO Div. Wildl. Res. Tech. Pap., no. 23, 11 pp.

Calaby, J. H., and C. White. 1967. The Tasmanian devil *(Sarcophilus harrisi)* in northern Australia in Recent times. Austral. J. Sci. 29:473–75.

Callister, D. J. 1991. A review of the Tasmanian brushtail possum industry. Traffic Bull. 12(3):49–58.

Caughley, G. 1984. The grey kangaroo overlap zone. Austral. Wildl. Res. 11:1–10.

Caughley, G., R. G. Sinclair, and G. R. Wilson. 1977. Numbers, distribution, and harvesting rate of kangaroos on the inland plains of New South Wales. Austral. Wildl. Res. 4:99–108.

Caughley, J. 1986. Distribution and abundance of the mountain pygmy-possum, *Burramys parvus* Broom, in Kosciusko National Park. Austral. Wildl. Res. 13:507–16.

Cerqueira, R. 1985. The distribution of *Didelphis* in South America (Polyprotodontia, Didelphidae). J. Biogeogr. 12:135–45.

Christensen, P. 1975. The breeding burrow of the banded ant-eater or numbat *(Myrmecobius fasciatus).* W. Austral. Nat. 13:32–34.

Christensen, P., K. Maisey, and D. H. Perry. 1984. Radiotracking the numbat, *Myrmecobius fasciatus,* in the Perup Forest of Western Australia. Austral. Wildl. Res. 11:275–88.

Clark, M. J., and W. E. Poole. 1967. The reproductive system and embryonic diapause in the female grey kangaroo, *Macropus giganteus.* Austral. J. Zool. 15:441–59.

Clemens, W. A., B. J. Richardson, and P. R. Baverstock. 1989. Biogeography and phylogeny of the Metatheria. *In* Walton and Richardson (1989), 527–59.

Close, R. L., J. D. Murray, and D. A. Briscoe. 1990. Electrophoretic and chromosome surveys of the taxa of short-nosed bandicoots within the genus *Isoodon. In* Seebeck et al. (1990), 19–27.

Cockburn, A., and K. A. Lazenby-Cohen. 1992. Use of nest trees by *Antechinus stuartii,* a semelparous lekking marsupial. J. Zool. 226:657–80.

Cole, J. R., and D. F. Gibson. 1991. Distribution of stripe-faced dunnarts *Sminthopsis macroura* and desert dunnarts *S. youngsoni* (Marsupialia: Dasyuridae) in the Northern Territory. Austral. Mamm. 14:129–31.

Colgan, D., and T. F. Flannery. 1992. Biochemical systematic studies in the genus *Petaurus* (Marsupialia: Petauridae). Austral. J. Zool. 40:245–56.

Colgan, D., T. F. Flannery, J. Trimble, and K. Aplin. 1993. Electrophoretic and morphological analysis of the systematics of the *Phalanger orientalis* (Marsupialia) species complex in Papua New Guinea and Solomon Islands. Austral. J. Zool. 41:355–78.

Collins, L. R. 1973. Monotremes and marsupials: a reference for zoological institutions. Smithson. Inst. Press, Washington, D.C., v + 323 pp.

Cooper, D. W., and P. A. Woolley. 1983. Confirmation of a new species of small dasyurid marsupial by electrophoretic analysis of enzymes and proteins. Austral. J. Zool. 31: 743–51.

Copley, P. B. 1983. Studies on the yellow-footed rock-wallaby, *Petrogale xanthopus* Gray (Marsupialia: Macropodidae). I. Distribution in South Australia. Austral. Wildl. Res. 10:47–61.

Copley, P. B., V. T. Read, A. C. Robinson, and C. H. S. Watts. 1990. Preliminary studies of the Nuyts Archipelago bandicoot *Isoodon obesulus nauticus* on the Franklin Islands, South Australia. In Seebeck et al. (1990), 345–56.

Corbet, G. B., and J. E. Hill. 1991. A world list of mammalian species. Natural History Museum Publ., London, and Oxford Univ. Press, viii + 243 pp.

———. 1992. The mammals of the Indomalayan region: a systematic review. Oxford Univ. Press, viii + 488 pp.

Corbett, L. K. 1975. Geographical distribution and habitat of the marsupial mole, *Notoryctes typhlops.* Austral. Mamm. 1:375–78.

Craig, S. A. 1985. Social organization, reproduction, and feeding behaviour of a population of yellow-bellied gliders, *Petaurus australis* (Marsupialia: Petauridae). Austral. Wildl. Res. 12:1–18.

———. 1986. A record of twins in the yellow-bellied glider (*Petaurus australis* Shaw) (Marsupialia: Petauridae) with notes on the litter size and reproductive strategy of the species. Victorian Nat. 103:72–75.

Crawley, M. C. 1973. A live-trapping study of Australian brush-tailed possums, *Trichosurus vulpecula* (Kerr), in the Orongorongo Valley, Wellington, New Zealand. Austral. J. Zool. 21:75–90.

Creighton, G. K. 1985. Phylogenetic inference, biogeographic interpretations, and the patterns of speciation in *Marmosa* (Marsupialia: Didelphidae). Acta Zool. Fennica 170:121–24.

Croft, D. B. 1980. Behaviour of red kangaroos, *Macropus rufus* (Desmarest, 1822) in northwestern New South Wales, Australia. Austral. Mamm. 4:5–58.

———. 1989. Social organization of the Macropodoidea. *In* Grigg, Jarman, and Hume (1989), 505–25.

———. 1991*a*. Home range of the euro, *Macropus robustus erubescens.* J. Arid Environ. 20:99–111.

———. 1991*b*. Home range of the red kangaroo *Macropus rufus.* J. Arid Environ. 20:83–98.

Crowcroft, P. 1977. Breeding of wombats *(Lasiorhinus latifrons)* in captivity. Zool. Garten 47:313–22.

Cuttle, P. 1982. Life history of the dasyurid marsupial *Phascogale tapoatafa. In* Archer (1982*a*), 13–22.

D

Da Silva, M. N. F., and A. Langguth. 1989. A new record of *Glironia venusta* from the lower Amazon, Brazil. J. Mamm. 70:873–75.

Dawson, L. 1982*a*. Taxonomic status of fossil devils (*Sarcophilus,* Dasyuridae, Marsupialia) from late Quaternary eastern Australian localities. *In* Archer (1982*a*), 517–25.

———. 1982*b*. Taxonomic status of fossil thylacines (*Thylacinus,* Thylacinidae, Marsupialia) from late Quaternary deposits in eastern Australia. *In* Archer (1982*a*), 527–36.

Dawson, L., and T. Flannery. 1985. Taxonomic and phylogenetic status of living and fossil kangaroos and wallabies of the genus *Macropus* Shaw (Macropodidae: Marsupialia), with a new subgeneric name for the larger wallabies. Austral. J. Zool. 33:473–98.

Deems, E. F., Jr., and D. Pursley, eds. 1978. North American furbearers. Internatl. Assoc. Fish and Wildl. Agencies, Univ. Maryland Press, College Park, x + 165 pp.

Delroy, L. B., J. Earl, I. Radbone, A. C. Robinson, and M. Hewett. 1986. The breeding and re-establishment of the brush-tailed bettong, *Bettongia penicillata,* in South Australia. Austral. Wildl. Res. 13:387–96.

De Vivo, M., and N. F. Gomès. 1989. First record of *Caluromysiops irrupta* Sanborn, 1951 (Didelphidae) from Brasil. Mammalia 53:310–11.

Dickman, C. R., and R. W. Braithwaite. 1992. Postmating mortality of males in the dasyurid marsupials, *Dasyurus* and *Parantechinus.* J. Mamm. 73:143–47.

Dickman, C. R., D. H. King, M. Adams, and P. R. Baverstock. 1988. Electrophoretic identification of a new species of *Antechinus* (Marsupialia: Dasyuridae) in south-eastern Australia. Austral. J. Zool. 36:455–63.

Dimpel, H., and J. H. Calaby. 1972. Further observations on the mountain pigmy possum *(Burramys parvus).* Victorian Nat. 89: 101–6.

Dixon, J. M. 1971. *Burramys parvus* Broom (Marsupialia) from Falls Creek area of the Bogong high plains, Victoria. Victorian Nat. 88:133–38.

———. 1978. The first Victorian and other records of the little pigmy possum *Cercartetus lepidus* (Thomas). Victorian Nat. 95:4–7.

———. 1988. Notes on the diet of three mammals presumed to be extinct: the pig-footed bandicoot, the lesser bilby and the desert rat kangaroo. Victorian Nat. 105:208–11.

Dressen, W. 1993. On the behaviour and social organization of agile wallabies, *Macropus agilis* (Gould, 1842) in two habitats of northern Australia. Z. Saugetierk. 58:201–11.

Dufty, A. C. 1991. Some population characteristics of *Perameles gunnii* in Victoria. Wildl. Res. 18:355–66.

Dwyer, P. D. 1977. Notes on *Antechinus* and *Cercartetus* (Marsupialia) in the New Guinea highlands. Proc. Roy. Soc. Queensland 88: 69–73.

———. 1983. An annotated list of mammals from Mt. Erimbari, Eastern Highlands Province, Papua New Guinea. Sci. New Guinea 10:28–38.

E

Eberhard, I. H. 1978. Ecology of the koala, *Phascolarctos cinereus* (Goldfuss) Marsupialia: Phascolarctidae, in Australia. *In* Montgomery (1978), 315–27.

Edmunds, R. M., J. W. Goertz, and G. Linscombe. 1978. Age ratios, weights, and reproduction of the Virginia opossum in northern Louisiana. J. Mamm. 59:884–85.

Edwards, D., and M. Westerman. 1992. DNA-DNA hybridisation and the position of Leadbeater's possum (*Gymnobelideus leadbeateri* McCoy) in the family Petauridae (Marsupialia: Diprotodontia). Austral. J. Zool. 40:563–71.

Edwards, G. P. 1989. The interaction between macropodids and sheep: a review. *In* Grigg, Jarman, and Hume (1989), 795–804.

Edwards, G. P., and E. H. M. Ealey. 1975. Aspects of the ecology of the swamp wallaby, *Wallabia bicolor* (Marsupialia: Macropodidae). Austral Mamm. 1:307–17.

Eisenberg, J. F. 1989. Mammals of the neotropics: the northern neotropics. Univ. Chicago Press, x + 449 pp.

Eisentraut, M. 1970. Beitrag zur Fortpflanzungsbiologie der Zwerbeutelratte *Marmosa murina* (Didelphidae, Marsupialia). Z. Saugetierk 35:159–72.

Eldridge, M. D. B., and R. L. Close. 1992. Taxonomy of rock wallabies, *Petrogale* (Marsupialia: Macropodidae). I. A revision of the eastern *Petrogale* with the description of three new species. Austral. J. Zool. 40:605–25.

———. 1993. Radiation of chromosome shuffles. Current Opinion in Genetics and Development 3:915–22.

Eldridge, M. D. B., P. G. Johnston, and P. S. Lowry. 1992. Chromosomal rearrangements in rock wallabies, *Petrogale* (Marsupialia: Macropodidae). VII. G-banding analysis of *Petrogale brachyotis* and *P. concinna:* species with dramatically altered karyotypes. Cytogenet. Cell Genet. 61:34–39.

Ellis, M. 1992. The mulgara, *Dasycercus cristicauda* (Krefft, 1867): a new dasyurid record for New South Wales. Austral. Zool. 28:57–58.

Ellis, M., P. Wilson, and S. Hamilton. 1991. The golden bandicoot, *Isoodon auratus* Ramsay 1887, in western New South Wales during European times. Austral. Zool. 27:36–37.

Emison, W. B., J. W. Porter, K. C. Norris, and G. J. Apps. 1975. Ecological distribution of the vertebrate animals of the volcanic plains—Otway Range area of Victoria. Victoria Fish. and Wild. Pap., no. 6, 93 pp.

———. 1978. Survey of the vertebrate fauna in the Grampians-Edenhope area of southwestern Victoria. Mem. Natl. Mus. Victoria 39:281–363.

Emmons, L. H. 1990. Neotropical rainforest mammals: a field guide. Univ. Chicago Press, xiv + 281 pp.

Enders, R. K. 1966. Attachment, nursing, and survival of young in some didelphids. Symp. Zool. Soc. London 15:195–203.

Ewer, R. F. 1968. A preliminary survey of the behaviour in captivity of the dasyurid marsupial, *Sminthopsis crassicaudata* (Gould). Z. Tierpsychol. 25:319–65.

F

Fadem, B. H., and R. S. Rayve. 1985. Characteristics of the oestrous cycle and influence of social factors in grey short-tailed opossums *(Monodelphis domestica).* J. Reprod. Fert. 73:337–42.

Fanning, F. D. 1982. Reproduction, growth, and development in *Ningaui* sp. (Dasyuridae, Marsupialia) from the Northern Territory. *In* Archer (1982*a*), 23–37.

Feiler, A. 1978*a*. Bemerkungen uber *Phalanger* der "orientalis-Gruppe" nach Tate (1945). Zool. Abhandl. (Dresden) 34:385–95.

———. 1978*b*. Über artliche Abgrenzung und innerartliche Ausformung bei *Phalanger maculatus* (Mammalia, Marsupialia, Phalangeridae). Zool. Abhandl. (Dresden) 35:1–30.

Fish, F. E. 1993. Comparison of swimming kinematics between terrestrial and semiaquatic opossums. J. Mamm. 74:275–84.

Fitzgerald, A. E. 1976. Diet of the opossum, *Trichosurus vulpecula* (Kerr) in the Orongorongo Valley, Wellington, New Zealand, in relation to food-plant availability. New Zealand J. Zool. 3:399–419.

———. 1978. Aspects of the food and nutrition of the brush-tailed opossum, *Trichosurus vulpecula* (Kerr, 1792), Marsupialia: Phalangeridae, in New Zealand. *In* Montgomery (1978), 289–303.

Flannery, T. F. 1983. Revision in the macropodid subfamily Sthenurinae (Marsupialia: Macropodoidea) and the relationships of the species of *Troposodon* and *Lagostrophus.* Austral. Mamm. 6:15–28.

———. 1987. A new species of *Phalanger* (Phalangeridae: Marsupialia) from montane western Papua New Guinea. Rec. Austral. Mus. 39:183–93.

———. 1989*a*. *Microhydromys musseri* n. sp., a new murid (Mammalia) from the Torricelli Mountains, Papua New Guinea. Proc. Linnean Soc. New South Wales 111:215–22.

———. 1989*b*. Phylogeny of the Macropodoidea: a study in convergence. *In* Grigg, Jarman, and Hume (1989), 1–46.

———. 1990*a*. *Echymipera davidi,* a new species of Perameliformes (Marsupialia) from Kiriwina Island, Papua New Guinea, with notes on the systematics of the genus *Echymipera. In* Seebeck et al. (1990), 29–35.

———. 1990*b*. Mammals of New Guinea. Robert Brown & Assoc., Carina, Queensland, iii + 439 pp.

———. 1992. Taxonomic revision of the *Thylogale brunii* complex (Macropodidae: Marsupialia) in Melanesia, with description of a new species. Austral. Mamm. 15:7–23.

———. 1993. Taxonomy of *Dendrolagus goodfellowi* (Macropodidae: Marsupialia) with description of a new subspecies. Rec. Austral. Mus. 45:33–42.

———. 1995. Mammals of the south-west Pacific and Moluccan Islands. Comstock/Cornell Univ. Press, Ithaca, 464 pp.

Flannery, T. F., M. Archer, and G. Maynes. 1987. The phylogenetic relationships of living phalangerids (Phalangeroidea: Marsupialia) with a suggested new taxonomy. *In* Archer (1987*a*), 477–506.

Flannery, T. F., P. Bellwood, P. White, A. Moore, B. Boeadi, and G. Nitihaminoto. 1995. Fossil marsupials (Macropodidae, Peroryctidae) and other mammals of Holocene age from Halmahera, North Moluccas, Indonesia. Alcheringa 19:17–25.

Flannery, T. F., and B. Boeadi. 1995. Systematic revision within the *Phalanger ornatus* complex (Phalangeridae: Marsupialia), with description of a new species and subspecies. Austral. Mamm. 18:35–44.

Flannery, T. F., B. Boeadi, and A. L. Szalay. 1995. A new tree-kangaroo *(Dendrolagus:* Marsupialia) from Irian Jaya, Indonesia, with notes on ethnography and the evolution of tree-kangaroos. Mammalia 59:65–84.

Flannery, T. F., and J. H. Calaby. 1987. Notes on the species of *Spilocuscus* (Marsupialia: Phalangeridae) from northern New Guinea and the Admiralty and St. Matthias Island groups. *In* Archer (1987*a*), 547–58.

Flannery, T. F., and P. Schouten. 1994. Possums of the world. GEO Productions, Chatswood, New South Wales, Australia, 240 pp.

Flannery, T. F., and L. Seri. 1990. *Dendrolagus scottae* n. sp. (Marsupialia: Macropodidae): a new tree-kangaroo from Papua New Guinea. Rec. Austral. Mus. 42:237–45.

Flannery, T. F., and F. Szalay. 1982. *Bohra paulae,* a new giant fossil tree kangaroo (Marsupialia: Macropodidae) from New South Wales, Australia. Austral. Mamm. 5:83–94.

Fleming, M. R. 1980. Thermoregulation and torpor in the sugar glider, *Petaurus breviceps* (Marsupialia: Petauridae). Austral. J. Zool. 28:521–34.

———. 1985. The thermal physiology of the mountain pygmy-possum *Burramys parvus* (Marsupialia: Burramyidae). Austral. Mamm. 8:79–90.

Fleming, M. R., and A. Cockburn. 1979. *Ningaui:* a new genus of dasyurid for Victoria. Victorian Nat. 96:142–45.

Fleming, M. R., and H. Frey. 1984. Aspects of the natural history of feathertail gliders *(Acrobates pygmaeus)* in Victoria. *In* Smith and Hume (1984), 403–8.

Fleming, T. H. 1972. Aspects of the population dynamics of three species of opossums in the Panama Canal Zone. J. Mamm. 53:619–23.

———. 1973. The reproductive cycles of three species of opossums and other mammals in the Panama Canal Zone. J. Mamm. 54:439–55.

Fletcher, M., C. J. Southwell, N. W. Sheppard,

G. Caughley, D. Grice, G. C. Grigg, and L. A. Beard. 1990. Kangaroo population trends in the Australian rangelands, 1980–1987. Search 21:28–29.

Fletcher, T. P. 1985. Aspects of reproduction in the male eastern quoll, *Dasyurus viverrinus* (Shaw) (Marsupialia: Dasyuridae), with notes on polyoestry in the female. Austral. J. Zool. 33:101–10.

Friend, G. R. 1990. Breeding and population dynamics of *Isoodon macrourus* (Marsupialia: Peramelidae): studies from the wet-dry tropics of northern Australia. *In* Seebeck et al. (1990), 357–65.

Friend, J. A. 1989. Myrmecobiidae. *In* Walton and Richardson (1989), 583–90.

Friend, J. A. 1990*a*. The numbat *Myrmecobius fasciatus* (Myrmecobiidae): history of decline and potential for recovery. Proc. Ecol. Soc. Austral. 16:369–77.

Friend, J. A. 1990*b*. Status of bandicoots in Western Australia. *In* Seebeck et al. (1990), 73–84.

Fry, E. 1971. The scaly-tailed possum *Wyulda squamicaudata* in captivity. Internatl. Zoo Yearbook 11:44–45.

Fuller, P. J., and A. A. Burbidge. 1987. Discovery of the dibbler, *Parantechinus apicalis*, on islands at Jurien Bay. W. Austral. Nat. 16: 177–81.

G

Ganslosser, U. 1984. On the occurrence of female coalitions in tree kangaroos (Marsupialia, Macropodidae, *Dendrolagus*). Austral. Mamm. 7:219–21.

Gardner, A. L. 1973. The systematics of the genus *Didelphis* (Marsupialia: Didelphidae) in North and Middle America. Spec. Publ. Mus. Texas Tech Univ., no. 4, 81 pp.

———. 1981. Review of *The mammals of Suriname*, by A. M. Husson. J. Mamm. 62:445–48.

Gardner, A. L., and G. K. Creighton. 1989. A new generic name for Tate's (1933) *Microtarsus* group of South American mouse opossums (Marsupialia: Didelphidae). Proc. Biol. Soc. Washington 102:3–7.

Gardner, A. L., and J. L. Patton. 1972. New species of *Philander* (Marsupialia: Didelphidae) and *Mimon* (Chiroptera: Phyllostomatidae) from Peru. Occas. Pap. Mus. Zool. Louisiana State Univ., no. 43, 12 pp.

Gaski, A. L. 1988. 'Roo update: a short history of U.S. kangaroo skin and product imports, 1984–1987. Traffic (U.S.A.) 8(2):1–5.

Geiser, F. 1986. Thermoregulation and torpor in the kultarr, *Antechinomys laniger* (Marsupialia: Dasyuridae). J. Comp. Physiol., ser. B, 156:751–57.

———. 1993. Hibernation in the eastern pygmy possum, *Cercartetus nanus* (Marsupialia: Burramyidae). Austral. J. Zool. 41: 67–75.

Geiser, F., M. L. Augee, H. C. K. McCarron, and J. Raison. 1984. Correlates of torpor in the insectivorous marsupial *Sminthopsis murina*. Austral. Mamm. 7:185–91.

Geiser, F., and R. V. Baudinette. 1988. Daily torpor and thermoregulation in the small dasyurid marsupials *Planigale gilesi* and *Ningaui yvonneae*. Austral. J. Zool. 36: 473–81.

Geiser, F., and L. S. Broome. 1991. Hibernation in the mountain pygmy possum *Burramys parvus* (Marsupialia). J. Zool. 223: 593–602.

Geiser, F., and P. Masters. 1994. Torpor in relation to reproduction in the mulgara, *Dasycercus cristicauda* (Dasyuridae: Marsupialia). J. Thermal Biol. 19:33–40.

Geiser, F., L. Matwiejczyk, and R. V. Baudinette. 1986. From ectothermy to heterothermy: the energetics of the kowari, *Dasyuroides byrnei* (Marsupialia: Dasyuridae). Physiol. Zool. 59:220–29.

Gemmell, R. T. 1988. The oestrous cycle length of the bandicoot *Isoodon macrourus*. Austral. Wildl. Res. 15:633–35.

———. 1990. Longevity and reproductive capability of *Isoodon macrourus* in captivity. *In* Seebeck et al. (1990), 213–17.

George, G. G. 1979. The status of endangered Papua New Guinea mammals. *In* Tyler (1979), 93–100.

———. 1987. Characterisation of the living species of cuscus (Marsupialia: Phalangeridae). *In* Archer (1987*a*), 507–26.

George, G. G., and G. M. Maynes. 1990. Status of New Guinea bandicoots. *In* Seebeck et al. (1990), 93–105.

George, G. G., and U. Schürer. 1978. Some notes on macropods commonly misidentified in zoos. Internatl. Zoo Yearbook 18:152–56.

Gilmore, D. P. 1977. The success of marsupials as introduced species. *In* Stonehouse and Gilmore (1977), 169–78.

Godfrey, G. K. 1969. Reproduction in a laboratory colony of the marsupial mouse *Sminthopsis larapinta* (Marsupialia: Dasyuridae). Austral. J. Zool. 17:637–54.

———. 1975. A study of oestrus and fecundity in a laboratory colony of mouse opossums *(Marmosa robinsoni)*. J. Zool. 175: 541–55.

Godfrey, G. K., and P. Crowcroft. 1971. Breeding the fat-tailed marsupial mouse *Sminthopsis crassicaudata* in captivity. Internatl. Zoo Yearbook 11:33–38.

Godin, A. J. 1977. Wild mammals of New England. Johns Hopkins Univ. Press, Baltimore, xii + 304 pp.

Goldingay, R. L. 1990. The foraging behaviour of a nectar feeding marsupial *Petaurus australis*. Oecologia 85:191–99.

———. 1992. Socioecology of the yellow-bellied glider *(Petaurus australis)* in a coastal forest. Austral. J. Zool. 40:267–78.

———. 1994. Loud calls of the yellow-bellied glider, *Petaurus australis:* territorial behaviour by an arboreal marsupial? Austral. J. Zool. 42:279–93.

Goldingay, R. L., and R. P. Kavanagh. 1990. Socioecology of the yellow-bellied glider, *Petaurus australis*, at Waratah Creek, NSW. Austral. J. Zool. 38:327–41.

———. 1993. Home-range estimates and habitat of the yellow-bellied glider *(Petaurus australis)* at Waratah Creek, New South Wales. Wildl. Res. 20:387–404.

Goodwin, H. A., and J. M. Goodwin. 1973. List of mammals which have become extinct or are possibly extinct since 1600. Internatl. Union Conserv. Nat. Occas. Pap., no. 8, 20 pp.

Gordon, G., L. S. Hall, and R. G. Atherton. 1990. Status of bandicoots in Queensland. *In* Seebeck et al. (1990), 37–42.

Gordon, G., and A. J. Hulbert. 1989. Peramelidae. *In* Walton and Richardson (1989), 603–24.

Gordon, G., and B. C. Lawrie. 1977. The rufescent bandicoot, *Echymipera rufescens* (Peters and Doria) on Cape York Peninsula. Austral. Wildl. Res. 5:41–45.

———. 1980. The rediscovery of the bridled nail-tailed wallaby, *Onychogalea fraenata* (Gould) (Marsupialia: Macropodidae) in Queensland. Austral. Wildl. Res. 7:339–45.

Gordon, G., D. G. McGreevy, and B. C. Lawrie. 1978. The yellow-footed rock-wallaby, *Petrogale xanthopus* Gray (Macropodidae), in Queensland. Austral. Wildl. Res. 5:295–97.

———. 1990. Koala population turnover and male social organization. *In* Lee, Handasyde, and Sanson (1990), 189–92.

Gordon, G., P. McRae, L. Lim, D. Reimer, and G. Porter. 1993. The conservation status of the yellow-footed rock-wallaby in Queensland. Oryx 27:159–68.

Grant, T. R. 1973. Dominance and association among members of a captive and free-ranging group of grey kangaroos *(Macropus giganteus)*. Anim. Behav. 21:449–56.

Grant, T. R., and P. D. Temple-Smith. 1987. Observations on torpor in the small marsupial *Dromiciops australis* (Marsupialia: Microbiotheriidae) from southern Chile. *In* Archer (1987*a*), 273–77.

Green, R. H. 1967. Notes on the devil *(Sarcophilus harrisi)* and the quoll *(Dasyurus viverrinus)* in north-eastern Tasmania. Rec. Queen Victoria Mus., no. 27, 13 pp.

———. 1972. The murids and small dasyurids in Tasmania. Parts 5, 6, and 7. Rec. Queen Victoria Mus., no. 46, 34 pp.

Green, R. H., and J. L. Rainbird. 1987. The common wombat *Vombatus ursinus* (Shaw, 1800) in northern Tasmania—Part 1. Breeding, growth, and development. Rec. Queen Victoria Mus., no. 91, 20 pp.

Grigg, G. C., L. A. Beard, G. Caughley, D. Grice, J. A. Caughley, N. Sheppard, M. Fletcher, and C. Southwell. 1985. The Australian kangaroo populations, 1984. Search 16:277–79.

Grimwood, I. R. 1969. Notes on the distribution and status of some Peruvian mammals. Spec. Publ. Amer. Comm. Internatl. Wildl. Protection, no. 21, v + 86 pp.

Groves, C. P. 1976. The origin of the mammalian fauna of Sulawesi (Celebes). Z. Saugetierk. 41:201–16.

———. 1982. The systematics of tree kangaroos (*Dendrolagus;* Marsupialia, Macropodidae). Austral. Mamm. 5:157–86.

———. 1987*a*. On the cuscuses (Marsupialia: Phalangeridae) of the *Phalanger orientalis* group from Indonesian territory. *In* Archer (1987*a*), 569–79.

———. 1987*b*. On the highland cuscuses (Marsupialia: Phalangeridae) of New Guinea. *In* Archer (1987*a*), 559–67.

Groves, C. P., and T. Flannery. 1989. Revision of the genus *Dorcopsis* (Macropodidae: Marsupialia). *In* Grigg, Jarman, and Hume (1989), 117–28.

———. 1990. Revision of the families and genera of bandicoots. *In* Seebeck et al. (1990), 1–11.

Grzimek, B., ed. 1975. Grzimek's animal life encyclopedia: mammals, I–IV. Van Nostrand Reinhold, New York, vols. 10–13.

———, ed. 1990. Grzimek's encyclopedia of mammals. McGraw-Hill, New York, 5 vols.

Guiler, E. R. 1961. Breeding season of the thylacine. J. Mamm. 42:396–97.

———. 1970*a*. Observations on the Tasmanian devil, *Sarcophilus harrisii* (Marsupialia: Dasyuridae). I. Numbers, home range, movements, and food in two populations. Austral. J. Zool. 18:49–62.

———. 1970*b*. Observations on the Tasmanian devil, *Sarcophilus harrisii* (Marsupialia: Dasyuridae). II. Reproduction, breeding, and growth of pouch young. Austral. J. Zool. 18:63–70.

———. 1971*a*. Food of the Potoroo (Marsupialia, Macropodidae). J. Mamm. 52:232–34.

———. 1971*b*. The husbandry of the potoroo *Potorous tridactylus*. Internatl. Zoo Yearbook 11:21–22.

———. 1971*c*. The Tasmanian devil, *Sarcophilus harrisii*, in captivity. Internatl. Zoo Yearbook 11:32–33.

———. 1982. Temporal and spatial distribution of the Tasmanian devil, *Sarcophilus harrisii* (Dasyuridae: Marsupialia). Pap. Proc. Roy. Soc. Tasmania 116:153–63.

Guiler, E. R., and D. A. Kitchener. 1967. Further observations on longevity in the wild potoroo, *Potorous tridactylus*. Austral. J. Sci. 30:105–6.

H

Hall, E. R. 1955. Handbook of mammals of Kansas. Univ. Kansas Mus. Nat. Hist. Misc. Publ., no. 7, 303 pp.

———. 1981. The mammals of North America. John Wiley & Sons, New York, 2 vols.

Hall, L. S. 1987. Syndactyly in marsupials—problems and prophecies. *In* Archer (1987*a*), 245–55.

Handasyde, K. A., I. R. McDonald, K. A. Than, J. Michaelides, and R. W. Martin. 1990. Reproductive hormones and reproduction in the koala. *In* Lee, Handasyde, and Sanson (1990), 203–10.

Handley, C. O., Jr. 1976. Mammals of the Smithsonian Venezuelan Project. Brigham Young Univ. Sci. Bull., Biol. Ser., 20(5):1–89.

Handley, C. O., Jr., and L. K. Gordon. 1979. New species of mammals from northern South America: mouse possums, genus *Marmosa* Gray. *In* Eisenberg (1979), 65–72.

Harding, H. R., F. N. Carrick, and C. D. Shorey. 1987. The affinities of the koala *Phascolarctos cinereus* (Marsupialia: Phascolarctidae) on the basis of sperm ultrastructure and development. *In* Archer (1987*a*), 353–64.

Harper, F. 1945. Extinct and vanishing mammals of the Old World. Spec. Publ. Amer. Comm. Internatl. Wildl. Protection, no. 12, xv + 850 pp.

Hart, R. P., and D. J. Kitchener. 1986. First record of *Sminthopsis psammophila* (Marsupialia: Dasyuridae) from Western Australia. Rec. W. Austral. Mus. 13:139–44.

Harvie, A. E. 1973. Diet of the opossum (*Trichosurus vulpecula* Kerr) on farmland northeast of Waverly, New Zealand. Proc. New Zealand Ecol. Soc. 20:48–52.

Hayssen, V., A. Van Tienhoven, and A. Van Tienhoven. 1993. Asdell's patterns of mammalian reproduction: a compendium of species-specific data. Comstock/Cornell Univ. Press, Ithaca, viii + 1023 pp.

Henry, S. 1984. Social organisation of the greater glider in Victoria. *In* Smith and Hume (1984), 221–28.

Henry, S., and S. A. Craig. 1984. Diet, ranging behaviour, and social organization of the yellow-bellied glider (*Petaurus australis* Shaw) in Victoria. *In* Smith and Hume (1984), 331–41.

Hershkovitz, P. 1976. Comments on generic names of four-eyed opossums (family Didelphidae). Proc. Biol. Soc. Washington 89:295–304.

———. 1981. *Philander* and four-eyed opossums once again. Proc. Biol. Soc. Washington 93:943–46.

———. 1992*a*. Ankle bones: the Chilean opossum *Dromiciops gliroides* Thomas, and marsupial phylogeny. Bonner Zool. Beitr. 43:181–213.

———. 1992*b*. The South American gracile opossums, genus *Gracilinanus* Gardner and Creighton, 1989 (Marmosidae, Marsupialia): a taxonomic review with notes on general morphology and relationships. Fieldiana Zool., n.s., no. 70, 56 pp.

———. 1995. The staggered marsupial third lower incisor: hallmark of cohort Didelphimorphia, and description of a new genus and species with staggered i3 from the Albian (lower Cretaceous) of Texas. Bonner Zool. Beitr. 45:153–69.

Heuvelmans, B. 1958. On the track of unknown animals. Hill & Wang, New York, 558 pp.

Hickman, V. V., and J. L. Hickman. 1960. Notes on the habits of the Tasmanian dormouse phalangers *Cercartetus nanus* (Desmarest) and *Eudromicia lepida* (Thomas). Proc. Zool. Soc. London 135:365–74.

Holmes, D. J. 1991. Social behavior in captive Virginia opossums, *Didelphis virginianus*. J. Mamm. 72:402–10.

Hope, J. H. 1981. A new species of *Thylogale* (Marsupialia: Macropodidae) from Mapala Rock Shelter, Jaya (Carstensz) Mountains, Irian Jaya (western New Guinea), Indonesia. Rec. Austral. Mus. 33:369–87.

How, R. A. 1976. Reproduction, growth, and survival of young in the mountain possum, *Trichosurus caninus* (Marsupialia). Austral. J. Zool. 24:189–99.

———. 1978. Population strategies of four species of Australian "possums." *In* Montgomery (1978), 305–13.

How, R. A., J. L. Barnett, A. J. Bradley, W. J. Humphreys, and R. W. Martin. 1984. The population biology of *Pseudocheirus peregrinus* in a *Leptospermum laevigatum* thicket. *In* Smith and Hume (1984), 261–68.

Howe, D. 1975. Observations on a captive marsupial mole, *Notoryctes typhlops*. Austral. Mamm. 1:361–65.

Hughes, R. L., L. S. Hall, K. P. Aplin, and M. Archer. 1987. Organogenesis and fetal membranes in the New Guinea pen-tailed possum, *Distoechurus pennatus* (Acrobatidae: Marsupialia). *In* Archer (1987*a*), 715–24.

Hulbert, A. J., G. Gordon, and T. J. Dawson. 1971. Rediscovery of the marsupial *Echymipera rufescens* in Australia. Nature 231:330–31.

Hume, I. D., P. J. Jarman, M. B. Renfree, and P. D. Temple-Smith. 1989. Macropodidae. *In* Walton and Richardson (1989), 679–715.

Humphreys, W. F., R. A. How, A. J. Bradley, C. M. Kemper, and D. J. Kitchener. 1984. The biology of *Wyulda squamicaudata* Alexander 1919. *In* Smith and Hume (1984), 162–69.

Hunsaker, D., II. 1977. Ecology of New World marsupials. *In* Hunsaker (1977*a*), 95–156.

Hunsaker, D., II, and D. Shupe. 1977. Behavior of New World marsupials. *In* Hunsaker (1977*a*), 279–347.

Husband, T. P., G. D. Hobbs, C. N. Santos, and H. J. Stillwell. 1992. First record of *Metachirus nudicaudatus* for northeast Brazil. Mammalia 56:298–99.

Husson, A. M. 1978. The mammals of Suriname. E. J. Brill, Leiden, xxxiv + 569 pp.

Hutchins, M., and R. Smith. 1990. Biology and status of wild tree kangaroos. *In* Roberts, M., and M. Hutchins, eds., The biology and management of tree kangaroos, Amer. Assoc. Zool. Parks Aquar. Marsupial and Monotreme Advisory Group Bull., no. 1, 1–6.

I

Ingleby, S. 1991*a*. Distribution and abundance of the northern nailtail wallaby, *Onychogalea unguifera* (Gould, 1841). Wildl. Res. 18:655–76.

———. 1991*b*. Distribution and status of the spectacled hare-wallaby, *Lagorchestes conspicillatus*. Wildl. Res. 18:501–19.

Izor, R. J., and R. H. Pine. 1987. Notes on the black-shouldered opossum, *Caluromysiops irrupta*. Fieldiana Zool., n.s., 39:117–24.

J

Jackson, H. H. T. 1961. Mammals of Wisconsin. Univ. Wisconsin Press, Madison, xii + 504 pp.

Johnson, C. N., and D. G. Crossman. 1991. Dispersal and social organization of the northern hairy-nosed wombat *Lasiorhinus krefftii*. J. Zool. 225:605–13.

Johnson, C. N., and K. A. Johnson. 1983. Behaviour of the bilby, *Macrotis lagotis* (Reid) (Marsupialia: Thylacomyidae) in captivity. Austral. Wildl. Res. 10:77–87.

Johnson, K. A. 1980. Spatial and temporal use of habitat by the red-necked pademelon, *Thylogale thetis* (Marsupialia: Macropodidae). Austral. Wildl. Res. 7:157–66.

———. 1989. Thylacomyidae. *In* Walton and Richardson (1989), 625–35.

Johnson, K. A., and R. I. Southgate. 1990. Present and former status of bandicoots in the Northern Territory. *In* Seebeck et al. (1990), 85–92.

Johnson, K. A., and D. W. Walton. 1989. Notoryctidae. *In* Walton and Richardson (1989), 591–602.

Johnson, P. M. 1978. Husbandry of the rufous rat-kangaroo *Aeprymnus rufescens* and brush-tailed rock wallaby *Petrogale penicillata* in captivity. Internatl. Zoo Yearbook 18:156–57.

———. 1979. Reproduction in the plain rock-wallaby, *Petrogale penicillata inornata* Gould, in captivity, with age estimation of the pouch young. Austral. Wildl. Res. 6:1–4.

———. 1980. Observations of the behaviour of the rufous rat-kangaroo, *Aeprymnus rufescens* (Gray), in captivity. Austral. Wildl. Res. 7:347–57.

———. 1993. Reproduction of the spectacled hare-wallaby, *Lagorchestes conspicillatus* Gould (Marsupialia: Macropodidae), in captivity, with age estimation of the pouch young. Wildl. Res. 20:97–101.

Johnson, P. M., and R. Strahan. 1982. A further description of the musky rat-kangaroo, *Hypsiprymnodon moschatus* Ramsay, 1876 (Marsupialia, Potoroidae), with notes on its biology. Austral. Zool. 21:27–46.

Johnson, P. M., and K. Vernes. 1994. Reproduction in the red-legged pademelon, *Thylogale stigmatica* Gould (Marsupialia: Macropodidae), and age estimation and development of pouch young. Wildl. Res. 21:553–58.

Johnson-Murray, J. L. 1987. The comparative myology of the gliding membranes of *Acrobates*, *Petauroides*, and *Petaurus* contrasted with the cutaneous myology of *Hemibelideus* and *Pseudocheirus* (Marsupialia: Phalangeridae) and with selected gliding Rodentia (Sciuridae and Anomaluridae). Austral. Wildl. Res. 35:101–13.

Johnston, P. G., and G. B. Sharman. 1976. Studies on populations of *Potorous* Desmarest (Marsupialia). I. Morphological variation. Austral. J. Zool. 24:573–88.

———. 1977. Studies on populations of *Potorous* Desmarest (Marsupialia). II. Electrophoretic, chromosomal, and breeding studies. Austral. J. Zool. 25:733–47.

Jones, B. A., R. A. How, and D. J. Kitchener. 1994. A field study of *Pseudocheirus occidentalis* (Marsupialia: Petauridae). II. Population studies. Wildl. Res. 21:189–201.

Jones, C. J., and F. Geiser. 1992. Prolonged and daily torpor in the feathertail glider, *Acrobates pygmaeus* (Marsupialia: Acrobatidae). J. Zool. 227:101–8.

Jones, F. W. 1923–25. The mammals of South Australia. A. B. Jones, Government Printer, Adelaide, 458 pp.

Jones, J. K., Jr. 1964. Distribution and taxonomy of mammals of Nebraska. Univ. Kansas Publ. Mus. Nat. Hist. 16:1–356.

Jones, J. K., Jr., H. H. Genoways, and J. D. Smith. 1974. Annotated checklist of mammals of the Yucatan Peninsula, Mexico. III. Marsupialia, Insectivora, Primates, Edentata, Lagomorpha. Occas. Pap. Mus. Texas Tech Univ., no. 23, 12 pp.

Jones, M. L. 1982. Longevity of captive mammals. Zool. Garten 52:113–28.

K

Kaufmann, J. H. 1974. Social ethology of the whiptail wallaby, *Macropus parryi*, in northeastern New South Wales. Anim. Behav. 22:281–369.

———. 1975. Field observations of the social behaviour of the eastern grey kangaroo, *Macropus giganteus*. Anim. Behav. 23:214–21.

Keast, A. 1982. The thylacine (Thylacinidae, Marsupialia): how good a pursuit carnivore? *In* Archer (1982*a*), 675–84.

Kelt, D. A., and D. R. Martínez. 1989. Notes on the distribution and ecology of two marsupials endemic to the Valdivian forests of southern South America. J. Mamm. 70:220–24.

Kemper, C. 1990. Status of bandicoots in South Australia. *In* Seebeck et al. (1990), 67–72.

Kennedy, M. 1992. Australasian marsupials and monotremes: an action plan for their conservation. IUCN (World Conservation Union), Gland, Switzerland, vii + 103 pp.

Kerle, J. A. 1984*a*. The behaviour of *Burramys parvus* Broom (Marsupialia) in captivity. Mammalia 48:317–25.

———. 1984*b*. Growth and development of *Burramys parvus* in captivity. *In* Smith and Hume (1984), 409–12.

Kerle, J. A., and C. J. Howe. 1992. The breeding biology of a tropical possum, *Trichosurus vulpecula arnhemensis* (Phalangeridae: Marsupialia). Austral. J. Zool. 40:653–65.

Kerle, J. A., G. M. McKay, and G. B. Sharman. 1991. A systematic analysis of the brushtail possum, *Trichosurus vulpecula* (Kerr, 1792) (Marsupialia: Phalangeridae). Austral. J. Zool. 39:313–31.

Kerle, J. C., and A. Borsboom. 1984. Home range, den tree use, and activity patterns in the greater glider, *Petauroides volans*. *In* Smith and Hume (1984), 229–36.

Kirkpatrick, T. H. 1965. Studies of Macropodidae in Queensland. 2. Age estimation in the grey kangaroo, the red kangaroo, the eastern wallaroo and the red-necked wallaby, with notes on dental abnormalities. Queensland J. Agric. Anim. Sci. 22:301–17.

———. 1967. The grey kangaroo in Queensland. Queensland Agric. J. 93:550–52.

———. 1968. Studies on the wallaroo. Queensland Agric. J. 94:362–65.

———. 1970*a*. The agile wallaby in Queensland. Queensland Agric. J. 96:169–70.

———. 1970*b*. The swamp wallaby in Queensland. Queensland Agric. J. 96:335–36.

Kirsch, J. A. W. 1968. Burrowing by the quenda, *Isoodon obesulus*. W. Austral. Nat. 10:178–80.

———. 1977*a*. The classification of marsupials. *In* Hunsaker (1977*a*), 1–50.

———. 1977*b*. The comparative serology of Marsupialia, and a classification of marsupials. Austral. J. Zool., Suppl. Ser., no. 52, 152 pp.

Kirsch, J. A. W., and J. H. Calaby. 1977. The species of living marsupials: an annotated list. *In* Stonehouse and Gilmore (1977), 9–26.

Kirsch, J. A. W., and W. E. Poole. 1972. Taxonomy and distribution of the grey kangaroos, *Macropus giganteus* Shaw and *Macropus fuliginosus* (Desmarest), and their subspecies (Marsupialia: Macropodidae). Austral. J. Zool. 20:315–39.

Kirsch, J. A. W., M. S. Springer, C. Krajewski, M. Archer, K. Aplin, and A. W. Dickerman. 1990. DNA/DNA hybridization studies of the carnivorous marsupials. I: The intergeneric relationships of bandicoots (Marsupialia: Perameloidea). J. Molecular Evol. 30:434–48.

Kirsch, J. A. W., and P. F. Waller. 1979. Notes on the trapping and behavior of the Caenolestidae (Marsupialia). J. Mamm. 60:390–95.

Kitchener, D. J. 1972. The importance of shelter to the quokka, *Setonix brachyurus* (Marsupialia), on Rottnest Island. Austral. J. Zool. 20:281–99.

———. 1981. Breeding, diet, and habitat preference of *Phascogale calura* (Gould, 1844) (Marsupialia: Dasyuridae) in the southern wheatbelt, Western Australia. Rec. W. Austral. Mus. 9:173–86.

———. 1988. A new species of false antechinus (Marsupialia: Dasyuridae) from the Kimberley, Western Australia. Rec. W. Austral. Mus. 14:61–71.

Kitchener, D. J., and N. Caputi. 1988. A new species of false antechinus (Marsupialia: Dasyuridae) from Western Australia, with remarks on the generic classification within the Parantechini. Rec. W. Austral Mus. 14:35–59.

Kitchener, D. J., S. Hisheh, L. H. Schmitt, and I. Maryanto. 1993. Morphological and genetic variation in *Aethalops alecto* (Chiroptera, Pteropodidae) from Java, Bali, and Lombok Is., Indonesia. Mammalia 57:255–72.

Kitchener, D. J., and G. Sanson. 1978. *Petrogale burbidgei* (Marsupialia, Macropodidae), a new rock wallaby from Kimberley, Western Australia. Rec. W. Austral. Mus. 6:269–85.

Kitchener, D. J., J. Stoddart, and J. Henry. 1983. A taxonomic appraisal of the genus *Ningaui* Archer (Marsupialia: Dasyuridae), including description of a new species. Austral. J. Zool. 31:361–79.

———. 1984. A taxonomic revision of the *Sminthopsis murina* complex (Marsupialia: Dasyuridae) in Australia, including descriptions of four new species. Rec. W. Austral. Mus. 11:201–48.

Krajewski, C., A. C. Driskell, P. R. Baverstock, and M. J. Braun. 1992. Phylogenetic relationships of the thylacine (Mammalia: Thylacinidae) among dasyuroid marsupials: evidence from cytochrome *b* DNA sequences. Proc. Roy. Soc. London, ser. B, 250:19–27.

Krajewski, C., J. Painter, L. Buckley, and M. Westerman. 1994. Phylogenetic structure of the marsupial family Dasyuridae based on cytochrome *b* DNA sequences. J. Mamm. Evol. 2:25–35.

L

Laurance, W. F. 1990. Comparative responses of five arboreal marsupials to tropical forest fragmentation. J. Mamm. 71: 641–53.

Laurie, E. M. O., and J. E. Hill. 1954. List of land mammals of New Guinea, Celebes, and adjacent islands, 1758–1952. British Mus. (Nat. Hist.), London, 175 pp.

Lazell, J. D., Jr., T. W. Sutterfield, and W. D. Giezentanner. 1984. The population of rock wallabies (genus *Petrogale*) on Oahu, Hawaii. Biol. Conserv. 30:99–108.

Lazenby-Cohen, K. A. 1991. Communal nesting in *Antechinus stuartii* (Marsupialia: Dasyuridae). Austral. J. Zool. 39:273–83.

Lazenby-Cohen, K. A., and A. Cockburn. 1988. Lek promiscuity in a semelparous mammal, *Antechinus stuartii* (Marsupialia: Dasyuridae)? Behav. Ecol. Sociobiol. 22:195–202.

Lee, A. K., A. J. Bradley, and R. W. Braithwaite. 1977. Corticosteroid levels and male mortality in *Antechinus stuartii*. *In* Stonehouse and Gilmore (1977), 209–22.

Lee, A. K., and F. N. Carrick. 1989. Phascolarctidae. *In* Walton and Richardson (1989), 740–54.

Lemke, T. O., A. Cadena, R. H. Pine, and J. Hernandez-Camacho. 1982. Notes on opossums, bats, and rodents new to the fauna of Colombia. Mammalia 46:225–34.

Leopold, A. S. 1959. Wildlife of Mexico. Univ. California Press, Berkeley, xiii + 568.

Lidicker, W. Z., Jr., and B. J. Marlow. 1970. A review of the dasyurid marsupial genus *Antechinomys* Krefft. Mammalia 34:212–27.

Lidicker, W. Z., Jr., and A. C. Ziegler. 1968. Report on a collection of mammals from eastern New Guinea including species keys for fourteen genera. Univ. California Publ. Zool. 87:i–v + 1–64.

Lindenmayer, D. B., and J. M. Dixon. 1992. An additional historical record of Leadbeater's possum, *Gymnobelideus leadbeateri* McCoy, prior to the 1961 rediscovery of the species. Victoria Nat. 109:217–18.

Lindenmayer, D. B., H. A. Nix, J. P. McMahon, M. F. Hutchinson, and M. T. Tanton. 1991. The conservation of Leadbeater's possum, *Gymnobelideus leadbeateri* (McCoy): a case study of the use of bioclimatic modelling. J. Biogeogr. 18:371–83.

Linscombe, G. 1994. U.S. fur harvest (1970–1992) and fur value (1974–1992) statistics by state and region. Louisiana Department of Wildlife and Fisheries, Baton Rouge, 29 pp.

Lobert, B., and A. K. Lee. 1990. Reproduction and life history of *Isoodon obesulus* in Victorian heathland. *In* Seebeck et al. (1990), 311–18.

Lowery, G. H., Jr. 1974. The mammals of Louisiana and its adjacent waters. Louisiana State Univ. Press, xxiii + 565 pp.

Lumsden, L. F., A. F. Bennett, and P. Robertson. 1988. First record of the paucident planigale, *Planigale gilesi* (Marsupialia: Dasyuridae), for Victoria. Victorian Nat. 105:81–87.

Lundie-Jenkins, G., L. K. Corbett, and C. M. Phillips. 1993. Ecology of the rufous hare-wallaby, *Lagorchestes hirsutus* Gould (Marsupialia: Macropodidae), in the Tanami Desert, Northern Territory. III. Interactions with introduced mammal species. Wildl. Res. 20:495–511.

Lyne, A. G. 1974. Gestation period and birth in the marsupial *Isoodon macrourus*. Austral. J. Zool. 22:303–9.

———. 1976. Observations on oestrus and the oestrous cycle in the marsupials *Isoodon macrourus* and *Perameles nasuta*. Austral. J. Zool. 24:513–21.

———. 1990. A brief review of bandicoot studies. *In* Seebeck et al. (1990), xxiii–xxix.

Lyne, A. G., and P. A. Mort. 1981. Comparison of skull morphology in the marsupial bandicoot genus *Isoodon*: its taxonomic implications and notes on a new species, *Isoodon arnhemensis*. Austral. Mamm. 4: 107–33.

M

Mackowski, C. M. 1986. Distribution, habitat, and status of the yellow-bellied glider, *Petaurus australis* Shaw (Marsupialia: Petauridae), in northeastern New South Wales. Austral. Mamm. 9:141–44.

Mahoney, J. A. 1981. The specific name of the honey possum (Marsupialia: Tarsipedidae: *Tarsipes rostratus* Gervais and Verreaux, 1842). Austral. Mamm. 4:135–38.

Main, A. R., and M. Yadav. 1971. Conservation of macropods in reserves in Western Australia. Biol. Conserv. 3:123–33.

Mansergh, I. 1984*a*. Ecological studies and conservation of *Burramys parvus*. *In* Smith and Hume (1984), 545–52.

———. 1984*b*. The mountain pygmy-possum *(Burramys parvus)* (Broom): a review. *In* Smith and Hume (1984), 413–16.

———. 1984*c*. The status, distribution, and abundance of *Dasyurus maculatus* (tiger quoll) in Australia, with particular reference to Victoria. Austral. Zool. 21:109–22.

Mansergh, I., and D. Scotts. 1990. Aspects of the life history and breeding of the mountain pygmy-possum, *Burramys parvus* (Marsupialia: Burramyidae), in alpine Victoria. Austral. Mamm. 13:179–91.

Marlow, B. J. 1961. Reproductive behavior of the marsupial mouse, *Antechinus flavipes* (Waterhouse) (Marsupialia), and the development of pouch young. Austral. J. Zool. 9:203–18.

Marples, T. G. 1973. Studies on the marsupial glider, *Schoinobates volans* (Kerr). IV. Feeding biology. Austral. J. Zool. 21:213–16.

Marshall, L. G. 1977. *Lestodelphys halli*. Mammalian Species, no. 81, 3 pp.

———. 1978*a*. *Chironectes minimus*. Mammalian Species, no. 109, 6 pp.

———. 1978*b*. *Dromiciops australis*. Mammalian Species, no. 99, 5 pp.

———. 1978*c*. *Glironia venusta*. Mammalian Species, no. 107, 3 pp.

———. 1978*d*. *Lutreolina crassicaudata*. Mammalian Species, no. 91, 4 pp.

———. 1980. Systematics of the South American marsupial family Caenolestidae. Fieldiana Zool., n.s., no. 5, 145 pp.

———. 1982. Systematics of the South American marsupial family Microbiotheriidae. Fieldiana Geol., n.s., no. 10, 75 pp.

———. 1984. Monotremes and marsupials. *In* Anderson and Jones (1984), 59–116.

———. 1987. Systematics of Itaboraian (middle Paleocene) age "opossum-like" marsupials from the limestone quarry at Sao Jose de Itaborai, Brazil. *In* Archer (1987*a*), 91–160.

Marshall, L. G., J. A. Case, and M. O. Woodburne. 1990. Phylogenetic relationships of the families of marsupials. Current Mamm. 2:433–506.

Martin, R., and K. Handasyde. 1990. Population dynamics of the koala *(Phascolarctos cinereus)* in southeastern Australia. *In* Lee, Handasyde, and Sanson (1990), 75–84.

Martin, R. W., and A. Lee. 1984. The koala, *Phascolarctos cinereus*, the largest marsupial folivore. *In* Smith and Hume (1984), 463–67.

Massoia, E. 1980. El estado sistemático de cuatro especies de cricetidos Sudamericanos y comentarios sobre otras especies congenéricas. Ameghiniana, Rev. Asoc. Paleontol. Argentina 17:280–87.

Massoia, E., and J. Foerster. 1974. Un mamifero nuevo para la Republica Argentina: *Caluromys lanatus lanatus* (Illiger), (Mammalia—Marsupialia—Didelphidae). IDIA, January–February 1974, 5–7.

Maynes, G. M. 1973. Reproduction in the parma wallaby, *Macropus parma* Waterhouse. Austral. J. Zool. 21:331–51.

———. 1974. Occurrence and field recognition of *Macropus parma*. Austral. Zool. 18:72–87.

———. 1977*a*. Breeding and age structure of the population of *Macropus parma* on Kawau Island, New Zealand. Austral. J. Ecol. 2:207–14.

———. 1977*b*. Distribution and aspects of the biology of the parma wallaby, *Macropus parma*, in New South Wales. Austral. Wildl. Res. 4:109–25.

———. 1982. A new species of rock wallaby, *Petrogale persephone* (Marsupialia: Macropodidae), from Proserpine, central Queensland. Austral. Mamm. 5:47–58.

———. 1989. Zoogeography of the Macropodoidea. *In* Grigg, Jarman, and Hume (1989), 47–66.

McCarthy, T. J. 1982. *Chironectes, Cyclopes, Cabassous*, and probably *Cebus* in southern Belize. Mammalia 46:397–400.

McCracken, H. E. 1990. Reproduction in the greater bilby, *Macrotis lagotis* (Reid)—a comparison with other perameloids. *In* Seebeck et al. (1990), 199–204.

McEvoy, J. S. 1970. Red-necked wallaby in Queensland. Queensland Div. Plant Industry Adv. Leaf., no. 1050, 4 pp.

McKay, G. M. 1982. Nomenclature of the gliding possum genera *Petaurus* and *Petauroides* (Marsupialia: Petauridae). Austral. Mamm. 5:37–39.

———. 1989. Family Petauridae. *In* Walton and Richardson (1989), 665–78.

McKenzie, N. L., and M. Archer. 1982. *Sminthopsis youngsoni* (Marsupialia: Dasyuridae), the lesser hairy-footed dunnart, a new species from arid Australia. Austral. Mamm. 5:267–79.

McLean, R. G., and S. R. Ubico. 1993. A first record of the water opossum *(Chironectes minimus)* from Guatemala. Southwestern Nat. 38:402–4.

McManus, J. J. 1970. Behavior of captive opossums, *Didelphis marsupialis virginiana*. Amer. Midl. Nat. 84:144–69.

———. 1974. *Didelphis virginiana*. Mammalian Species, no. 40, 6 pp.

McNee, A., and A. Cockburn. 1992. Specific identity is not correlated with behavioural and life-history diversity in *Antechinus stuartii* sensu lato. Austral. J. Zool. 40:127–33.

McQuade, L. R. 1984. Taxonomic relationship of the greater glider *Petauroides volans* and lemur-like possum *Hemibelideus lemuroides*. *In* Smith and Hume (1984), 303–10.

Medellín, R. A., G. Cancino Z., A. Clemente M., and R. O. Guerrero V. 1992. Noteworthy records of three mammals from Mexico. Southwestern Nat. 37:427–28.

Menkhorst, P. W., and J. H. Seebeck. 1990. Distribution and conservation status of bandicoots in Victoria. *In* Seebeck et al. (1990), 51–60.

Menzies, J. I. 1990. Notes on spiny bandicoots, *Echymipera* spp. (Marsupialia: Peramelidae) from New Guinea and description of a new species. Sci. New Guinea 16:86–98.

Menzies, J. I., and J. C. Pernetta. 1986. A taxonomic revision of cuscuses allied to *Phalanger orientalis* (Marsupialia: Phalangeridae). J. Zool., ser. B, 1:551–618.

Merchant, J. C. 1976. Breeding biology of the agile wallaby, *Macropus agilis* (Gould) (Marsupialia: Macropodidae), in captivity. Austral. Wildl. Res. 3:93–103.

———. 1989. Lactation in macropodoid marsupials. *In* Grigg, Jarman, and Hume (1989), 355–66.

Miller, G. S., Jr., and N. Hollister. 1922. A new phalanger from Celebes. Proc. Biol. Soc. Washington 35:115–16.

Miller, S. D., J. Rottmann, K. J. Raedeke, and R. D. Taber. 1983. Endangered mammals of Chile: status and conservation. Biol. Conserv. 25:335–52.

Minta, S. C., T. W. Clark, and P. Goldstraw. 1989. Population estimates and characteristics of the eastern barred bandicoot in Victoria, with recommendations for population monitoring. *In* Clark, T. W., and J. H. Seebeck, eds., Management and conservation of small populations, Chicago Zoological Society, 47–75.

Mitchell, P. 1990*a*. The home ranges and social activity of koalas—a quantitative analysis. *In* Lee, Handasyde, and Sanson (1990), 171–87.

———. 1990*b*. Social behaviour and communication of koalas. *In* Lee, Handasyde, and Sanson (1990), 151–70.

Mitchell, P., and R. Martin. 1990. The structure and dynamics of koala populations—French Island in perspective. *In* Lee, Handasyde, and Sanson (1990), 97–108.

Moors, P. J. 1975. The urogenital system and notes on the reproductive biology of the female rufous rat-kangaroo, *Aepyprymnus rufescens* (Gray) (Macropodidae). Austral. J. Zool. 23:355–61.

Morcombe, M. K. 1967. The rediscovery after eighty-three years of the dibbler *Antechinus apicalis* (Marsupialia, Dasyuridae). W. Austral. Nat. 10:103–11.

Morton, S. R. 1978*a*. An ecological study of *Sminthopsis crassicaudata* (Marsupialia: Dasyuridae). I. Distribution, study areas, and methods. Austral. Wildl. Res. 5:151–62.

———. 1978*b*. An ecological study of *Sminthopsis crassicaudata* (Marsupialia: Dasyuridae). II. Behaviour and social organization. Austral. Wildl. Res. 5:163–82.

———. 1978*c*. An ecological study of *Sminthopsis crassicaudata* (Marsupialia: Dasyuridae). III. Reproduction and life history. Austral. Wildl. Res. 5:183–211.

———. 1978*d*. Torpor and nest-sharing in free-living *Sminthopsis crassicaudata* (Marsupialia) and *Mus musculus* (Rodentia). J. Mamm. 59:569–75.

Morton, S. R., and T. C. Burton. 1973. Observations on the behaviour of the macropodid marsupial *Thylogale billardieri* (Desmarest) in captivity. Austral. Zool. 18:1–14.

Morton, S. R., C. R. Dickman, and T. P. Fletcher. 1989. Dasyuridae. *In* Walton and Richardson (1989), 560–82.

Morton, S. R., and A. K. Lee. 1978. Thermoregulation and metabolism in *Planigale maculata* (Marsupialia: Dasyuridae). J. Thermal Biol. 3:117–20.

Morton, S. R., J. W. Wainer, and T. P. Thwaites. 1980. Distributions and habitats of *Sminthopsis leucopus* and *S. murina* (Marsupialia: Dasyuridae) in southeastern Australia. Austral. Mamm. 3:19–30.

Muir, B. G. 1985. The dibbler *(Parantechinus apicalis)* found in Fitzgerald River National Park, Western Australia. W. Austral. Nat. 16:48–51.

Musser, G. G., and H. G. Sommer. 1992. Taxonomic notes on specimens of the marsupials *Pseudocheirus schlegelii* and *P. forbesi* (Diprotodontia, Pseudocheiridae) in the American Museum of Natural History. Amer. Mus. Novit., no. 3044, 16 pp.

N

Nagy, K. A., R. S. Seymour, A. K. Lee, and R. Braithwaite. 1978. Energy and water budgets in free-living *Antechinus stuartii* (Marsupialia: Dasyuridae). J. Mamm. 59:60–68.

Nelson, J. E., and A. Goldstone. 1986. Reproduction in *Peradorcas concinna* (Marsupialia: Macropodidae). Austral. Wildl. Res. 13:501–5.

Newsome, A. E. 1971*a*. Competition between wildlife and domestic livestock. Austral. Vet. J. 47:577–86.

———. 1971*b*. The ecology of red kangaroos. Austral. Zool. 16:32–50.

———. 1975. An ecological comparison of the two arid-zone kangaroos of Australia, and their anomalous prosperity since the introduction of ruminant stock to their environment. Quart. Rev. Biol. 50:389–424.

Nicholls, D. G. 1971. Daily and seasonal movements of the quokka, *Setonix brachyurus* (Marsupialia), on Rottnest Island. Austral. J. Zool. 19:215–26.

Novak, M., M. E. Obbard, J. G. Jones, R. Newman, A. Booth, A. J. Satterthwaite, and G. Linscombe. 1987. Furbearer harvests in North America, 1600–1984. Ontario Ministry Nat. Res., Toronto, xvi + 270 pp.

O

O'Connell, M. A. 1983. *Marmosa robinsoni.* Mammalian Species, no. 203, 6 pp.

Olds, T. J., and L. R. Collins. 1973. Breeding Matschie's tree kangaroo *Dendrolagus matschiei* in captivity. Internatl. Zoo Yearbook 13:123–25.

P

Pacheco, V., H. de Macedo, E. Vivar, C. Ascorra, R. Arana-Cardó, and S. Solari. 1995. Lista anotada de los mamíferos Peruanos. Conservation International Occas. Pap., no. 2, 35 pp.

Paddle, R. N. 1993. Thylacines associated with the Royal Zoological Society of New South Wales. Austral. Zool. 29:97–101.

Painter, J., C. Krajewski, and M. Westerman. 1995. Molecular phylogeny of the marsupial genus *Planigale* (Dasyuridae). J. Mamm. 76:406–13.

Parker, S. A. 1971. Notes on the small black wallaroo *Macropus bernardus* (Rothschild, 1904) of Arnhem Land. Victorian Nat. 88: 41–43.

Partridge, J. 1967. A 3,300 year old thylacine (Marsupialia: Thylacinidae) from the Nullarbor Plain, Western Australia. J. Roy. Soc. W. Austral. 50:57–59.

Patterson, B. D., and M. H. Gallardo. 1987. *Rhyncholestes raphanurus.* Mammalian Species, no. 286, 5 pp.

Patterson, B. D., P. L. Meserve, and B. K. Lang. 1990. Quantitative habitat associations of small mammals along an elevational transect in temperate rainforests of Chile. J. Mamm. 71:620–33.

Payne, O. 1995. Koalas out on a limb. Natl. Geogr. 187(4):36–59.

Pearson, D. J. 1992. Past and present distribution and abundance of the black-footed rock-wallaby in the Warburton region of Western Australia. Wildl. Res. 19:605–22.

Pearson, D. J., and A. C. Robinson. 1990. New records of the sandhill dunnart, *Sminthopsis psammophila* (Marsupialia: Dasyuridae) in South and Western Australia. Austral. Mamm. 13:57–59.

Pearson, O. P. 1983. Characteristics of a mammalian fauna from forests in Patagonia, southern Argentina. J. Mamm. 64:476–92.

Pemberton, D., and D. Renouf. 1993. A field study of communication and social behaviour of the Tasmanian devil at feeding sites. Austral. J. Zool. 41:507–26.

Perrers, C. 1965. Notes on a pigmy possum, *Cercartetus nanus* Desmarest. Austral. Zool. 13:126.

Phillips, B. 1990. Koalas: the little Australians we'd all hate to lose. Australian National Parks and Wildlife Service, Canberra, ix + 104 pp.

Phillips, C. J., and J. K. Jones, Jr. 1968. Additional comments on reproduction in the woolly opossum *(Caluromys derbianus)* in Nicaragua. J. Mamm. 49:320–21.

———. 1969. Notes on reproduction and development in the four-eyed opossum, *Philander opossum,* in Nicaragua. J. Mamm. 50:345–48.

Pine, R. H. 1972. A new subgenus and species of murine opossum (genus *Marmosa*) from Peru. J. Mamm. 53:279–82.

———. 1973. Anatomical and nomenclatural notes on opossums. Proc. Biol. Soc. Washington 86:391–402.

———. 1975. A new species of *Monodelphis* (Mammalia: Didelphidae) from Bolivia. Mammalia 39:320–22.

———. 1976. *Monodelphis umbristriata* (A. de Miranda-Ribeiro) is a distinct species of opossum. J. Mamm. 57:785–87.

———. 1977. *Monodelphis iheringi* (Thomas) is a recognizable species of Brazilian opossum (Mammalia: Marsupialia: Didelphidae). Mammalia 41:235–37.

———. 1979. Taxonomic notes on "*Monodelphis dimidiata itatiayae* (Miranda-Ribeiro)," *Monodelphis domestica* (Wagner), and *Monodelphis maraxina* Thomas (Mammalia: Marsupialia: Didelphidae). Mammalia 43:495–99.

———. 1981. Reviews of the mouse opossums *Marmosa parvidens* Tate and *Marmosa invicta* Goldman (Mammalia: Marsupialia: Didelphidae) with description of a new species. Mammalia 45:56–70.

Pine, R. H., and J. P. Abravaya. 1978. Notes on the Brazilian opossum *Monodelphis scalops* (Thomas) (Mammalia: Marsupialia: Didelphidae). Mammalia 42:379–82.

Pine, R. H., P. L. Dalby, Jr., and J. O. Matson. 1985. Ecology, postnatal development, morphometrics, and taxonomic status of the short-tailed opossum, *Monodelphis dimidiata,* an apparently semelparous annual marsupial. Ann. Carnegie Mus. 54:195–231.

Pine, R. H., and C. O. Handley, Jr. 1984. A review of the Amazonian short-tailed opossum *Monodelphis emiliae.* Mammalia 48:239–45.

Pine, R. H., S. D. Miller, and M. L. Schamberger. 1979. Contributions to the mammalogy of Chile. Mammalia 43:339–76.

Poole, W. E. 1973. A study of breeding in grey

kangaroos, *Macropus giganteus* Shaw and *M. fuliginosus* (Desmarest), in central New South Wales. Austral. J. Zool. 21:183–212.

———. 1975. Reproduction in the two species of grey kangaroos, *Macropus giganteus* Shaw and *M. fuliginosus* (Desmarest). II. Gestation, parturition, and pouch life. Austral. J. Zool. 23:333–53.

———. 1976. Breeding biology and current status of the grey kangaroo, *Macropus fuliginosus fuliginosus,* of Kangaroo Island, South Australia. Austral. J. Zool. 24:169–87.

———. 1977. The eastern grey kangaroo, *Macropus giganteus,* in south-east South Australia: its limited distribution and need of conservation. CSIRO Div. Wildl. Res. Tech. Pap., no. 31, 15 pp.

———. 1978. Management of kangaroo harvesting in Australia. Austral. Natl. Parks Wildl. Serv. Occas. Pap., no. 2, 28 pp.

———. 1979. The status of the Australian Macropodidae. *In* Tyler (1979), 13–27.

———. 1984. Management of kangaroo harvesting in Australia (1984). Austral. Natl. Parks Wildl. Serv. Occas. Pap., no. 9, 25 pp.

Poole, W. E., S. M. Carpenter, and N. G. Simms. 1990. Subspecific separation in the western grey kangaroo, *Macropus fuliginosus:* a morphometric study. Austral. Wildl. Res. 17:159–68.

Poole, W. E., and P. C. Catling. 1974. Reproduction in the two species of grey kangaroos, *Macropus giganteus* Shaw and *M. fuliginosus* (Desmarest). I. Sexual maturity and oestrus. Austral. J. Zool. 22:277–302.

Poole, W. E., J. T. Wood, and N. G. Simms. 1991. Distribution of the tammar, *Macropus eugenii,* and the relationships of populations as determined by cranial morphometrics. Wildl. Res. 18:625–39.

Q

Quin, D. G., A. P. Smith, S. W. Green, and H. B. Hines. 1992. Estimating the home ranges of sugar gliders *(Petaurus breviceps)* (Marsupialia: Petauridae), from grid-trapping and radiotelemetry. Wildl. Res. 19:471–87.

R

Rageot, R. 1978. Observaciones sobre el monito del monte. Chile Min. Agric. Corp. Nac. For. Dept. Tec. IX—Reg. Interp.—V. Silvestre, 16 pp.

Read, D. G. 1984. Reproduction and breeding season of *Planigale gilesi* and *P. tenuirostris* (Marsupialia: Dasyuridae). Austral. Mamm. 7:161–73.

Reading, R. P., P. Myronuik, G. Backhouse, and T. W. Clark. 1992. Eastern barred bandicoot reintroductions in Victoria, Australia. Species 19:29–31.

Redford, K. H., and J. F. Eisenberg, eds. 1992. Mammals of the neotropics: the southern cone. Univ. Chicago Press, ix + 430 pp.

Reig, O. A., J. A. W. Kirsch, and L. G. Marshall. 1987. Systematic relationships of the living and neocenozoic American "opossum-like" marsupials (suborder Didelphimorphia), with comments on the classification of these and of the Cretaceous and Paleogene New World and European metatherians. *In* Archer (1987*a*), 1–89.

Renfree, M. B. 1980. Embryonic diapause in the honey possum *Tarsipes spencerae.* Search 11:81.

Renfree, M. B., E. M. Russell, and R. D. Wooller. 1984. Reproduction and life history of the honey possum, *Tarsipes rostratus. In* Smith and Hume (1984), 427–37.

Richardson, B. J., and G. B. Sharman. 1976. Biochemical and morphological observations on the wallaroos (Macropodidae: Marsupialia) with a suggested new taxonomy. J. Zool. 179:499–513.

Ride, W. D. L. 1970. A guide to the native mammals of Australia. Oxford Univ. Press, Melbourne, xiv + 249 pp.

Rigby, R. G. 1972. A study of the behaviour of caged *Antechinus stuartii.* Z. Tierpsychol. 31:15–25.

Robinson, F. J., A. C. Robinson, C. H. S. Watts, and P. R. Baverstock. 1978. Notes on rodents and marsupials and their ectoparasites collected in Australia in 1974–75. Trans. Roy. Soc. S. Austral. 102:59–70.

Robinson, N. A., N. D. Murray, and W. B. Sherwin. 1993. VNTR loci reveal differentiation between and structure within populations of the eastern barred bandicoot *Perameles gunnii.* Molecular Biol. 2: 195–207.

Robinson, N. A., W. B. Sherwin, and P. R. Brown. 1991. A note on the status of the eastern barred bandicoot, *Perameles gunnii,* in Tasmania. Wildl. Res. 18:451–57.

Roig, V. G. 1991. Desertification and distribution of mammals in the southern cone of South America. *In* Mares, M. A., and D. J. Schmidly, eds., Latin American mammalogy: history, biodiversity, and conservation, Univ. Oklahoma Press, Norman, 239–79.

Rose, R. W. 1978. Reproduction and evolution in female Macropodidae. Austral. Mamm. 2:65–72.

———. 1986. The habitat, distribution, and conservation status of the Tasmanian bettong, *Bettongia gaimardi* (Desmarest). Austral. Wildl. Res. 13:1–6.

———. 1987. Reproductive biology of the Tasmanian bettong (*Bettongia gaimardi:* Macropodidae). J. Zool. 212:59–67.

———. 1989. Reproductive biology of the rat-kangaroos. *In* Grigg, Jarman, and Hume (1989), 307–15.

Rose, R. W., and D. J. McCartney. 1982. Reproduction of the red-bellied pademelon, *Thylogale billardierii* (Marsupialia). Austral. Wildl. Res. 9:27–32.

Rosenthal, M. A. 1975. Observations on the water opossum or yapok *Chironectes minimus* in captivity. Internatl. Zoo Yearbook 15:4–6.

Rounsevell, D. E., and S. J. Smith. 1982. Recent alleged sightings of the thylacine (Marsupialia, Thylacinidae) in Tasmania. *In* Archer (1982*a*), 233–36.

Russell, E. M. 1974*a*. The biology of kangaroos (Marsupialia-Macropodidae). Mamm. Rev. 4:1–59.

———. 1974*b*. Recent ecological studies on Australian marsupials. Austral. Mamm. 1: 189–211.

———. 1979. The size and composition of groups in the red kangaroo, *Macropus rufus.* Austral. Wildl. Res. 6:237–44.

Russell, E. M., and M. B. Renfree. 1989. Tarsipedidae. *In* Walton and Richardson (1989), 769–82.

S

Sarich, V. M., J. M. Lowenstein, and B. J. Richardson. 1982. Phylogenetic relationships of *Thylacinus cynocephalus* (Marsupialia) as reflected in comparative serology. *In* Archer (1982*a*), 703–9.

Scarlett, N. 1969. The bilby, *Thylacomys lagotis,* in Victoria. Victorian Nat. 86:292–94.

Schmitt, L. H., A. J. Bradley, C. M. Kemper, D. J. Kitchener, W. F. Humphreys, and R. A. How. 1989. Ecology and physiology of the northern quoll, *Dasyurus hallucatus* (Marsupialia, Dasyuridae), at Mitchell Plateau, Kimberley, Western Australia. J. Zool. 217: 539–58.

Schwartz, C. W., and E. R. Schwartz. 1959. The wild mammals of Missouri. Univ. Missouri Press, Columbia, vi + 341 pp.

Scott, M. P. 1987. The effect of mating and agonistic experience on adrenal function and mortality of male *Antechinus stuartii* (Marsupialia). J. Mamm. 68:479–86.

Seebeck, J. H. 1981. *Potorous tridactylus* (Kerr) (Marsupialia: Macropodidae): its distribution, status, and habitat preferences in Victoria. Austral. Wildl. Res. 8:285–306.

———. 1992*a*. Breeding, growth, and development of captive *Potorous longipes* (Marsupialia: Potoroidae); and a comparison with *P. tridactylus.* Austral. Mamm. 15:37–45.

———. 1992*b*. Sub-fossil potoroos in south-eastern Australia; with a record of *Potorous longipes* from New South Wales. Victorian Nat. 109:173–76.

Seebeck, J. H., A. F. Bennett, and D. J. Scotts. 1989. Ecology of the Potoroidae—a review. *In* Grigg, Jarman, and Hume (1989), 67–88.

Seebeck, J. H., P. R. Brown, R. L. Wallis, and C. M. Kemper, eds. 1990. Bandicoots and bilbies. Surrey Beatty & Sons, Chipping Norton, New South Wales, xxix + 392 pp.

Seebeck, J. H., and P. G. Johnston. 1980. *Potorous longipes* (Marsupialia: Macropodidae); a new species from eastern Victoria. Austral. J. Zool. 28:119–34.

Seebeck, J. H., and R. W. Rose. 1989. Potoroidae. *In* Walton and Richardson (1989), 716–39.

Serena, M., and T. R. Soderquist. 1988. Growth and development of pouch young of wild and captive *Dasyurus geoffroii* (Marsupialia: Dasyuridae). Austral. J. Zool. 36:533–43.

———. 1989*a*. Nursery dens of *Dasyurus geoffroii* (Marsupialia: Dasyuridae), with notes on nest building behaviour. Austral. Mamm. 12:35–36.

———. 1989*b*. Spatial organization of a riparian population of the carnivorous marsupial *Dasyurus geoffroii*. J. Zool. 219:373–83.

———. 1990. Occurrence and outcome of polyoestry in wild western quolls, *Dasyurus geoffroii* (Marsupialia: Dasyuridae). Austral. Mamm. 13:205–8.

Sharman, G. B., R. L. Close, and G. M. Maynes. 1990. Chromosome evolution, phylogeny, and speciation of rock wallabies (*Petrogale:* Macropodidae). Austral. J. Zool. 37: 351–63.

Sharman, G. B., C. E. Murtagh, P. M. Johnson, and C. M. Weaver. 1980. The chromosomes of a rat-kangaroo attributable to *Bettongia tropica* (Marsupialia: Macropodidae). Austral. J. Zool. 28:59–63.

Sharman, G. B., and P. E. Pilton. 1964. The life history and reproduction of the red kangaroo *(Megaleia rufa)*. Proc. Zool. Soc. London 142:29–48.

Sherwin, W. B., and P. R. Brown. 1990. Problems in the estimation of the effective population size of the eastern barred bandicoot *Perameles gunnii* at Hamilton, Victoria. *In* Seebeck et al. (1990), 367–74.

Shield, J. 1968. Reproduction of the quokka, *Setonix brachyurus*, in captivity. J. Zool. 155:427–44.

Short, J., S. D. Bradshaw, J. Giles, R. I. T. Prince, and G. R. Wilson. 1992. Reintroduction of macropods (Marsupialia: Macropodidae) in Australia—a review. Biol. Conserv. 62:189–204.

Short, J., and G. Milkovits. 1990. Distribution and status of the brush-tailed rock-wallaby in south-eastern Australia. Austral. Wildl. Res. 17:169–79.

Short, J., and B. Turner. 1991. Distribution and abundance of spectacled hare-wallabies and euros on Barrow Island, Western Australia. Wildl. Res. 18:421–29.

———. 1992. The distribution and abundance of the banded and rufous hare-wallabies, *Lagostrophus fasciatus* and *Lagorchestes hirsutus*. Biol. Conserv. 60:157–66.

———. 1993. The distribution and abundance of the burrowing bettong (Marsupialia: Macropodidae). Wildl. Res. 20:525–34.

Simonetta, A. M. 1979. First record of *Caluromysiops* from Colombia. Mammalia 43:247–48.

Simons, E. L., and T. M. Bown. 1984. A new species of *Peratherium* (Didelphidae: Polyprotodonta): the first African marsupial. J. Mamm. 65:539–48.

Simpson, G. G. 1945. The principles of classification and a classification of the mammals. Bull. Amer. Mus. Nat. Hist. 85:i–xvi + 1–350.

Smith, A. 1982. Is the striped possum (*Dactylopsila trivirgata;* Marsupialia, Petauridae) an arboreal anteater? Austral. Mamm. 5:229–34.

———. 1984*a*. Demographic consequences of reproduction, dispersal, and social interaction in a population of Leadbeater's possum. *In* Smith and Hume (1984), 359–73.

———. 1984*b*. Diet of Leadbeater's possum, *Gymnobelideus leadbeateri* (Marsupialia). Austral. Wildl. Res. 11:265–73.

Smith, M. J. 1971. Breeding the sugar-glider *Petaurus breviceps* in captivity; and growth of pouch young. Internatl. Zoo Yearbook 11:26–28.

———. 1973. *Petaurus breviceps*. Mammalian Species, no. 30, 5 pp.

Smith, M. J., B. K. Brown, and H. J. Frith. 1969. Breeding of the brush-tailed possum, *Trichosurus vulpecula* (Kerr), in New South Wales. CSIRO Wildl. Res. 14:181–93.

Smith, M. J., and R. A. How. 1973. Reproduction in the mountain possum, *Trichosurus caninus* (Ogilby), in captivity. Austral. J. Zool. 21:321–29.

Smith, M. J., and G. C. Medlin. 1982. Dasyurids of the northern Flinders Ranges before pastoral development. *In* Archer (1982*a*), 563–72.

Smith, R. F. C. 1969. Studies on the marsupial glider, *Schoinobates volans* (Kerr). Austral. J. Zool. 17:625–36.

Soderquist, T. R. 1993. Maternal strategies of *Phascogale tapoatafa* (Marsupialia: Dasyuridae). I. Breeding seasonality and maternal investment. Austral. J. Zool. 41:549–66.

———. 1994. Anti-predator behaviour of the brush-tailed phascogale *(Phascogale tapoatafa)*. Victorian Nat. 111:22–24.

———. 1995. Spatial organization of the arboreal carnivorous marsupial *Phascogale tapoatafa*. J. Zool. 237.

Soderquist, T. R., and L. Ealey. 1994. Social interactions and mating strategies of a solitary carnivorous marsupial, *Phascogale tapoatafa*, in the wild. Wildl. Res. 21:527–42.

Soderquist, T. R., and A. Lill. 1995. Natal dispersal and philopatry in the carnivorous marsupial *Phascogale tapoatafa* (Dasyuridae). Ethology 99:297–312.

Soriano, P. 1987. On the presence of the short-tailed opossum *Monodelphis adusta* (Thomas) in Venezuela. Mammalia 51:321–24.

Southgate, R. I. 1990. Distribution and abundance of the greater bilby *Macrotis lagotis* Reid (Marsupialia: Peramelidae). *In* Seebeck et al. (1990), 293–302.

Springer, M. S., J. A. W. Kirsch, K. Aplin, and T. Flannery. 1990. DNA hybridization, cladistics, and the phylogeny of phalangerid marsupials. J. Molecular Evol. 30:298–311.

Stodart, E. 1977. Breeding and behaviour of Australian bandicoots. *In* Stonehouse and Gilmore (1977), 179–91.

Strahan, R. 1975. Status and husbandry of Australian monotremes and marsupials. *In* Martin (1975*b*), 171–82.

———, ed. 1983. The Australian Museum complete book of Australian mammals. Angus & Robertson, London, xxi + 530 pp.

Streilen, K. E. 1982*a*. Ecology of small mammals in the semiarid Brazilian Caatinga. I. Climate and faunal composition. Ann. Carnegie Mus. 51:79–106.

———. 1982*b*. The ecology of small mammals in the semiarid Brazilian Caatinga. III. Reproductive biology and population ecology. Ann. Carnegie Mus. 51:251–69.

———. 1982*c*. The ecology of small mammals in the semiarid Brazilian Caatinga. V. Agonistic behavior and overview. Ann. Carnegie Mus. 51:345–69.

Suckling, G. C. 1984. Population ecology of the sugar glider, *Petaurus breviceps*, in a system of fragmented habitats. Austral. Wildl. Res. 11:49–75.

Sunquist, M. E., S. N. Austad, and F. Sunquist. 1987. Movement patterns and home range in the common opossum *(Didelphis marsupialis)*. J. Mamm. 68:173–76.

Szalay, F. S. 1982. A new appraisal of marsupial phylogeny and classification. *In* Archer (1982*a*), 621–40.

———. 1993. Metatherian taxon phylogeny: evidence and interpretation from the cranioskeletal system. *In* Szalay, Novacek, and McKenna (1993*a*), 216–42.

———. 1994. Evolutionary history of the marsupials and an analysis of osteological characters. Cambridge. Univ. Press, xii + 481 pp.

T

Tate, G. H. H. 1933. A systematic revision of the marsupial genus *Marmosa,* with a discussion of the adaptative radiation of the murine opossums *(Marmosa).* Bull. Amer. Mus. Nat. Hist. 66:1–250.

———. 1945. Results of the Archbold Expeditions. No. 52. The marsupial genus *Phalanger.* Amer. Mus. Novit., no. 1283, 30 pp.

Taylor, A. C., W. B. Sherwin, and R. K. Wayne. 1994. Genetic variation of microsatellite loci in a bottlenecked species: the northern hairy-nosed wombat *Lasiorhinus krefftii.* Molecular Ecol. 3:277–90.

Taylor, J. M., J. H. Calaby, and T. D. Redhead. 1982. Breeding in wild populations of the marsupial-mouse *Planigale maculata sinualis* (Dasyuridae, Marsupialia). *In* Archer (1982*a*), 83–87.

Taylor, J. M., and B. E. Horner. 1970. Gonadal activity in the marsupial mouse, *Antechinus bellus,* with notes on other species of the genus (Marsupialia: Dasyuridae). J. Mamm. 51:659–68.

Taylor, R. J. 1992. Seasonal changes in the diet of the Tasmanian bettong *(Bettongia gaimardi),* a mycophagous marsupial. J. Mamm. 73:408–14.

———. 1993*a*. Habitat requirements of the Tasmanian bettong *(Bettongia gaimardi),* a mycophagous marsupial. Wildl. Res. 20: 699–710.

———. 1993*b*. Home range, nest use, and activity of the Tasmanian bettong, *Bettongia gaimardi.* Wildl. Res. 20:87–95.

Thomas, O. 1921. A new genus of opossum from southern Patagonia. Ann. Mag. Nat. Hist., ser. 9, 8:136–39.

Thompson, V. D. 1995. Queensland koala *(Phascolarctos adustus)* and Victorian koala *(Phascolarctos cinereus victor):* North American regional studbook. Zoological Society of San Diego, 73 pp.

Thornback, J., and M. Jenkins. 1982. The IUCN mammal red data book. Part 1: Threatened mammalian taxa of the Americas and the Australasian zoogeographic region (excluding Cetacea). Internatl. Union Conserv. Nat., Gland, Switzerland, xl + 516 pp.

Troy, S., and G. Coulson. 1993. Home range of the swamp wallaby, *Wallabia bicolor.* Wildl. Res. 20:571–77.

Turner, K. 1970. Breeding Tasmanian devils, *Sarcophilus harrisii,* at Westbury Zoo. Internatl. Zoo. Yearbook 10:65.

Tyndale-Biscoe, C. H. 1968. Reproduction and post-natal development in the marsupial *Bettongia lesueur* (Quoy and Gaimard). Austral. J. Zool. 16:577–602.

———. 1973. Life of marsupials. American Elsevier, New York, viii + 254 pp.

Tyndale-Biscoe, C. H., and R. B. MacKenzie. 1976. Reproduction in *Didelphis marsupialis* and *D. albiventris* in Colombia. J. Mamm. 57:249–65.

Tyndale-Biscoe, C. H., and R. F. C. Smith. 1969. Studies on the marsupial glider, *Schoinobates volans* (Kerr). II. Population structure and regulatory mechanisms. J. Anim. Ecol. 38:637–49.

U

USDI (United States Department of the Interior). 1995. Removal of three kangaroos from the List of Endangered and Threatened Wildlife. Federal Register 60:12887–906.

V

Van Deusen, H. M., and J. K. Jones, Jr. 1967. Marsupials. *In* Anderson and Jones (1967), 61–86.

Van Deusen, H. M., and K. Keith. 1966. Range and habitat of the bandicoot, *Echymipera clara,* in New Guinea. J. Mamm. 47: 721–23.

Van Dyck, S. 1980. The cinnamon antechinus, *Antechinus leo* (Marsupialia: Dasyuridae), a new species from the vine-forests of Cape York Peninsula. Austral. Mamm. 3:5–17.

———. 1982*a*. The relationships of *Antechinus stuartii* and *A. flavipes* (Dasyuridae, Marsupialia) with special reference to Queensland. *In* Archer (1982*a*), 723–66.

———. 1982*b*. The status and relationships of the Atherton antechinus, *Antechinus godmani* (Marsupialia: Dasyuridae). Austral. Mamm. 5:195–210.

———. 1985. *Sminthopsis leucopus* (Marsupialia: Dasyuridae) in north Queensland rainforest. Austral. Mamm. 8:53–60.

———. 1986. The chestnut dunnart, *Sminthopsis archeri* (Marsupialia: Dasyuridae), a new species from the savannahs of Papua New Guinea and Cape York Peninsula, Australia. Austral. Mamm. 9:111–24.

———. 1987. The bronze quoll, *Dasyurus spartacus* (Marsupialia: Dasyuridae), a new species from the savannahs of Papua New Guinea. Austral. Mamm. 11:145–56.

Van Dyck, S., J. C. Z. Woinarski, and A. J. Press. 1994. The Kakadu dunnart, *Sminthopsis bindi* (Marsupialia: Dasyuridae), a new species from the stony woodlands of the Northern Territory. Mem. Queensland Mus. 37:311–23.

Van Gelder, R. G. 1977. Mammalian hybrids and generic limits. Amer. Mus. Novit., no. 2635, 25 pp.

Vose, H. M. 1973. Feeding habits of the western Australian honey possum, *Tarsipes spenserae.* J. Mamm. 54:245–47.

W

Wainer, J. W. 1976. Studies of an island population of *Antechinus minimus* (Marsupialia, Dasyuridae). Austral. Zool. 19:1–7.

Waithman, J. 1979. A report on a collection of mammals from southwest Papua, 1972–1973. Austral. Zool. 20:313–26.

Wakefield, N. A. 1970*a*. Notes on Australian pigmy-possums (*Cercartetus,* Phalangeridae, Marsupialia). Victorian Nat. 87:11–18.

———. 1970*b*. Notes on the glider-possum, *Petaurus australis* (Phalangeridae, Marsupialia). Victorian Nat. 87:221–36.

———. 1971. The brush-tailed rock-wallaby *(Petrogale penicillata)* in western Victoria. Victorian Nat. 88:92–102.

Wallis, I. R., P. J. Jarman, B. E. Johnson, and R. W. Liddle. 1989. Nest sites and use of nests by rufous bettongs *Aepyprymnus rufescens. In* Grigg, Jarman, and Hume (1989), 619–23.

Ward, G. D. 1978. Habitat use and home range of radio-tagged opossums *Trichosurus vulpecula* (Kerr) in New Zealand lowland forest. *In* Montgomery (1978), 267–87.

Ward, S. J. 1990*a*. Life history of the eastern pygmy possum, *Cercartetus nanus* (Burramyidae: Marsupialia), in south-eastern Australia. Austral. J. Zool. 38: 287–304.

———. 1990*b*. Life history of the feathertail glider, *Acrobates pygmaeus* (Acrobatidae: Marsupialia) in south-eastern Australia. Austral. J. Zool. 38:503–17.

———. 1990*c*. Reproduction in the western pygmy-possum, *Cercartetus concinnus* (Marsupialia: Burramyidae), with notes on reproduction of some other small possum species. Austral. J. Zool. 38:423–38.

Ward, S. J., and M. B. Renfree. 1988*a*. Reproduction in females of the feathertail glider, *Acrobates pygmaeus* (Marsupialia). J. Zool. 216:225–39.

———. 1988*b*. Reproduction in males of the feathertail glider, *Acrobates pygmaeus* (Marsupialia). J. Zool. 216:241–51.

Wells, R. T. 1978. Field observations of the hairy-nosed wombat, *Lasiorhinus latifrons* (Owen). Austral. Wildl. Res. 5:299–303.

———. 1989. Vombatidae. *In* Walton and Richardson (1989), 755–68.

Westerman, M. 1991. Phylogenetic relationships of the marsupial mole, *Notoryctes typhlops* (Marsupialia: Notoryctidae). Austral. J. Zool. 39:529–37.

Westerman, M., A. H. Sinclair, and P. A. Woolley. 1984. Cytology of the feathertail possum *Distoechurus pennatus. In* Smith and Hume (1984), 423–25.

Wilson, B. A., and A. R. Bourne. 1984. Reproduction in the male dasyurid *Antechinus minimus maritimus* (Marsupialia: Dasyuridae). Austral. J. Zool. 32:311–18.

Wilson, D(on). E., and D. M. Reeder, eds. 1993. Mammal species of the world: a taxonomic and geographic reference. Smithsonian Inst. Press, Washington, D.C., xviii + 1206 pp.

Winkel, K., and I. Humphrey-Smith. 1987. Diet of the marsupial mole, *Notoryctes typhlops* (Stirling 1889) (Marsupialia: Notoryctidae). Austral. Mamm. 11:159–61.

Winter, J. W. 1979. The status of endangered Australian Phalangeridae, Petauridae, Burramyidae, Tarsipedidae, and the Koala. *In* Tyler (1979), 45–59.

———. 1984. Conservation studies of tropical rainforest possums. *In* Smith and Hume (1984), 469–81.

Wodzicki, K., and J. E. C. Flux. 1971. The parma wallaby and its future. Oryx 11:40–47.

Wood, D. H. 1970. An ecological study of *Antechinus stuartii* (Marsupialia) in a southeast Queensland rain forest. Austral. J. Zool. 18: 185–207.

Woodburne, M. O., R. H. Tedford, and M. Archer. 1987. New Miocene ringtail possums (Marsupialia: Pseudocheiridae) from South Australia. *In* Archer (1987*a*), 639–79.

Woolley, P. A. 1971*a*. Maintenance and breeding of laboratory colonies of *Dasyuroides byrnei* and *Dasycercus cristicauda*. Internatl. Zoo Yearbook 11:351–54.

———. 1971*b*. Observations on the reproductive biology of the dibbler, *Antechinus apicalis* (Marsupialia: Dasyuridae). J. Roy Soc. W. Austral. 54:99–102.

———. 1977. In search of the dibbler, *Antechinus apicalis* (Marsupialia: Dasyuridae). J. Roy. Soc. W. Austral. 59:111–17.

———. 1980. Further searches for the dibbler, *Antechinus apicalis* (Marsupialia: Dasyuridae). J. Roy. Soc. W. Austral. 63:47–52.

———. 1982*a*. Observations on the feeding and reproductive status of captive feathertailed possums, *Distoechurus pennatus* (Marsupialia: Burramyidae). Austral. Mamm. 5: 285–87.

———. 1982*b*. Phallic morphology of the Australian species of *Antechinus* (Dasyuridae, Marsupialia): a new taxonomic tool? *In* Archer (1982*a*), 767–81.

———. 1984. Reproduction in *Antechinomys laniger* ("spenceri" form) (Marsupialia: Dasyuridae): field and laboratory investigations. Austral. Wildl. Res. 11:481–89.

———. 1988. Reproduction in the Ningbing antechinus (Marsupialia: Dasyuridae): field and laboratory observations. Austral. Wildl. Res. 15:149–56.

———. 1989. Nest location by spool-and-line tracking of dasyurid marsupials in New Guinea. J. Zool. 218:689–700.

———. 1990. Mulgaras, *Dasycercus cristicauda* (Marsupialia: Dasyuridae): their burrows, and records of attempts to collect live animals between 1966 and 1979. Austral. Mamm. 13:61–64.

———. 1991*a*. Reproduction in *Dasykaluta rosamondae* (Marsupialia: Dasyuridae): field and laboratory observations. Austral. J. Zool. 39:549–68.

———. 1991*b*. Reproduction in *Pseudantechinus macdonnellensis* (Marsupialia: Dasyuridae): field and laboratory observations. Wildl. Res. 18:13–25.

———. 1991*c*. Reproductive pattern of captive Boullanger Island dibblers, *Parantechinus apicalis* (Marsupialia: Dasyuridae). Wildl. Res. 18:157–63.

———. 1992. New records of the Julia Creek dunnart, *Sminthopsis douglasi* (Marsupialia: Dasyuridae). Wildl. Res. 19:779–84.

Woolley, P. A., S. A. Raftopoulos, G. J. Coleman, and S. M. Armstrong. 1991. A comparative study of circadian activity patterns of two New Guinean dasyurid marsupials, *Phascolosorex dorsalis* and *Antechinus habbema*. Austral. J. Zool. 39:661–71.

Woolley, P., and A. Valente. 1982. The dibbler, *Parantechinus apicalis* (Marsupialia: Dasyuridae): failure to locate populations in four regions in the south of Western Australia. Austral. Mamm. 5:241–45.

Z

Ziegler, A. C. 1972. Additional specimens of *Planigale novaeguineae* (Dasyuridae: Marsupialia) from Territory of Papua. Austral. Mamm. 1:43–45.

———. 1977. Evolution of New Guinea's marsupial fauna in response to a forested environment. *In* Stonehouse and Gilmore (1977), 117–38.

———. 1981. *Petaurus abidi*, a new species of glider (Marsupialia: Petauridae) from Papua New Guinea. Austral. Mamm. 4:81–88.

Index

The scientific names of orders, families, and genera that have titled accounts in the text are in boldface type, as are the page numbers on which such accounts begin. Other scientific names and vernacular names appear in roman.